ECHOES
of the
ANCIENT
SKIES

ECHOES
of the
ANCIENT
SKIES

the astronomy
of lost civilizations

by DR. E.C. KRUPP

DOVER PUBLICATIONS, INC.
Mineola, New York

Bibliographical Note

This Dover edition, first published in 2003, is an unabridged republication of the 1994 Oxford University Press paperback edition of the work originally published in 1983 by Harper & Row, Publishers, Inc., New York.

Library of Congress Cataloging-in-Publication Data

Krupp, E. C. (Edwin C.), 1944–
    Echoes of the ancient skies : the astronomy of lost civilizations / E. C. Krupp.
    p. cm.
    Originally published: New York, Harper & Row, c1983.
    Includes bibliographical references and index.
    ISBN 0-486-42882-6 (pbk.)
    1. Astronomy, Ancient. 2. Archæology. 3. Civilization—History. I. Title.

QB16.K78 2003
520'.93—dc21

2002041788

Manufactured in the United States of America
Dover Publications, Inc., 31 East 2nd Street, Mineola, N.Y. 11501

for my grandmother,
Eleanor Olander,
who, still a florist after all these years,
sees renewal in every season
and cosmic order in every bloom

Foreword
to the 1994 edition

When *Echoes of the Ancient Skies* was first published in 1983, we had begun to understand how intricately astronomy was woven into many different aspects of ancient and traditional culture, and that was the theme of the book. Now, more than a decade later, much additional research has confirmed that premise and revealed much more detail about the part the sky plays in people's affairs. Recognizing the continuing value of the material in this book, Oxford University Press has brought it back into print, allowed me to make a few corrections, and provided room for an update on new developments in the study of ancient astronomy.

Archaeoastronomy—the interdisciplinary study of ancient, prehistoric, and traditional astronomy and its cultural context—became truly international in 1981, when the First "Oxford" International Conference on Archaeoastronomy was held in Oxford, England. Research from that conference was reported in the first edition of this book, but since then three more "Oxford" conferences have been organized—in Mérida (Yucatán, Mexico–1986), in St. Andrews (Scotland–1990), and Stara Zagora (Bulgaria–1993). Accordingly, archaeoastronomy has evolved. It is much more anthropologically oriented, and archaeoastronomists are more interested in understanding how astronomy affects society and culture than in identifying astronomical alignments (although these remain an important element of research).

It is easy to see, then, why the First International Conference on Ethnoastronomy, which took place in 1983 at the Smithsonian's National Museum of Natural History and National Air and Space Museum, in Washington, D.C., was as significant as the first "Oxford" meeting. The value of ethnographic and ethnohistoric data was demonstrated there by research from all over the world.

The cultural dimension of astronomy became even clearer in two traveling museum exhibits, "Star Gods of the Ancient Americas" (1983–1984) and "Skywatchers of Ancient California" (1983–1985). Traditional peoples use their symbolic vocabularies, with striking redundancy, to disclose what they believe are the fundamental principles of their lives. These two exhibits put on public display an abundance of objects with astronomical imagery and objects used in celestially tempered rituals. These artifacts restated for us the function of astronomical concepts in ideology, which happily coincided with a reawakening of anthropological interest in the function of ideology in the development of complex societies.

Regional symposia also established their value. In October 1983, the Maxwell Museum of Anthropology and the University of New Mexico sponsored "Astronomy and Ceremony in the Prehistoric Southwest" in Albuquerque. A month later, the First Western Regional Conference on Archaeoastronomy explored California Indian astronomies at California State University, Northridge. American Indians were invited to and participated at both conferences.

In 1984 the Universidad Nacional Autónoma de México hosted a symposium with international participation: "Arqueoastronomía y Etnoastronomía en Mesoamérica." Partially inspired by the success of the first ethnoastronomy conference in Washington, D.C., it concentrated attention on the astronomical traditions of central Mexico and in Maya territory, and consolidated research. An international conference on rock art and archaeoastronomy took place in Little Rock, Arkansas, in 1984. Three conferences, emphasizing the astronomical traditions of Asia, were held in Hyderabad, India, at the Birla Planetarium. The Historical Astronomy Division of the American Astronomical Society included sessions on archaeoastronomy at least once a year. Specialized programs were also scheduled for meetings of the International Congress of Americanists and for the International Astronomical Union.

Scholars in Poland organized a national symposium in 1982, and a variety of meetings in Europe followed, including Tobukhin in 1988 and Venice in 1989. The "Time and Astronomy at the Meeting of Two Worlds" conference held in Frombork, Poland, in 1992, was timed to coincide with the year of the five hundredth anniversary of the arrival of Columbus in the New World.

A new European journal, *Astronomie et Sciences Humaines* (Astronomy and the Human Sciences), was inaugurated by the Astronomical Observatory of Strasbourg at the University of Strasbourg in 1988. In 1990 the Europeans also instituted an annual regional meeting.

All of this activity confirms that the value of understanding ancient astronomical traditions goes beyond the study of the development of astronomical knowledge and techniques. Astronomy is important because it allows at least partial entrance into the belief systems of our ancestors. In this way, we see how they saw the world and their place in it. We see how their view of the universe—their cosmovision—affected the way they adapted to the forces that framed and altered their lives.

Although the fundamental information in this book retains its value, there have been significant developments since it was first published. For example, a more detailed examination by T. P. Ray of the architecture of Newgrange was published in *Nature* in 1989. Ray showed that the alignment of this prehistoric passage grave in Ireland was more accurately planned and more astronomically precise than previously thought. Winter solstice sunlight originally entered in the chamber just as the sun rose above the horizon more than five thousand years ago. The beam of sunlight was narrower than it is today, and its path bisected the back chamber. Martin Brennan, despite criticism of his methods and many of his conclusions, recognized that vertical grooves on decorated stones on the front and back of Newgrange seemed to define the same line that coincided with winter solstice sunrise. All of this reinforces the claim that Newgrange is the oldest monumental structure known to have an astronomical alignment.

Several other chambered passage graves and other tombs in northwestern Europe also are aligned astronomically, for example, winter solstice sunset at Clava in Scotland, winter solstice sunset at Maes Howe (Orkney, Scotland), winter solstice sunrise at La Roche-aux-Fées in Brittany (France), and winter solstice sunrise and the major standstill southern moonrise at Gavr'inis, near Carnac, Brittany. Although the complicated site at Brainport Bay (or Minard), Scotland, is not a chambered tomb, it combines characteristics associated with prehistoric ritual with a well-defined alignment of upright stones that appear to be targeted on the horizon for summer solstice sunrise.

In 1991 the remains of an unprecedented neolithic structure were uncovered at a gravel quarry in Godmanchester, near Cambridge, England. Believed to be a temple, it was contrived with earthworks and wood posts almost five thousand years ago. Its main axis coincides with the sunrise at the beginning of May and the beginning of August. These dates were singled out in the Celtic calendar much later as key dates of seasonal change, and Alexander Thom has argued that the same dates were observed in neolithic and bronze age Britain. Other alleged alignments at the site include solstice and equinox sunrises and sunsets and the northern and southern major and minor standstills of the moon.

Alexander Thom's precise prehistoric observatories continued to command attention and controversy during the last decade, but this interpretation encountered a challenge that has prompted many to abandon the idea. Astronomers Bradley Schaefer and William Liller have argued through practical observation that day-to-day changes in atmospheric refraction introduce significant and unpredictable variations in the horizon positions of the sun, moon, planets, and stars. Because the variation at higher latitudes can be much larger than subtle effects such as the small gravitational perturbation on inclination of the moon's orbit, precise alignments are impossible. This analysis has been disputed, but it has diminished some support for the Thom picture of prehistoric astronomy. Despite that, rigorous statistical studies by Clive Ruggles suggest to him that less precise alignments on the lunar standstills may well exist in alignments of standing stones toward the horizon. This is an interesting development, for Anthony Aveni has maintained that there is not textual evidence anywhere for an interest in the lunar standstills. He disputes that the lunar extremes were observed by anyone.

New fieldwork on the Great Pyramid and the Sphinx has provided compelling new information, and some of it is related to astronomy. Zahi Hawass, Director General of the Giza Plateau and Saqqara Archaeological Sites for the Egyptian Antiquities Organization, agrees that the Great Pyramid was intended not only to contain the body of the pharaoh Khufu but also to facilitate the transformation of his spirit and to launch his soul to the sky. Independent scholar John Charles Deaton has found additional evidence for this in the names of other pyramids. Hawass believes Khufu identified himself as the sun and that the boat found in a pit adjacent to the Great Pyramid was the counterpart of either the boat that ferried the sun through the hours of the day or the vessel that transported the sun through the dangerous hours of the night.

Remote sensing has provided new information about the interior of the Great Pyramid. For example, the shafts that extend from the so-called Queen's Chamber continue, like those from the King's Chamber, to the outer surface of the structure.

Finally, Egyptologist Mark Lehner has identified a number of astronomical alignments at Giza and believes the summer solstice sun was intended to create the "akhet" symbol when it set between the Great Pyramid and Khafre's Pyramid. The equinox may also account for arrangement of Khafre's Pyramid with respect to the Sphinx, which symbolizes the Horus, the son of Re, the sun god. Khafre, who was the son of Khufu, seems to have associated himself with Horus and portrayed himself as the Sphinx. The Temple of the Sphinx has two niches, one on the east wall and one on the west. They were probably dedicated to the rising and setting sun. The temples twenty-four pillars appear to symbolize the hours of the day.

In 1988 the results of my 1982 fieldwork on Egyptian solar chapels were published. Measurements confirmed the conclusions Gerald Hawkins reached at Karnak and described in Chapter 10 and also added new details. Winter Solstice sunrise alignment was also found at the solar sanctuary in Hatshepsut's mortuary temple at Deir el-Bahri, and these sanctuaries were linked with ancient Egyptian beliefs about the passage of Re through the netherworld and the transformation of the soul of the deceased pharaoh. The nearly cardinal alignment of the Sun Temple of Neuserre' at Abu Ghurab suggested that a change from cardinal to solstitial orientation of solar temples may have been adopted between the Old and New Kingdoms.

The explosive expansion of our ability to read Maya hieroglyphic writing has reinforced with dramatic new details the rather general commentary in Chapter 3 about celestial events and royal power. In 1970, when Erich von Däniken's romantic and misinformed account of ancient astronauts, *Chariots of the Gods?*, was published, the glyph for Pacal, the dead Maya king depicted on the Palenque sarcophagus (described in Chapter 5), was still unrecognized. Inspired by the absence of translated text, von Däniken insisted that the portrait of the dying king was really an astronaut rocketing into space. In just three years, however, the names of kings and references to dynastic events became readable on inscribed monuments at one site after another. We have lived through the golden age of decipherment of Maya writing, and the inscriptions confirm that astronomical metaphor and celestial events helped define the ritual landscape for Maya rulers. For example, transfers of royal power appear to have been timed by summer solstice at Yaxchilán, and Carolyn Tate has reported that the main doorway of Temple 33 allowed the light of the summer solstice rising sun to fall upon a statue of Bird-Jaguar IV, the Yaxchilán king who built it.

In another case Dieter Dütting and Anthony Aveni have associated a Palenque inscription with the conjunction of Jupiter, Saturn, and Mars with the moon on 23 July 690 A.D. Chan-Bahlum, Pacal's son and successor, dedicated the Cross Group on that day, and it is thought that Pacal's heir saw in the late July sky a reenactment of the primordial birth of the three ancestor gods of the Palenque dynasty (the three planets) by the First Mother (the moon). With that in mind, he consecrated his accession monument in the same way the First Father originally dedicated the dome of the sky after he had built it millennia before.

The Maya even went to war by the sky, and the planet Venus was the trigger. Linda Schele, Mary Miller, and other Mayanists have discussed the Venus war regalia of the Maya and the timing of raids and captures with appearances of Venus, particularly as an evening "star." John Carlson, Director of the Center for Archaeoastronomy, spotlighted this activity in his *National Geographic* article, "America's Ancient Skywatchers" (March 1990). Carlson also explored the role of Venus-regulated warfare and ritual sacrifice in central Mexico at Cacaxtla and Teotihuacán. Its presence there confirms that Venus warfare was widely established by the first millennium A.D.

Research in North America, particularly in the Southwest and in California, has delivered more sites with astronomical connotations and more information about symbolic and practical astronomy among North America's prehistoric peoples. Anna Sofaer and her collaborators have identified other petroglyphs on Chaco Canyon's Fajada Butte that interact seasonally with midday light and shadow. Robert and Ann Preston investigated light-and-shadow effects and line-of-sight alignments of nineteen rock art sites, most in Arizona. In Utah Jesse Warner and his associates have reported numerous light-and-shadow events on Fremont Culture rock art, some more convincing than others. John Rafter, Arlene Benson, and others continued to explore the astronomical potential of California sites and have integrated ethnographic data on California Indian ritual and myth with their studies of rock art. I have been conducting a similar analysis of California material, particularly in Chumash and Luiseño territory. A winter solstice wedge of sunlight, somewhat like that seen at Burro Flats, falls upon a set of concentric rings at Condor Cave.

Some specialists remain skeptical about light-and-shadow effects. It is hard to demonstrate that native peoples exploited this technique. It is not so different, however, from observations of a beam of light passing through an aperture and onto the wall of a Pueblo building, and we have ethnographic descriptions of that. In California we even know the names of two Chumash shamans who went into the mountains at the time of winter solstice to paint on the rocks. Whether they were interested in solstitial lighting on the pictographs is not mentioned, but the connection between California Indian shamanism, rock art, and winter solstice ritual is clear.

J. McKim Malville expanded the study of prehistoric Pueblo astronomy with fieldwork at Yellow Jacket, where a line of pointed monoliths was directed toward the summer solstice sunrise over the mountain horizon. Malville also believes the major standstill northern moonrise was watched between the two natural rock pillars by the people of Chimney Rock Pueblo, the most eastern Chaco outlier. Anna Sofaer, Rolf Sinclair, and Michael P. Marshall have offered an intriguing symbolic explanation for Chaco Canyon's Great North Road. They associate this peculiar feature of the Anasazi landscape with Pueblo cosmology and the journey of the spirits from the "middle of the world" to the mythic ancestral land in the far north. Sofaer and Sinclair have also collaborated with Joey B. Donahue in a study of solar and lunar orientations in major buildings of Chaco Canyon and Chaco outliers.

A substantive collection of research papers on the Nazca lines of Peru's south coastal desert, edited by Anthony F. Aveni, appeared in 1990. Along with the briefer reports by Johann Reinhard, it persuades us that the geoglyphs were oriented to mountains, to the flow of water, and sometimes to celestial targets. Certainly some of them appear to have been ritual pathways, the use of which may have been seasonally and calendrically determined. Also in Peru, further studies of Inca astronomy were undertaken by Mariusz Ziólkowski and Robert Sadowski of Poland, Arnold Lebeuf of France, and David S. P. Dearborn, Raymond E. White, and Katharina J. Schreiber of the United States. The last three discovered the December solstice sunrise operation of Intimachay, a cave below the main ruins of Machu Picchu.

Important ethnographic accounts of astronomical and cosmological traditions of North American Indians have been published by Claire R. Farrer (Mescalero Apache) and Trudy Griffin-Pierce (Navajo). In South America's Amazon Basin, Peter G. Roe looked carefully at the astronomy and cosmology of the Shipibo. Bororo cosmology was described independently by both Jon Christopher Crocker and Stephen Michael Fabian.

In Asia, J. McKim Malville evaluated the astronomical possibilities at sites in India, including the Sun Temple at Konorak and Vijayanagara, the fourteenth-century capital of the medieval Hindu kings. I have continued to collect information in China and have published a report on the astronomical and seasonal characteristics of the suburban altars of imperial Beijing and how they displayed and legitimized the authority of the Chinese emperor. A far more detailed and comprehensive synthesis of the sacred cosmological dimensions of Beijing was written by Jeffrey F. Meyer. His book, *The Dragons of Tiananmen*, appeared in 1991. The earliest textual reference to the names of the twenty-eight reference stars, or *hsiu*, in traditional astronomy was found in 1979, on the lid of a ceramic box recovered from a tomb dated to 433 B.C., but information on the find was not available in English until 1984.

The Chinese announced the discovery in April 1987 of what may be the oldest star map in the world painted on the ceiling of a small Western Han tomb in Xi'an. The tomb was constructed at least as early as 86 B.C., and so the map is probably older than Egypt's famous "circular zodiac" of Dendera, usually dated to about 30 B.C., International financial assistance and conservation expertise are badly needed to preserve this remarkable relic. I was fortunate to see it in 1992, but without help, the world's oldest star map will crumble away.

Interest in astronomy's place in Hellenistic religion was revived in 1989 by David Ulansey's book, *The Origin of the Mithraic Mysteries*. Mithraism's chief competitor, Christianity, also incorporated astronomical and cosmological symbolism during its early development, and *Jesus Christ, Sun of God* (1993) by David Fideler, documents much of that tradition.

After *Echoes of the Ancient Skies*, I turned some of my attention to celestial mythology and wrote *Beyond the Blue Horizon: Myths and Legends of the Sun, Moon, Stars, and Planets* (1990). No comprehensive cross-cultural study of this material had been undertaken in more than a century.

The passage of time harvests the lives of people, and in the last ten years the interdisciplinary conversation that is archaeoastronomy lost three pioneers. Alexander Thom, whose megalithic lunar observatories set some experts on their ears and lit a fire under others, died on 7 November 1985. There would be a lot less California Indian astronomy without D. Travis Hudson, who, with Ernest Underhay, integrated data from John Peabody Harrington's unpublished ethnographic notes on the Chumash. Sadly, Travis Hudson took his own life on 6 July 1985. Finally, Horst Hartung, well known as an architect, as an expert on the planning of Maya ceremonial centers, and as a principal collaborator with Anthony Aveni on Mesoamerican projects, died on 18 July 1990.

For those who would like to track down the details of some of the developments I have outlined here, along with other research for which I had no space to mention, here are some highlights added in the last decade to the shelf on ancient astronomy.

Abhayankar, K. D. and B. G. Sidharth *Treasures of Ancient Indian Astronomy*. Delhi: Ajanta Publications, 1993.

Aveni, Anthony F. *Conversing with the Planets*. New York: Times Books, 1992.

———., *Empires of Time*. New York: Basic Books, 1989.

———., ed. *The Lines of Nazca*. Philadelphia: The American Philosophical Society, 1990.

———., ed. *New Directions in American Archaeoastronomy*. Oxford: B.A.R., 1988.

———., ed. *The Sky in Mayan Literature*. New York: Oxford University Press, 1992.

————., ed. *World Archaeoastronomy*. Cambridge: Cambridge University Press, 1989.

Aveni, Anthony F., and Gordon Brotherston. *Calendars in Mesoamerica and Peru and Native American Computations of Time*. Oxford: B.A.R., 1983.

Benson, Arlene, and Tom Hoskinson, eds. *Earth and Sky: Papers from the Northridge Conference on Archaeoastronomy*. Thousand Oaks, Calif.: Slo'w Press, 1985.

Broda, Johanna, Stanislaw Iwaniszewski, and Lucrecia Maupomé, eds. *Arqueoastronomía y Etnoastronomía en Mesoamérica*. Mexico City: Universidad Nacional Autónoma de México, 1991.

Burl, Aubrey. *Prehistoric Astronomy and Ritual*. Aylesbury, U.K.: Shire Publications, 1983.

Carlson, John B. "America's Ancient Skywatchers." *National Geographic Magazine* (March 1990): 76-107.

————., *Venus-regulated Warfare and Ritual Sacrifice in Mesoamerica: Teotihuacán and the Cacaxtla "Star Wars" Connection*. College Park, Md.: Center for Archaeoastronomy, 1991.

Carlson, John B., and W. James Judge. *Astronomy and Ceremony in the Prehistoric Southwest*. Albuquerque: Maxwell Museum of Anthropology, 1987.

Crocker, John Christopher. *Vital Souls*. Tucson: The University of Arizona Press, 1985.

Fabian, Stephen Michael. *Space-Time of the Bororo of Brazil*. Gainesville: University Press of Florida, 1992.

Farrer, Claire R. *Living Life's Circle*. Albuquerque: University of New Mexico Press, 1991.

Fideler, David. *Jesus Christ, Sun of God*. Wheaton, Ill.: Quest Books, 1993.

Griffin-Pierce, Trudy. *Earth Is My Mother, Sky Is My Father*. Albuquerque: University of New Mexico Press, 1992.

Hadingham, Evan. *Early Man and the Cosmos*. New York: Walker & Company, 1984.

Iwaniszewski, Stanislaw, ed. *Readings in Archaeoastronomy*. Warsaw: State Archaeological Museum and Department of Historical Anthropology, Institute of Archaeology, Warsaw University, 1992.

Krupp, E. C. *Beyond the Blue Horizon: Myths and Legends of the Sun, Moon, Stars, and Planets*. New York: HarperCollins, 1990; Oxford University Press, 1991.

Malville, J. McKim, and Claudia Putnam. *Prehistoric Astronomy in the Southwest*. Boulder, Colo.: Johnson Publishing Company, 1989.

McCoy, Ron. *Archaeoastronomy: Skywatching in the Native American Southwest*. Flagstaff: The Museum of Northern Arizona Press, 1992.

Meyer, Jeffrey F. *The Dragons of Tiananmen: Beijing as a Sacred City*. Columbia: University of South Carolina Press, 1991.

O'Neil, W. M. *Early Astronomy from Babylon to Copernicus*. Sydney, Australia: Sydney University Press, 1986

Roe, Peter G. *The Cosmic Zygote*. New Brunswick, N. J.: Rutgers University Press, 1982.

Ruggles, C. L. N., ed. *Archaeoastronomy in the 1990s*. Loughborough, U.K.: Group D Publications, 1993.

————., ed. *Astronomy and Cultures*. Niwot: University Press of Colorado, 1993.

————., ed. *Records in Stone: Papers in Memory of Alexander Thom*. Cambridge: Cambridge University Press, 1988.

Schiffman, Robert A., ed. *Visions of the Sky: Archaeological and Ethnological Studies of California Indian Astronomy*. Coyote Press Archives of California Prehistory No. 16. Salinas, Calif.: Coyote Press, 1988.

Swarup, G., A. K. Bag, and K. S. Shukla. *History of Oriental Astronomy*. Cambridge: Cambridge University Press, 1987.

Ulansey, David. *The Origins of the Mithraid Mysteries*. New York: Oxford University Press, 1989.

Williamson, Ray A. *Living the Sky: the Cosmos of the American Indian*. Boston: Houghton Mifflin Company, 1984.

Williamson, Ray A. and Claire R. Farrer. *Earth & Sky: Visions of the Cosmos in Native American Folklore*. Albuquerque: University of New Mexico Press, 1992.

Worthen, Thomas D. *The Myth of Replacement: Stars, Gods, and Order in the Universe*. Tucson: The University of Arizona Press, 1991.

Ziólkowski, Mariusz S., and Robert M. Sadowski, eds. *Time and Calendars in the Inca Empire*. Oxford: B.A.R., 1989.

Contents

Acknowledgments

Those who in some way helped pull this book together seem as numerous as the stars in the night sky. Those whose assistance, influence, or advice propelled the manuscript forward comprise a catalog of first magnitude friends. They deserve credit for the book's successes, but for its shortcomings, no blame.

Dr. Anthony F. Aveni, Von Del Chamberlain, Dr. Ing. Horst Hartung, and Dr. Johannes Wilbert read parts or all of the preliminary text. Their comprehensive and freely given suggestions saved me—it is certain—from embarrassment and error.

The scholars mentioned already and many others as well have provided me with encouragement and information through their published work, through illuminating correspondence, and through many engaging conversations. Professor Aubrey Burl, Dr. John Eddy, Professor Owen Gingerich, Professor Ronald Hicks, Professor H. B. Nicholson, Dr. Gerardo Reichel-Dolmatoff, Professor Edward H. Schafer, Professor Alexander Thom, and Dr. A. S. Thom are among those who have shared their experience and ideas with me. I would never have been able to visit some of the more obscure archaeological sites with astronomical components had I not secured the direction and help many of these specialists provided. Tony Aveni taught me how to measure alignments, and both he and Horst Hartung helped me push to new limits the frontiers of where you can physically take a chartered bus in Mexico.

In the same breath, I want to acknowledge the enthusiasm and invitations of my colleagues in California who, as archaeologists or rock art specialists, let me join them in their on-site investigations. They include Arlene Benson, Bob Cooper, Ken Hedges, Dr. Tom Hoskinson, Dr. Travis Hudson, Vernon Hunter, Dan Larson, John Rafter, John Romani, Gwen Romani, Larry Spanne, and the late Beverly Trupe.

Through the expertise and commitment of Mr. Chang Ze min, our National Guide in the People's Republic of China, and the other representatives of China International Travel Service, our efforts to reach important archaeological and astronomical sites were successful. Also, I am grateful to several scholars in China whose interest in ancient Chinese astronomy enhanced my own. Wang De-chang and the other members of the Historical Research Group at Purple Mountain Observatory, Dr. Xia Nai, Director of the Institute of Archaeology in Beijing, and Mr. Bian Depei, editor of *Amateur Astronomer* magazine and affiliated with the Beijing Planetarium, have all contributed to my knowledge of China's extensive astronomical tradition.

Dr. Ray Williamson, the organizer of the "Archaeoastronomy in the Americas" conference held in Santa Fe in 1979, saw to it that I was able to get to that watershed meeting. For that and for his willingness to share his ideas and work I thank him.

I owe a debt to Dr. John Carlson, Director of the Center for Archaeoastronomy, whose effective work keeps us all informed of what is going on in the field. Dr. Carlson has been a steady source of information, support, and friendship.

With uncompromising persistence, Mr. Roger Bingham has forced me to verbalize and then clarify some of the fundamental themes in this book, and while doing so he has shared his own unique insights into these matters.

This book cuts across the boundaries between academic disciplines, and I am certain that my undergraduate experience at Pomona College, where a liberal arts tradition fosters an appreciation for the connections between the sciences and the humanities, is responsible for making me believe the interdisciplinary path could be followed. Dr. George O. Abell, my advisor in graduate school at U.C.L.A., is keenly aware of the human side of the scientific enterprise. Even though his own research interests do not extend to archaeoastronomy, he has continuously encouraged my efforts in this area as well as in public education. I have enjoyed his humor and perspective since high school, and any science I do reflects his influence.

U.C.L.A. Extension's Department of the Sciences has permitted me to organize field study tours on a regular basis to some very unusual—and sometimes unlikely—places. Dr. Robert Barrett, in fact, was personally responsible for prompting me to go to the People's Republic of China, and Dr. Eve Haberfield has often done the hard work—far from the comfort of home and fire—to get us wherever we were going and back again as well.

I am indebted to the management of the Department of Recreation and Parks, City of Los Angeles, for permitting me to use illustrations from Griffith Observatory's extensive collection of diagrams, plans, and planetarium artwork, much of which has also appeared in the Observatory's monthly magazine, the *Griffith Observer*. Credit for most of these illustrations goes to Griffith Observatory employees: Joseph Bieniasz, Lois Cohen, Helen Jorjorian, and George Ewell Peirson.

The generosity of the Friends of the Observatory and its president, Mrs. Debra Griffith, helps make Observatory exhibits and programs possible—including those on archaeoastronomy. Two of our trips to Mexico were cosponsored by the Friends.

The text was typed in part by Edward Helms and in large measure by Susan Coleman. For their speed and accuracy, I am thankful, especially when I recall the quality of my handwriting, itself a demonstration of the forces of chaos.

Hugh Rawson's original decision to secure this book for the publisher was, of course, indispensable. His expert advice, as an editor and writer of many years' standing, helped tighten up the book's original concept and refine portions of the text at a critical early stage. Lawrence Peel Ashmead, Executive Editor at Harper & Row, took on the project from there and shepherded it with genuine interest and care. It was he who had contracted for my first book, and I sincerely appreciate his continuing concern for my work. Finally, Craig D. Nelson imposed a salutary editorial discipline on the book. He scrutinized it in detail, bringing his influence and sound critical judgment to the project. He is, in large measure, responsible for the satisfying shape in which the book now finds itself. I count myself fortunate to have enjoyed the dividends of his dedication and professionalism.

Jane Jordan Browne, of Multimedia Product Development, Inc., demonstrated her loyalty once again by making sure that no obstacle—even any that I created—kept this book from making its way through the uncertain realm of publishing. Her sharp editorial eye examined the original proposal, the manuscript itself, and the revisions, and when she blinked, the cold beam of common sense and plain talk fell upon the words I had written. She is a vital partner in these efforts. There wouldn't be a book here without her.

The photographic and artistic contributions of Robin Rector Krupp, my wife, are evident throughout this book, but I am even more beholden to her for defending the barricades to my library for more than a year, filtering out even family and close friends, and letting me work. This book is in part about social cohesion, and it is ironic that it had to be written while shunning the company of others. My parents, Florence and Edwin Krupp, and my in-laws, Margaret and Robert Rector, are among those others. Their patience, understanding, and approval have kept the whole process from souring.

My son, Ethan, has come to believe it is normal to spend every weeknight and weekend writing. Although I am grateful for this perspective of his, I shall try to modify this vision of reality in the months to come.

ECHOES
of the
ANCIENT
SKIES

I

The Lights We See

The way people look at the universe has a lot to do with how they behave. And the sky is what used to tell us about the big picture—about what really makes the world the way it is.

In this age of urbanization and artificial light, it is difficult to appreciate the paramount importance of the sky to our ancestors. Digital watches and desk calendars are readily available; there is no need to watch the sky to tell the time of day or the year. And under the lights of our cities, we can scarcely see anything overhead—the night is diluted. Most of the stars are fainter than the background of scattered light. For city dwellers, the night sky is preserved only under the dome of the local planetarium. We have struggled—successfully—to shelter ourselves from the elements, and we have managed to shut out the sky. In the process, we also have removed ourselves from one of the fundamental components of our culture.

For most of the history of humankind, going back to stone age times, the sky has served as a tool. Just as the hands of the first people grasped the flints they crafted, so their brains grasped the sky. The regularity of the motions of celestial objects enabled them to orient themselves in time and space. And just as their culture was partly a product of the tools they made with their hands—axes and arrowheads, needles and spear-throwers—so it was also shaped by their perceptions of the sky. From the sky they gained—and we, their descendants, have inherited—a profound sense of cyclic time, of order and symmetry, and of the predictability of nature. In this awareness lie not only the foundations of science but of our view of the universe and our place in it.

The sky was a very practical tool: It helped people survive. We are so used to the concept of time—so oppressed by time—that we take it for granted. It seems as straightforward as the calendar on the wall. There, before us, is an array of days to come and days just past. Mentally, we place ourselves somewhere among the orderly sequence of numbered days. By doing so, we can plan the future and evaluate the past. This consciousness of time permits complicated undertakings. We can interrupt the pattern of our personal lives and engage in planned, joint enterprises. Organized, cooperative groups have an evolutionary advantage, and the essence of social cohesion—effective human interaction—demands the invention of a common system of reference. Timekeeping and the calendar depend on reliable, repetitive celestial cycles for meaning and measure.

Our sense of location—of the organization of the landscape—also has helped us survive, and it, too, depends upon the sky. Directions on land derive their meaning from celestial phenomena—from the steadiness of the pole star and from the regular changes in the point at which the sun rises along the horizon. In this we are not so different from our fellow creatures. Honeybees, we know from the work of Austrian Nobel prizewinner Karl von Frisch, use the position of the sun and its polarized ultraviolet light to find their way from hive to flower and from flower to hive. Pigeons depend on the sun and their own internal clocks to find their way back to roost (magnetic particles in the tissue of their brains, in tune with the earth's magnetic field, are part of a backup navigation system). Were our habitat restricted to the ground, and were our eyes ant-high, the pattern of the trees above us might satisfy our need for references, as it seems to do for foraging ants who manage to find their way back to their nest through a maze of obstacles. But we wander the earth, and it is the sky that engages our brains.

For our ancestors, what went on in the sky was metaphor. It meant something. It was both the symbol of the principles that they felt ordered their lives and the force behind those principles. There was power in the sky. The tides resonated with the phases of the moon; the seasons fell into place in concert with the sun and stars; the world and its inhabitants followed the seasons. Modern, urbanized peoples have lost this sense of coherence between what goes on in the sky and in their lives, but some traditional peoples still have it. The Desana Indians of Colombia even describe the sky as a brain, its two hemispheres divided by the Milky Way. Their brains, they say, are in resonance with the sky. This integrates them into the world and gives them a sense of their role in the cosmos.

The perception of the Desana was common to many ancient peoples. They also sensed that the sky orders our psyches and our societies, and they expressed the bond between brain and sky in their works: in calendars and clocks; in star maps and almanacs; in gods and in myths; in ceremony, costume, and dance; in temples and in tombs. Sometimes they symbolized this bond on the ceiling, sometimes on the floor. They embedded it in the layout of their cities. They incorporated it on playing boards for games. They carved it on boundary stones that commemorated a royal grant of land to a loyal subject. They wove it into the protocol of kingship and social organization. Some used the sky to assess the state of the world. Others looked up into the darkness to prognosticate the future.

Our place in the universe can be known only by knowing the universe. Its structure, its creation, and its ultimate fate are deduced from the clues overhead. Ancient astronomers at genuine observatories kept vigils with the night and looked for meaning and understanding. Today their modern counterparts continue the same quest, and this old tradition of skywatching still gives us perspective, still tells us what and where and when in the cosmos we are. We perceived order in the sky and stitched it to earth. But this should not really surprise us. After all, the sky also is the mirror of our mind's own eye.

Looking Through the Eyes of Our Ancestors

Most of us have lost touch with the sky, but we are reminded of our old heritage, now and then, when the colors of sunset recapture us and we stop and watch the last gleam of sun slip behind the dark silhouette of a distant horizon. Alarm clocks awake us now instead of the morning songs of the birds, but it is still possible to experience the sense of renewal our ancestors found in the dawn. All we have to do is rise before sunup and wait for the first warm beams of light to spill over the landscape. By traveling outside our cities we can see the same stars people have watched for at least tens of thousands of years. Few of us have jobs and life-styles that permit us to live with the sky, as our ancestors did, but even a glimpse lets us feel what they felt.

Thousands of stars powder the sky. Some that are especially bright draw our attention, and the even brighter planets seem to stand apart from the many other stars around them. The smoky trail of the Milky Way bridges the sky from one side to the other, like the white ghost of a vast rainbow. The night sky is rich, beautiful, and mysterious.

To really know the sky, however, you have to keep watching it. A glance won't take it in, and it does change—from hour to hour, day to day, month to month, year to year, and in even longer cycles of appalling spans of time. If we take the trouble to notice them, the shorter cycles can be sensed just as our ancestors sensed them.

From one simple cycle, the earth's daily rotation, time is metered and directions are set. The cycle begins in the morning when the sun rises. By this we mean it crosses the horizon, that boundary between the earth spread out around us and the sky stretched out above. The original meaning for *horizon* is "boundary" or "limit," and our sense of territory, or bounded space, may owe something to the perception that the earth ends at the "edge" of the sky. As far as our eyes are concerned it does not matter if we know the earth to be round or if we think it flat. At any particular spot we are surrounded by the rim of the horizon.

Eventually the sun returns to the horizon—once it disappears the sky grows darker, and within an hour stars begin to shine overhead. By watching them closely, we see

The sun's first gleam upon an irregular desert horizon in California renews the sky's daily life. *(Robin Rector Krupp)*

that most of them do the same thing the sun did. They rise, they pass over the world, and they set. Those that were near the position of the setting sun follow it below the horizon in the early evening. Others, just rising when the sun went down, may be up the entire night. All the stars still in the sky at dawn vanish in the twilight as the sun brings day back to the world.

This reliable pattern of day and night is the sky's first cycle in the passage of time. Day and night are apportioned by the journey of the sun and stars across the sky. It is a parade animated by the spinning earth. We stand on the earth but we don't sense its motion. Instead we see it reflected in the sky. Our planet rotates from west to east, and to us, it looks like the pageant is rolling from east to west.

East is the realm of risings. Settings occur in the west. These directions acquire meaning because of the celestial events that define them, and these events have, in turn, symbolic meaning of their own. When we speak of east in a general sense—and not *due east*—we refer to the half of the horizon where celestial objects can appear. They are, in a sense, "born" there, and we associate birth, creation, and life with the east. East in Latin is *orient,* a word which derives from the verb "to rise." West, on the other hand, is *occident,* and similarly related to the verb "to fall." Ancient peoples equated settings of the sun and the other celestial objects to their "deaths," and we still speak of "sunset years" as a metaphor for old age. For many cultures the west was the land of the dead, and in World War I a soldier killed in action was said to have "gone west." The widely read novel *The Lord of the Rings* concludes as the two main characters, Frodo and Bilbo, in old age and at the end of an era, depart their homelands for "the West."

Some stars are placed so that they never descend beneath the horizon; throughout the day and night, they follow circular courses having a common center, a spot that never moves. In our era, in the northern hemisphere, an almost motionless star in Ursa Minor nearly occupies that spot. It is Polaris. The name refers to the north celestial pole, the center of the circular paths followed by the stars that never set. Just as the earth spins around its "pole," the sky appears to turn around this unique spot, and the stars that complete circles around it are called circumpolar stars. If we face

When we face due north in the northern hemisphere, we see the stars near the north celestial pole trace circular paths around it. The pole itself is stationary and therefore special, and it makes the direction we must face special, too. A dashed north-south line seems to step up to greet us, and the dashed arc of the meridian stretches up the sky from cardinal north, through the north celestial pole, and on across the zenith overhead. If we turn completely around, we face south and see the meridian arch down to meet the cardinal line on the ground at the south point. The paths of stars in this direction nearly cross the sky horizontally, in contrast with the circular trails of the stars in the north.   *(Robin Rector Krupp)*

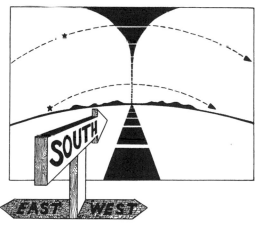

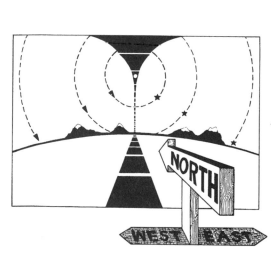

the north celestial pole, the stars turn counterclockwise around it, but below the earth's equator, in the southern hemisphere, we see stars moving clockwise in rings around a similar spot, the south celestial pole. No bright object points it out, but the daily movement of the stars would make it noticed.

Pole, in the sense we have used it here, derives from the word for "stake," and the concept behind the word is a pole that reaches to the canopy of the sky, supports it, and acts as the pivot of the sky's daily rotation. It is a cosmic axis and is described in the mythologies of various peoples as a mountain, as an actual pole, as a tree, or as some other sky-piercing staff. In any event, the pole of the sky is a special place, a motionless reference in a moving sky.

By following an imaginary line from the steady beacon of Polaris straight down to the horizon, we locate the direction north. It is because Polaris defines this direction for us in the northern hemisphere that it is also called the North Star. And once we've found north, the other three cardinal directions, south, east, and west, are automatically defined. Between the cardinal compass points are the intercardinal directions, northeast, northwest, southeast, and southwest, in the center of each quarter arc of horizon.

Seeing Seasons in the Sun and Stars

Time slides through the seasons, and this too shows up in the sky. This cycle is a long one. It takes a year, the length of time in which the earth orbits the sun. Again, we don't sense the motion directly but follow it by observing daily changes in the positions of the sun and the stars until, after a year's time, they return to their original starting places.

During each circuit of the sun, the earth spins around about 365¼ times. There are, therefore, 365¼ days in a year, and shifts in the sun's daily path measure out the annual cycle. The measurement is made this way: From any place we care to stand, we can notice the direction in which sunrise occurs. If we are in the northern hemisphere, once every year, in winter, the sun comes up far to the southeast, as far south of east as we will ever see it from our chosen sunwatching station. Although the sunrise seems to reappear in the same spot on several successive mornings, it gradually edges north until—half a year later, in summer—it occurs far to the northeast, at its northern limit. For several days, the point of sunrise lingers there, but it eventually reverses its movements again and returns toward the south, parceling out the second half of the year until the sunrise is back at its southern limit and winter returns. This annual seasonal cycle, which is mirrored by the moving point of sunset along the western half of the horizon, then begins once more. Thus, by the passage of the sun, time is regulated in an orderly sequence of days and years.

When the sunrise occurs at its southern extreme, in the southeast quarter of the horizon, we say it is the day of the winter solstice. Both the event and the date bear

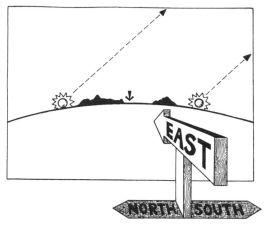

East is the direction of risings, and the two paths shown here belong to the sun at summer (left) and winter (right) solstice. At this location, the sun never appears to rise farther to the north (left) or farther to the south (right) than the places here where it first appears above the horizon. The sun rises midway between these two extremes, due east, at the equinoxes. This annual pattern of sunrise is mirrored in the west by the sunsets. Again, the extremes of the sun's annual motion are indicated by the two descending paths of winter (left and south) and summer (right and north) solstice. Equinox sunsets take place due west. *(Robin Rector Krupp)*

this name, and solstice literally means "sun stand still," an acknowledgment that sunrise (and sunset) dawdles for a while before reversing its course along the eastern horizon. Similarly, the northern limit is reached on the summer solstice. Although these limits of sunrise occur in the general direction of northeast and southeast, they do not necessarily coincide with those intercardinal directions. The exact positions of the limits depend upon latitude. The farther you travel from the earth's equator, the farther north and south the solstices occur.

Midway between its northern and southern extremes, the sun rises due east and sets due west. This occurs twice a year, once when the sunrise is headed north and again six months later when the sun returns south. Both events are called the equinox—in spring the vernal equinox and in fall the autumnal equinox.

Equinox means "equal night," and on either equinox, the duration in hours, say, of daylight is equal to the length of night. At the winter solstice, the nights are long and the days are short. By contrast, short nights and long days prevail in summer. The sun is up longer in summer because it follows a longer path that arcs high through the sky. The winter sun passes low over the southern sky. Its path is short, and the daylight hours are few. At the equinoxes the sun's course falls halfway between these two extremes, higher than the winter solstice sun and lower than the summer.

It is the height of the sun's course in the sky, not the distance of the earth from the sun, that determines whether it is winter or summer. In fact, it is in January, during winter in the northern hemisphere, that the earth comes closest to the sun (about 3 percent closer than in July, when it is farthest away). Seasonal changes in temperature result, instead, from variations in how directly the earth is heated. This depends, in turn, on the angle at which sunlight strikes our spot on the earth. A low winter sun means the sunlight arrives from a low angle, hits the earth at a slant, and spreads widely over it. Heating is less efficient, and the weather is colder. In summer the high path of the sun provides more direct sunlight and more intense heating.

The height of the sun's path varies from season to season because the axis of the earth's rotation is tilted in space. On one side of the earth's orbit, the sun shines more directly on the northern hemisphere, and the sun appears to rise and set in the north. On the orbit's other side, the tilted earth now exposes its southern hemisphere to the most direct sunlight. Below the equator it is summer, but in the north the sun appears to rise and set in the south. Winter is in the air. All of the seasonal changes we experience are products of this simple arrangement of the earth in space.

If we continue looking at the sky with our ancestors' eyes, we shall see that there are also gradual, nightly changes in the stars as well as in the daytime path of the sun across the sky. At first glance, the night sky can be bewildering to anyone who is unfamiliar with its geography. It seems as though the stars are innumerable, but this is not so. About 8,000 stars can be seen with the unaided eye over the entire sky, both the half we see overhead and the half obstructed by the earth below our feet. At any moment, this number is greatly reduced, of course, because half the sky can't be seen at all and because stars low in the sky are obscured by the atmosphere. Perhaps as many as 2,500 can be seen at one time under the best conditions. This is still quite a few, but a little practice soon makes them familiar. They are not all the same, and they are not strewn uniformly upon the sky. The brighter ones are like landmarks, and in combination with fainter stars around them they seem to form distinctive arrangements and shapes. These patterns, or "pictures," are called constellations.

Some constellations, like the Big Dipper and Orion, are well known, and a few— the Big Dipper, for example—even look like what they are named. Orion, by contrast, looks more like an hourglass, but it is supposed to represent a hunter. A bright red star, Betelgeuse, at the upper left corner of the hourglass, marks the hunter's right shoulder. A diagonal to the lower right reaches Rigel, at his left knee. At the waist of the hourglass three stars in a line, closely and evenly spaced, and of comparable brightness, form the famed belt of Orion. With fainter light, another three celestial lamps seem to hang in a line from the belt and provide Orion his sword. To the right of Orion are the stars of Taurus, the bull, and Aldebaran, the bull's red eye, is nearly in line with the belt. Almost on the same line on the hunter's other flank is Sirius, the brightest star of the sky.

The stars of Orion and their neighbors are conspicuous occupants of the winter sky because, at this point in the earth's circuit around the sun, our planet's night side faces their direction. As the earth proceeds further along in its orbit, however, Orion and its neighbors will appear higher and higher in the sky each night at sunset. By spring, Orion will be high overhead at sunset, and it will set halfway through the night. Other stars then appear at the horizon as the sun goes down, and they rule the sky by remaining up until dawn. Leo, the lion, is among the constellations of spring, but after a while, he, too, is replaced by the stars of summer, when the earth's night side faces directly away from Orion.

Vega, Deneb, and Altair, three of the brightest stars, are set like jewels at the corners of a triangle that spans the width of the Milky Way. They pass overhead throughout the summer night, and farther to the south, Antares, another bright star nearly as red as the planet Mars, burns in the heart of Scorpius, the scorpion. The scorpion's claws stretch into the constellation of Libra, the scales, to the west of the scorpion's head. Its body curls and hooks back to the east and ends in a starry stinger. Scorpius looks a bit like a scorpion, and just as winter is heralded by Orion, Scorpius signals the summer nights.

Summer slips by as well, however, and when the fall comes other stars take their places on the celestial stage. Pegasus, the winged horse, looks more like a square—a star at each corner—than a flying horse, but it is a distinctive pattern in the autumn sky. Perseus, the hero, follows soon behind, and through the fall the entire cast in his legend—Cassiopeia, the queen, Cepheus, the king, Andromeda, the chained lady, and Cetus, the sea monster—all take their celestial bows.

In a few short months it is winter again. We know, because Orion is rising in the east as the night begins. In a steady, seasonal cycle the stars tell us where in the year we are. At the same time, of course, the sun seems to move against the background stars. A different set of stars is behind it when it rises, and so, although we can't actually see the stars in sunlight, the changes we notice before sunrise or after sunset tell us the sun's position among the stars is shifting.

Arching across the sky during all seasons of the year is the Milky Way. On any single night, only part of the Milky Way can be seen. What is visible varies with the latitude and season. From most locations on earth, at most times of the year, it bridges heaven in a direction slightly skewed to the east-west rotation of the sky. The Milky Way is made up of stars, of course, but the stars that comprise it are too distant to see individually. They are far more numerous than the relatively nearby stars that we organize into constellations. All of the stars, including those of the Milky Way, are in the same huge galaxy of stars, gas, and dust. Most of the Milky Way galaxy is flattened into a disk. Bright arms of luminous objects and interstellar material spiral out from its center. We can't really see the whirling pattern, however, for we ourselves are in the disk, near the rim, in one of the outer turns of a spiral arm. Inner coils of the arms block our view of the distant center, where most of the Galaxy is really concentrated in a huge bulge of stars. All around us, however, intervening stars in the Galaxy's disk blend into a river of pale light. It circles entirely around the sky and is our inside, edge-on view of our own Galaxy.

Our name for this ribbon of light is clear enough. It looks like a path or trail, and it is, of course, milky white. The Greeks sometimes said it spilled across the sky when Hercules, as a baby, sucked too vigorously at Hera's breast. Others have called it a river, a serpent, a chain linked to heaven, or the path of the dead.

Precession's Celestial Promenades

For all practical purposes, the configurations of the stars and the form of the Milky Way do not change. Tens of thousands of years will have to pass before the Big Dipper looks like something else.

The stars very slowly shift the seasons of their appearances, however. In 13,000 years or so, Scorpius will herald winter, and Orion will rule the summer sky. These gradual changes reflect another celestial cycle—precession.

Precession is the product of gravity. The sun and the moon, pulling on the earth's equatorial bulge, cause our planet to wobble. This means that each time the earth returns to a particular place in its orbit around the sun—to the spot it occupies at the vernal equinox, say—its axis is pointing at a slightly different place in the sky. In fact, from one moment to the next, the axis is adjusting, but the effect is very small and is only measurable cumulatively, after several years. For the ancients it took centuries to see the shift. The pole, then, slips around a circle through the northern stars. Sometimes one star is the pole star and sometimes another; during some centuries there is no pole star at all.

The wobble of the earth is a bit like that of a top, but what takes only a moment for a top is a grand and stately cycle when mirrored in the apparent movements of the constellations. The entire cycle takes nearly 26,000 years. During this period the sun occupies different constellations—at an equinox or solstice, say—depending upon where in the cycle we are. For example, in our era, Pisces, the fishes, is the home of the vernal equinox sun. It began to appear there at that time of year about 2,000 years ago, when Aries, the ram, was abandoned. Before Aries, the vernal equinox occurred in Taurus. Still centuries ahead of us is the so-called Age of Aquarius, when the vernal equinox sun will shine among the stars of the water bearer. In 26,000 years, then, all twelve constellations of the zodiac, the ring of constellations through which the sun passes in a single year, stand their turns at the vernal equinox.

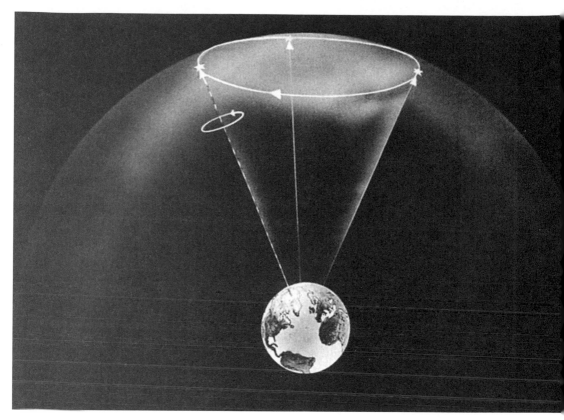

The spinning earth experiences an extra gravitational tug by the sun and moon upon its equatorial bulge. This force acts to pull the earth's axis upright, but because the planet is spinning, the axis swivels. This motion is called precession, and it takes about 26,000 years to complete one cycle. During this time the north celestial pole shifts its position through the background stars of the northern sky, and the constellations slip from one season to the next, until they mark the same times of the year they did at the start of the precessional cycle. *(Griffith Observatory)*

Wanderers Through the Stars

A few of the "stars" don't behave like stars. They move among them. Constellations are fixed patterns, but the planets, or "wanderers," add variety to the unchanging pictures in the sky. The ancients knew just five: Mercury, Venus, Mars, Jupiter, and Saturn. Seen from earth, they are luminous nomads in a wilderness of stars. Since the invention of the telescope, three more planets have been discovered, and rocket-powered space probes have given us closeup looks at several. Of the five original wanderers, Mercury and Venus have orbits inside that of the earth, while the others are more distant from the sun. All, however, stay close to the path the sun follows through the stars.

From the vantage of the unaided eye, the five wanderers are starlike objects that differ from each other in brightness and in the periods and patterns of their cycles through the sky. Venus is the brightest of them, so bright, in fact, it can sometimes be seen in the full daylight if one knows where to look. Mercury is the faintest, elusive enough to evade the notice of most people.

The interior orbits of Venus and Mercury give them a distinctive sequence of appearances and disappearances. Venus, for example, may be aligned with the sun in our line of sight but on the near side of its orbit. We cannot see it in the sun's glare, and it remains invisible for about a week. When its travels carry it far enough west of the sun, it rises before the sun and appears in the east in the morning sky. It remains a "morning star," roams farther and farther west of the sun, and rises earlier and earlier. Eventuallly, however, the planet moves back toward the dazzling sunlight, now on the side farthest from earth. After 263 days—almost nine months—as a morning star, Venus again disappears in conjuction with the sun. This conjuction is much longer, about seven weeks because Venus is farther from us. Once more it escapes the glare, but now it is east of the sun and appears in the west after sunset. Its tour as an "evening star" lasts about nine months and mirrors the morning star segement of its celestial cycle. The entire cycle takes 584 days, or about 19½ months. Mercury completes a similar circuit in much less time—116 days, or about four months. Its orbit is smaller, and from our viewpoint it strays less when it moves to either side of the sun.

Both planets share another distinctive trait: They are never seen in the west at dawn, only in the east. They are never seen in the east at sunset, only in the west. When they are morning stars they are in the sun's eastern realm. Similarly, as evening stars, they accompany the sun in the west. Neither is ever up all night. The other three bright planets experience no such restrictions on their movements.

Of the outer planets, Jupiter is the brightest. Mars is next. Saturn, though least bright, still outshines Mercury. Sometimes one of the outer planets may be in line with the sun on the far side of its orbit. Like Venus in its longer-lasting conjunction, the planet Jupiter, say, is invisible. And it too emerges as a morning star. Instead of seeming to reach the end of a tether to the sun, however, this outer planet continues on its orbital path until it is opposite the sun. At this time it rises when the sun sets, and sets when the sun and rises. It is up all night. This is something the inner planets cannot do. Jupiter (or Saturn, or Mars) then continues through the stars and eventually approaches the sun from the east. Now it is an evening star, edging ever closer to conjunction, when it vanishes once more.

The outer planets also share a distinctive behavior: Sometimes they seem to move backward in what astronomers call "retrograde motion." As we orbit the sun, we move faster than Mars, Jupiter, and Saturn. Somtimes we overtake them, and they appear, for awhile, to move east to west among the background stars, from night to night, rather than west to east, the normal direction of their "wandering" movement. Although the planets' circuits are not so direct as the sun's and involve loops of backward motion now and then, the paths they follow also pass through the zodiac.

Jupiter takes nearly 12 years to complete a journey through these stars. Saturn requires about 29½ years. Mars arrives at the same point in the zodiac after 687 days—almost two years.

Marking Time by Moonlight

Another celestial wanderer is the moon. Its cycles can be as complex as those of the planets, but paradoxically, it also acts as a handy regulator of time. The moon's reliable rhythm of phases, waxing full and waning to naught, subdivides the year for us, bundling the 365 days into 12 or so convenient packages of time, the months. The very word "month" derives from the word "moon," and both are rooted in a word that means "to measure." It is likely the moon was our first means of measuring the passage of the days. It is conspicuous—large, bright, fast-moving, quickly changing. All our calendars originate in its regular, cyclical phases.

The moon, of course, glows with no fire of its own; it shines by the reflected light of the sun. Because of this, its shape changes as it orbits the earth. When it is in the same direction as the sun, its dark side faces us, and we see no moon at all; this is called "new moon." In a day or so, after the moon moves east of the sun, a thin crescent shows up in the west after sunset; from our vantage point, it is the right-hand edge that is lit. Each day the moon moves quite a distance through the background stars and fills out as it goes. About a week after new moon, it is half lit; we call this "quarter moon," for the moon is a quarter through its cycle. Now it rises at about noon and sets about midnight. Another week passes and the moon, now opposite the sun, is full. It rises with sunset and sets at sunrise. It is bright and up all night.

Then the moon begins to wane. A week after being full it is half lit again—now the left half as we face it from earth. This moon rises at about midnight and sets at about noon. After another week, the moon has slimmed to a thin crescent in the morning sky. At last visibility it rises a little before the sun, and the next day it is gone. New moon has returned after 29½ days.

The moon's pattern is dependable for the short run, but it complicates the calendar in other ways. Ideally, we would keep track of the date in terms of the sun and the seasons. One year of 365¼ days is not exactly equivalent, however, to any even multiple of months, or cycles of the moon. Trying to coordinate the moon's time with the sun's time has guaranteed employment for calendar-keepers in more than one ancient civilization.

In a single month the moonrise will oscillate between two extremes on the eastern horizon, just as the sunrise does in a year. The positions of the moonrise limits may be inside the solstitial extremes, in the same direction as the solstices, or outside them. There are restrictions on the distance the moonrise limits can reach on either side of the solstices, and what the moon will actually do in any particular month depends on

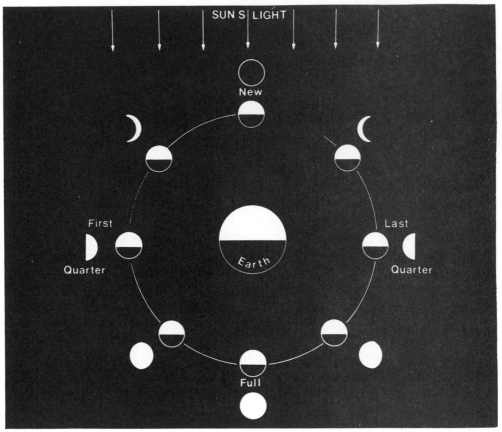

Because the moon shines only by reflected sunlight, its appearance is governed by its position with respect to the sun. As the moon orbits the earth, half of it is always lit, but the amount of sunlit disk we see from earth varies. In this diagram, the outer ring of moons, in various phases, shows what we see when the moon occupies those stations in its orbit. *(Griffith Observatory)*

the cycle by which they vary. It takes 18.6 years to go through the whole range, and although most of us are unaware of this change in the positions of extreme moonrise, simply watching the moon can make experts of us all.

Order and Chaos

Despite the regularity in the sky, unusual things happen there as well. A comet may arrive unexpectedly and linger a few weeks in the nighttime sky. Two or more planets may appear together in temporary conjunction and then gradually depart each other's company as they continue their respective courses. From time to time the moon crosses in front of a planet or a brighter star and occults it from view for an hour or so. The full moon itself may be obscured by passing through the earth's shadow. It first turns an eerie coppery red as it is cut by the dark disk projected upon its face. Even more disturbing, the moon can eclipse the sun, darkening the landscape in broad daylight and infusing the sky with an atmosphere of strangeness and uncertainty. Most of these infrequent celestial events were long regarded as calamities or, at least, as unsettling omens. All are departures from the normal behavior of the sky, and in that resides their impact. They intrude upon the regular celestial patterns and seem to threaten the cosmic order.

Conjunctions, occultations, and eclipses can be predicted, but it was difficult in ancient times to do so. Even though the old astronomer-priests might have sensed that even these events follow their own celestial rhythms, sorting out the cycles was a real challenge. From written records in Mesopotamia and China we know our ancestors were sometimes successful and sometimes taken by surprise. They predicted the next appearance of the first crescent moon and extrapolated the position of Venus from its past behavior. Now and then, something totally unpredictable might happen, however. A supernova—the appearance of a bright star where none had been seen before—could make maps of the sky and tables of the stars obsolete. Dark spots might mar the sun's shining face. Records of both phenomena are included among ancient Chinese observations.

Although chaos can intrude in the guise of a comet or an eclipse, the threat always passes. Order is restored. And much of what normally takes place in the sky is dependable and assists the brain in organizing its perceptions of the world. We reason that order must be an integral aspect of the universe. This is really an assumption.

Where would we be without the sky? Probably our brains would have sought symmetry and order and cyclical time in other phenomena—rock crystals, certain flowers, perhaps, and maybe tides, were there an unseen moon to move them. We have to work hard, however, to identify other repositories of cosmic order. The sky,

When the moon lines up exactly between the earth and the sun, the moon's disk totally eclipses the sun. Although the sun's bright disk is obscured, the outer halo of gases—or corona—provides a spectacular display while day is turned into night. *(Ivan Dryer, total solar eclipse, February 26, 1980, from Suryapet, India)*

by contrast, is obvious. Its usefulness to our brains is shown not only by the antiquity of astronomy itself but by the penetration of celestial imagery into virtually all aspects of the ancients' lives. For example, the god Osiris meant many different things to the ancient Egyptians. His rituals and myths have woven into them several threads from the sky, and these celestial connections linked the various aspects of Osiris together. We understand him best by understanding his celestial symbolism.

First and foremost, Osiris was the ruler of the dead. Depicted as a mummy, he presided over the judgment of souls and offered new life, through resurrection, to those who earned immortality by their proper conduct in this life. Osiris shows up, therefore, in numerous funerary papyri, in tomb paintings, on coffins, and on temple walls. Inscriptions identify him with a variety of titles, and they tell us he was much more than a god of the underworld and funeral rites. He was, as well, a god of kingship and the pharaoh's vitality, a personification of the land's fertility, the spirit or force of the cycle of vegetation, a god of life-sustaining water, and, in fact, the Nile itself. His most appropriate title is "Lord of Everything." He was also the sun, the moon, and the stars of Orion.

The myth of Osiris involves his own death and resurrection, a theme that echoes the daily cycle of the sun's death at sunset and its rebirth at dawn. As ruler of the dead, Osiris was "First of the Westerners." A fertility ceremony, recorded on the walls at Dendera, 417 miles upriver from Cairo, involved Osiris and 365 lighted lamps. Certainly this refers to the annual cycle of 365¼ days and links Osiris with the solar year. His temples were said to be surrounded by 365 trees, planted in his honor.

Osiris is connected even more closely to the moon, but to see how, we have to know his myth. There is no single source for it, but numerous partial versions exist, preserved from many different eras of Egyptian civilization. Plutarch, the Greek historian and biographer, collected material from Egyptian sources in the first century after Christ and retold the story of Osiris in his treatise, *On Isis and Osiris.* Despite the late date of this composition, many traditions of considerable antiquity seem to be preserved. At least parts of Plutarch's version are verified in the hieroglyphic inscriptions that remain.

Fathered by Geb, the earth, and born of Nut, the sky, Osiris became king of Egypt and brought civilization to the land and its people. He taught the people how to plant and harvest grain, how to lay out fields and measure their boundaries, and how to irrigate the land with canals and dams. Laws, religion, and urban life were all gifts of Osiris. He was Egypt's organizer and its spirit. He made Egypt what it was. Clearly, he is connected with order, and we should not be surprised to find him associated with the sky.

Success in Egypt made a missionary out of Osiris, but when he returned from his travels to his sister and consort, Isis, he was killed by conspirators led by his brother, Set.

Set was a troublesome figure among the Egyptian gods. He was associated with the

Osiris, a god of continuity and renewal of life, is portrayed as a mummy, in the tomb of Sennedjem, to emphasize his association with resurrection. *(E. C. Krupp)*

northern stars of the Big Dipper—what the Egyptians called the constellation of Meskhetiu, the "bull's leg and haunch." As the adversary of Osiris, he stood for sterility, the desert, mindless force, and violence. In the Osirian cycle, Set is the personification of chaos.

At a banquet, Set and his co-conspirators tricked Osiris into reclining inside a box, built like a coffin, exactly to his measure. Swarming around him, they closed the lid, nailed it down, and dumped it in the Nile. According to Plutarch, Osiris was slain— suffocated in the box—on the seventeenth day of the month of Athyr, when the sun was in Scorpius, in the twenty-eighth year of his reign.

The numbers are significant. Although the moon completes its phases in 29½ days, the number 28 was used symbolically for this interval. And an intricate logic links the moon with the Nile through the number 28. At Aswan, the traditional southern limit of Egypt, a nilometer measured the height of the river's maximum rise as 28 cubits. This peak occurred in the fall, in the month of Athyr, just as the Nile was about to decline. Osiris was slain on the seventeenth day of Athyr. Full moon was counted as the month's fifteenth day, but the moon looks full the day before and the day after. By day 17, it is obvious the moon has started to wane. So, the death of Osiris marked the "death," or fall, of the Nile from its annual flood height and coincided with the day of the moon's monthly "death." Osiris, the god of the Nile and Egypt's life, was a moon god because the river, the land, and the moon wax and wane, cycle after cycle.

Other lunar symbolism also threads through the Osiris myth. His casket floated downstream, out of the Nile, into the Mediterranean. Isis followed it, found his body at Byblos, in Syria, and carried it back to Egypt. Although unable to revive him, she managed through magic to conceive a child by him. She hid his body in the thickets of the Delta marshes and nursed, in secret, the newborn son of Osiris. One night, while hunting by moonlight, Set chanced upon the body of Osiris. He tore it into 14 parts and scattered them up and down the Nile. Once again Isis set out to retrieve the body of her husband, now dismembered and dispersed throughout the land. With patience and perseverance she sought the far-flung parts of Osiris; all but his sexual member were retrieved. The symbol of his vitality remained lost in the Nile.

The 14 pieces of the body of Osiris sound like the 14 days of the waning, or "dying" moon, and on the main ceiling of the Dendera temple are inscriptions and pictorial reliefs that leave no doubt. In one panel, an eye, installed in a disk, is transported in a boat. The eye, we know, was a symbol of the sun or moon. Thoth, the ibis-headed scribe god of wisdom and knowledge, pilots the barge. Thoth was closely associated with the moon and counted the days and seasons. The text for this panel refers to the period after full moon, and 14 gods accompany the eye in the disk.

Next to the portrayal of the waning moon, another carved panel represents the 14 days of the waxing moon. A staircase with 14 steps, a god on each, leads up to the same eye and disk, and hieroglyphics verify the god's association with days of the growing moon. Osiris, it is written, is "luminous," as the god of the moon.

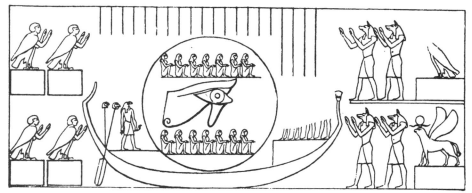

The far north panel of one of the registers on the ceiling of the Temple of Hathor at Dendera illustrates the 14 days of the waning moon as 14 seated figures that accompany the celestial eye in the lunar disk. *(E. A. W. Budge,* The Gods of the Egyptians)

On the same register that portrays the waning moon, the middle panel represents the 14 days of the waxing moon as 14 gods, each positioned upon a stairway that ascends to the disk of the moon. *(E. A. W. Budge,* The Gods of the Egyptians)

Finally, a third, adjacent panel shows Osiris in a boat with Isis and her sister Nephthys. Goddesses of the four cardinal directions support the sign of heaven, on which the boat floats, and the inscription says Osiris *is* the moon.

The myth of Osiris continues with his resurrection. Isis embalmed and mummified him, and through her help he was restored to everlasting life. In numerous tombs and temples this aspect of his myth is incorporated into one more celestial identity—the constellation Orion. Orion and Sirius, the night's counterpart to Isis, are often shown sailing in celestial boats, usually with Orion in the lead looking back at his wife, who follows him in the nightly journey from east to west. The sky is the pattern for the rest of the myth as well. Just as the dead Osiris floated out of Egypt, followed by Isis, Orion eventually leaves the nighttime sky, followed by Sirius a little later.

Osiris is identified unambiguously with the moon in the third and southernmost panel from the ceiling of the main hall at Dendera. He is joined in a celestial boat by the goddesses Isis and Nephthys, and the boat is sailing upon a symbol for the sky, itself supported by four goddesses. The accompanying texts say that Osiris has stepped into the full moon and that he is the moon. *(E. A. W. Budge,* The Gods of the Egyptians*)*

Osiris sails in his celestial boat upon the ceiling of the New Kingdom tomb of Senmut and turns his head away from his consort Isis, who follows him, as Sirius follows Orion across the southern sky. *(E. C. Krupp)*

Sirius was the key calibrator of the Egyptian calendar. The earth's orbital motion eventually puts the sun in the same direction as Sirius, and even the night's brightest star is lost in the glare of day. After disappearing from the night sky, however, Sirius eventually reappears in the dawn, before the sun comes up. The first time this occurs each year is called the star's *heliacal rising,* and on this day Sirius remains visible for only a short time before the sky gets too bright to see it. In ancient Egypt this annual reappearance of Sirius fell close to the summer solstice and coincided with the time of the Nile's inundation. Isis, as Sirius, was "the mistress of the year's beginning," for the Egyptian new year was set by this event. New Year's ceremony texts at Dendera say Isis coaxes out the Nile and causes it to swell. The metaphor is astronomical, hydraulical, and sexual, and it parallels the function of Isis in the myth. Sirius revives the

New life, in the form of a crop of grain, sprouts from the body of the dead Osiris and completes another cycle in the circuit of cosmic order. Osiris embodies the principle of rebirth and resurrection and is associated with everything that follows the pattern: the sun, the moon, the stars, the river, the plants, and the soul. He is the "Lord of Everything." *(E. A. W. Budge,* Egyptian Religion. *By permission of University Books, Inc., Secaucus, N.J. 07094.)*

Nile just as Isis revived Osiris. Her time in hiding from Set is when Sirius is gone from the night sky. She gives birth to her son Horus, as Sirius gives birth to the new year, and in texts Horus and the new year are equated. She is the vehicle for renewal of life and order. Shining for a moment, one morning in summer, she stimulates the Nile and starts the year.

In Egypt, Orion, like Sirius, is absent from the night sky for 70 days. This period is equated with the time Osiris spent in the transitional underworld. It is also equal to the stipulated period for the process of mummification. Orion's heliacal rising occurred a few weeks before the reappearance of Sirius and the Nile flood. Egyptian agriculture depended on the Nile. Without the river there would be no Egypt—only desert, the sterile domain of Set. With the summer's flood, life and water returned to the soil. Seeds were planted and nourished by the Nile. For Egyptians, the three seasons of farming and the seasons of the Nile were the same, and they named them accordingly. Each marked an important change in the landscape: Inundation, Emergence, and Low Water. There is a simple cycle here—birth, growth, death—and, with the next annual flood and new year, rebirth. This cycle is the essence of Egypt. It is paralleled by the myth. It is played out in the sky.

What we see in the lights overhead is the itinerary of cosmic order. Because it governs everything, it is reflected in the entire world. It is the core of our consciousness. It defines what is sacred and makes the sky the domain of gods.

2
The Skies We Watch

Archaeologists a thousand years from now—following a lead in a fragmentary reference recorded on an ancient electronic laser disk for home video replay—might set out in search of the lost observatory of Palomar Mountain. Their pilgrimage through the urban highrise sprawl of southern California would be rewarded, perhaps, with the discovery of the foundations of several buildings, a number of them round. In the midst of the largest of these enigmatic circular structures they might find the eroded, rusting ruins of the mount for the legendary 200-inch telescope, although the cagelike tube and huge glass mirror had long since disappeared. But for its mountaintop location and the north-south pole-oriented mounting, it would be hard to recognize the site as an ancient observatory. Guided by even limited knowledge of the astronomical techniques of their twentieth-century ancestors, however, the archaeologists of the future would stand a fair chance of identifying Palomar as the home of one of the greatest observatories of the ancient world. But without a written record it would be hard for them to fill in the details. Who observed at Palomar? What were their techniques? What precision did they seek? What did they want to know?

We of the twentieth century encounter the same problem when we try to understand the goals and procedures of the astronomers of antiquity. Ample evidence for ancient and prehistoric astronomy exists, but most of it is indirect. Celestially aligned architecture and celestially timed ceremonies tell us our ancestors watched the sky accurately and systematically, but signs of the astronomers themselves—their instruments, records, and techniques—are much harder to find.

We know that the calendar and ancient skylore were incorporated into rituals, temples, and tombs. This symbolic, applied astronomy is very helpful to us, for it immerses us into the belief systems of our ancestors. Although the religious applications of astronomy we encounter may not provide us with genuine observatories and astronomical treatises, they do give us an inkling of what was actually observed and how it may have been measured. And we can also understand quite well what cultural role astronomy played in the past. But what of the astronomers themselves? Did they see themselves as scientists rooting out nature's secrets through precise measurement and rigorous proof? Or perhaps they were exploitive theocrats who used the sky to maintain their power. Actually neither profile is really true.

No doubt the old skywatchers sought knowledge and commanded authority, even cynically at times, but in general they shared the beliefs of their times. They were participants in their societies, not outsiders. They were public servants. Their lineage leads back to the shamans, medicine men, and calendar priests who provided services their people needed. Astronomical observations were fixtures in the framework of the sacred life of the community. This is why, in fact, so many signs of the ancient astronomers are cultural applications of their professional activities. But like the shaman, who enjoys a personal relationship with the sacred, the ancient astronomers engaged in a direct experience with cosmic mystery every time they observed the sky. The quest for accuracy, precision, and understanding are not alien to this experience, but in fact a consequence of it. It is only when we equate these interests with the modern scientific method that we confuse the picture and mistakenly immerse our ancestors into our belief system. Just because we regard accurate, precise observation as a component of objective, scientific thought, we should not assume it fulfills the same function in other belief systems or that it can't exist in them at all.

Eyes on the Sky

A full understanding of the role of astronomy in culture includes, then, knowing who was responsible for making astronomical observations and how they went about their business. This tells us something about the extent of the investment in astronomy and helps us understand the motivations of the society that supports the specialist sky observer. Direct evidence of astronomical observation is difficult to verify, however. We've got to figure out how to prove that an astronomically oriented site is an observatory, how to interpret an astronomical record when we read one, and how to recognize an astronomical technique when we encounter it. This is rarely easy, but one ancient Egyptian ceremony—the "Stretching of the Cord"—echoes what must have been an actual astronomical procedure for laying out a temple.

When New Kingdom and Ptolemaic Egyptians began construction on a new temple, the foundation ceremonies included the "Stretching of the Cord." This ritual required the pharaoh to establish the basic reference line for the temple's orientation and plan. Texts and wall reliefs indicate that the pharaoh accomplished this task with the assistance of the goddess Seshat.

Seshat was associated with writing, record keeping, and, of course, the layout of

With one hand grasping a wand or stake and the other holding a mallet, both the pharaoh and the goddess Seshat participate in the "Stretching of the Cord" ceremony. A ring of cord loops both stakes at knee level in this relief on the walls of the Temple of Amun-Re at Karnak. Seshat's distinctive headdress—a seven-pointed star canopied by upturned cow horns—is supported above her head on a rod. *(Robin Rector Krupp)*

temples. Not surprisingly, she was closely affiliated with Thoth, the scribe god whose responsibilities included the calendar and the maintenance of world and social order. Seshat's emblems suggest other celestial connotations, too. Her close-fitting leopard skin is a counterpart of the leopard skin cloak of the *sem*-priest, who performed the ceremony of the Opening of the Mouth. The leopard is nocturnal. The adze with which the priest touched the mouth of the deceased pharaoh to restore life to his *ka* symbolized the Big Dipper. These connections with stars of the night sky suggest that the leopard's spotted pelt stood for the starry sky. In fact, a linen replica of such a garment—with a wooden leopard head and silver claws attached to it—was found among the treasures of Tutankhamun, and each of the leopard's "spots" is actually a star of solid gold. Maya priests dressed in the similarly spotted pelts of the jaguar, and for them, too, the pattern stood for the stars and the night sky.

Usually Seshat was portrayed with a seven-pointed star (although some have likened it to a seven-petaled flower) supported by a rod balanced upright upon her head.

Like a canopy over her star hangs what may be a pair of upturned horns of a cow or bull. This emblem was also the hieroglyph for her name. Both the horns and the seven points of the star seem to have something to do with the Big Dipper. We already know that the Bull's Thigh, or Meskhetiu, was the Big Dipper, and the Dipper contains seven bright stars. It is certain the Egyptians associated the number seven with the Big Dipper because several portrayals of Meskhetiu—at Dendera, Edfu, Esna, and Philae—surround the picture of the bull's leg with seven stars.

Also at Dendera, inside the Temple of Hathor, we read in an inscription that describes the Stretching of the Cord that the pharaoh

> ... stretches the rope in joy. With his glance toward the *ak* of the Bull's Thigh constellation, he establishes the temple of the Mistress of Dendera, as took place there before.

And the pharaoh himself describes what he is doing:

> Looking to the sky at the course of the rising stars, recognizing the *ak* of the Bull's Thigh constellation, I establish the corners of the temple of Her Majesty.

Upriver 110 miles similar inscriptions on the walls of the Temple of Horus at Edfu accompany portrayals of the pharaoh and Seshat driving stakes into the ground with mallets. A loop of rope that links both stakes represents the stretched cord. In one text the pharaoh says:

> I have grasped the stake along with the handle of the mallet. I take the measuring cord in the company of Seshat. I consider the progressive movement of the stars. My eye is fixed upon the Bull's Thigh constellation. I count off time, scrutinize the clock, and establish the corners of thy temple.

It is unlikely—despite the pharaoh's declaration—that he actually set out the foundation line for the temple. Instead, his activity in the Stretching of the Cord was ceremonial, more like the cutting of the year's first furrow by the Chinese emperor or the placement of a cornerstone at the foundation ceremonies for our own public buildings. What the pharaoh did, however, is carefully specified. This procedure required observation of a certain star at a certain time and, probably, in a certain position. Very likely his actions mimicked the actual procedure used by the Egyptian architects to lay out the reference line.

Seshat's symbol, the seven-pointed star poised upon a rod or wand, is consistent with the idea of sighting along the line to the Big Dipper established by the "Stretching of the Cord." The texts mention the *ak* of the Big Dipper, but we don't know what *ak* means. Most likely it refers to a particular position and orientation of the Big Dipper in its circular course around the pole. Even without all the details, however, the Stretching of the Cord implies a standardized technique for accurately surveying in a directed reference line. This means there was a tradition of astronomical observation to back it up.

Omens in the Sky

Bronze age China provides an entirely different kind of evidence for systematic observations of the sky. Shang dynasty oracle-bone inscriptions include references to stars and eclipses, but there is nothing scientific about these texts themselves. They are omens in which astronomical phenomena play a part in assessing the state of the kingdom:

> [On day] Kuei-Wei, divined. Cheng inquired: No ill fortune in the ensuing *hsün*? Three days later, on the evening-night of Yi-Yu, the moon was eclipsed— [so it was] heard. In the eighth moon.

The syntax here may seem a little confusing, but this is the standard form in which these divinations were recorded. The first part specifies the day, "Kuei-Wei," in the 60-day ritual cycle on which the omen was cast. An individual named Cheng asked of his ancestors the question that followed concerning the possibility of ill fortune in the subsequent ten-day interval, or *hsün*. The last statement is a comment added later that mentions a lunar eclipse occurring in the evening of the day Yi-Yu. This and similar commentaries were recorded between 1400 and 1200 B.C., and from them we can tell that eclipses were observed, reported, dated, recorded, and interpreted. Also, the calendar itself was clearly in place by the time these oracles were read. Even though we have no observatory archives to verify ongoing, systematic astronomy, the divination texts imply something like that was going on.

Recording the Sky

In ancient Mesopotamia the textual evidence for systematic observation—and even for mathematical astronomy—is more explicit, but most of the tablets are much more recent than the Shang oracle bones. With the exception of the Venus tables of Ammiza-duga, which probably originated in the seventeenth century B.C., most of the surviving Mesopotamian astronomical texts were written between 650 and 50 B.C. These clay tablets with cuneiform writing are called astronomical diaries, and they are the unmistakable observations of specialists: professional astronomer-scribes.

A typical diary entry begins with a statement on the length of the previous month. It might have been 29 or 30 days. Then, the present month's first observation—the time between sunset and moonset on the day of the first waxing crescent—is given, followed by similar information on the times between moonsets and sunrises and between moonrises and sunsets, at full moon. At the end of the month, the interval between the rising of the last waning crescent moon and sunrise is recorded.

When a lunar or solar eclipse took place, its date, time, and duration were noted along with the planets visible, the star that was culminating, and the prevailing wind at the time of the eclipse. Significant points in the various planetary cycles were all tabulated, and the dates of the solstices, equinoxes, and significant appearances of Sirius were provided.

The Babylonian astronomers used a set of 30 stars as references for celestial position, and their astronomical diaries detailed the locations of the moon and planets with respect to the stars. Reports of bad weather or unusual atmospheric phenomena—like rainbows and haloes—found their way into the diaries, too. Finally, various events of local importance (fires, thefts, and conquests), the amount of rise or fall in the river at Babylon, and the quantity of various commodities that could be purchased for one silver shekel filled out the diligent astronomer's report.

By the sixth century B.C., Neo-Babylonian astronomers were computing in advance the expected time intervals between moonrise or moonset and sunrise or sunset for various days in the months ahead. These calculations were based on systematic observations. Later, when combined with numerical tabulations of the monthly movement of the sun, the position of sun and moon at new moon, the length of daylight, half the length of night, an eclipse warning index, the rate of the moon's daily motion through the stars, and other related information, these computations enabled reasonably detailed and accurate predictions of what the moon would do and when it would do it.

Planets received similar attention, but because their movements were not uniform, the Mesopotamian astronomers had to devise mathematical techniques that would take variations in motion into account. As Jupiter, for example, makes its way through the zodiac in almost exactly 12 years, each year it more or less moves into a different zone, or constellation. Each year it also is seen in opposition to the sun—rising at sunset, setting at sunrise—but because Jupiter's motion is not uniform, it won't reach opposition on the same date each year. The Babylonians expressed this a little differently than we do and preferred to specify the position of Jupiter at each opposition, rather than the date. The effect is the same, however, and their tables show that they compensated for Jupiter's nonuniform motion by increasing its shift in position by the same amount for each opposition in one half of the 12-year cycle and by decreasing the shift by the same amount each time during the other half. When the shift of position is plotted through the successive oppositions of the planet, a zigzag line results.

Of course, the Babylonians never developed completely accurate representations of nonuniform motion, but in the later dynasties of Mesopotamia, and especially in the Seleucid period (301–164 B.C.) following the death of Alexander the Great, Babylonian (during this period called Chaldean) astronomers approximated the cyclical accelerations and decelerations of the moon and planets with the "zigzag functions." They did this numerically, not graphically, but the technique worked well enough for their purposes.

Despite the extensive written record of Babylonian astronomy, we have very little knowledge of the instruments used in ancient Mesopotamia, and we know even less about the observatories which must have existed. A clay "astrolabe" from Assyria is on display in the British Museum in London. Actually, a true astrolabe is used to measure the angular height of a celestial object, and the Assyrian devices look more like diagrams of the zones of the sky. They seem to be tables of handy astronomical information, designed to guide the astronomer in keeping time. Apart from a few limited references to an instrument used for measuring transits, the gnomon (or shadow stick), and the water clock, this is the complete inventory of our knowledge of Babylonian astronomical instruments.

It should not surprise us, however, that the astronomical instruments and observatories of ancient civilizations are hard to find. It is unlikely there were many of them, and the observatories that still exist may be hard to recognize for what they are. The actual equipment probably disappeared long ago, and the walls that housed the ancient astronomers may be all that remains today. If such observatories were incorporated into temples or palaces, they might be even harder to recognize. When we find a structure with astronomical alignments, it is not always easy to tell if the structure was used ritually or for actual observations or both.

A Circle for the Sun

In southern Illinois, Cahokia's Sun Circle is a good example of an astronomical monument whose purpose is unclear. No one even knew it existed until 1961 when Dr. Warren L. Wittry, an archaeologist now conducting site surveys on grants from the Illinois Department of Transportation, discovered a confusing collection of large, bathtub-shaped pits in the course of salvage excavation. These pits are located about 3,000 feet west of Monks Mound, Cahokia's largest earth "pyramid," and gradually Wittry realized they were arranged on several circular arcs—perhaps as many as four. The pits once contained timbers, probably tall posts, judging from the 18-inch-diameter postholes. Three of the arcs intersect, suggesting several stages of building. Although none of them has been completely excavated, by 1964 Wittry had discovered 16 postholes on the arc he arbitrarily called Circle 2. They spanned the compass from due north to the direction of winter solstice sunrise in the southeast.

Some aspects of Circle 2's design reflect considerable care. The postholes are evenly spaced in intervals about 27½ feet wide. Wittry thought an entire circle with a diameter of 410 feet had been laid out, but he was unable to excavate other portions of the site until 1977. With a grant from the Illinois Department of Conservation, 11 more postholes were found. Most matched the spacing, and all fit the diameter of the part of the circle uncovered earlier. Unfortunately, in 1961 tons of dirt were gouged from the west side of the circle for road fill before the importance of the area was recognized. Much of the rest of the circle is, therefore, lost. Wittry believes that the original ring held 47 posts and at least three of them marked significant astronomical events.

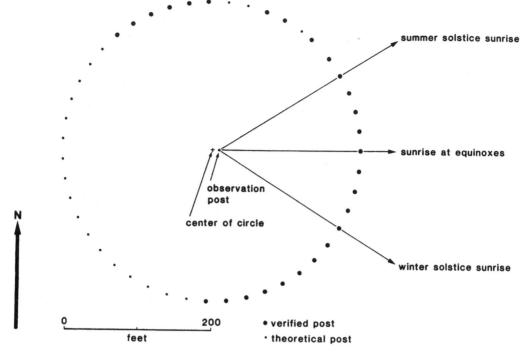

summer solstice sunrise

sunrise at equinoxes

observation post

center of circle

winter solstice sunrise

N

0   200

feet

● verified post

· theoretical post

Positions of known posts in Circle 2 at Cahokia combine with a posthole slightly offset from the circle's true center to provide observational sightlines to the equinox and solstice sunrises. *(Griffith Observatory, from Warren Wittry)*

Through measurement and calculation Wittry ascertained that sunrises for the solstices and the equinoxes aligned with posts on the east half of the circle. The spacing of the posts and the size of the circle required, however, a backsight 5 feet east of the circle's center. From the true center the alignments would be off. A posthole—as large as those on the ring itself—was found at the properly offset spot.

Geometric design, careful measurement, and astronomical alignments seem to be part of the plan in Cahokia's Sun Circle, but the purpose of the Sun Circle is still in doubt. Wittry believes the progress of sunrise along the eastern horizon may have been systematically observed by a "sun priest" from a platform on top of the offset central pole. Indians removed the posts centuries ago, however, and even their height is in doubt. There is, fortunately, a hint in the bathtub shape of the pits. This type of pit is convenient for sliding and propping very tall and heavy posts into place—Wittry guessed that the posts might have been about 30 feet high. He got his first opportunity to see a Cahokia sunrise—as the sun priest may have seen it in about A.D. 1000—on the 1977 autumnal equinox. Wittry had convinced a local utility, the Union Electric Company, to provide and install three posts—each 28 feet high—in the solstice and equinox pits. Before sunup on September 23 he balanced himself near the top of the observation post, also erected for the occasion in the offset pit near the center of the ring, and watched the sun come up in a notch formed by the south slope of Monks Mound and the bluffs on the Mississippi River 7 miles farther east. The equinox post was silhouetted in the sunrise on line with the notch. Wittry also confirmed the winter solstice sunrise line on several cold and bitter mornings the following December.

Was the Sun Circle a genuine astronomical observatory? If so, its size creates some problems. At a distance of 205 feet from the center, the Sun Circle's posts are fairly close to the sunwatcher's station. Like Stonehenge, with its Heel Stone only 256 feet from its center, and the Bighorn Medicine Wheel, with its central cairn but 36 feet northeast of the summer solstice sunrise backsight, Cahokia's Sun Circle will not provide a precise determination of the actual day of the solstice. The backsight and foresight are too close together. It is possible, of course, that the Cahokian sunwatcher had no interest in precision, but use of more distant horizon features, particularly at the equinoxes, would have allowed precise calibration of a solar calendar.

We know very little about these prehistoric circles of upright posts, but John White, a colonist in Sir Walter Raleigh's settlement in Virginia, included a seven-post circle in an Indian dance scene he painted. Most of the participants shown in this sixteenth-century eyewitness account are on the ring's perimeter, but three women occupy a spot near the center. The surfaces at the tops of the posts are carved with faces. In another illustration, Jacques le Moyne de Morgues recorded a group of Timucua Indians gathered inside a circle of tall posts. De Morgues was a member of a colony of French Huguenots, and the Indians in this picture he drew in 1564 in northern Florida seem to be engaged in warfare ritual. Parts from the dismembered bodies of their enemies hang from the tops of the posts.

Both of these early records of Indian life might convince us that astronomical observation was the last thing that took place in the post circle at Cahokia, but one detail permits serious consideration of that idea. The rings of the Virginia and Florida Indians include no central post; Cahokia's does. The Sun Circle's offset central post appears to be deliberately placed and favors the astronomical alignments.

The only other clue was discovered by Wittry in a fire pit a few feet inside the circle and in front of the winter solstice sunrise post. There he found a fragment of a ceramic beaker with a distinctive incised design. At the center is a cross, probably symbolic of the cardinal directions. A circle surrounds the cross, but it opens into a leg—something like the stroke that turns an *O* into a *Q*—on the lower right. If we imagine this as an intercardinal indicator, it could symbolize the southeast. A similar leg extends to the lower left, but the circular border does not open into it. The beaker may have been used—and broken—in a ritual that Wittry believes was connected with the southeast pole and the winter solstice sunrise.

More recent study has led Wittry to consider the idea that the bright star Capella may have been used to lay out the circle. Alignments with its rising and setting have

Measurements and modern observations allow us to reconstruct what a prehistoric Caho-kia sunwatcher may have observed at equinox sunrise from atop the offset central post in Circle 2. Monks Mound is profiled in shadow as the sun first appears, about midway up the sloping ramp on the south side of the mound and in line with a post on the circumfer-ence of Circle 2. *(Griffith Observatory)*

been proposed but await detailed measurement and calculation. Even if they prove to be real, however, the purpose of the Sun Circle will remain uncertain. It could have been used as an observatory, but precise determinations of the solstice appear to be outside its astronomical capacities. Other formal aspects of its design link it with ceremony and sacred space. An entire circle of posts is not needed, after all, to keep track of solstices and equinoxes, and, of course, the sun will never be seen outside the limits of the solstice markers. But even these qualms fail to diminish the apparent importance of the "central" post and the possibility of horizon observations of the sun. The Sun Circle may not have been a device for measuring the movement of the sun, but perhaps it *was* the place where the sun's actual travels were ritually observed.

A Platform for the Sun

In Scotland Professor Alexander Thom has interpreted certain prehistoric monuments as high-precision solar observatories. If they were used as he proposes, they pinpointed the solstices to the exact day. This considerably exceeds the precision that seems possible at Cahokia.

The megalithic alignments are there, but how can we tell they were intended? How can we know if they were meant to be precise? Because these sites probably date back to the Early Bronze Age (about 1700 B.C.), no "owner's manuals" are available to tell us what they did. However, one of these prehistoric sites provides the closest thing to an experimental test of a megalithic observatory we are likely to encounter. The monument is known as Kintraw, and it is located on a fairly large and level section of the steep slopes overlooking the head of Loch Craignish, an arm of the sea that cuts into Argyll (now the County of Strathclyde) from the Sound of Jura on the west coast of Scotland.

There really isn't much at Kintraw: a small, low cairn about 21½ feet across, a much larger—though now badly ruined—cairn with a diameter of 49 feet and height

This photograph of the standing stone at Kintraw, taken in 1973, before it fell over, also shows a portion of the large cairn behind it. *(E. C. Krupp)*

of 8 feet, and a single standing stone, or menhir, about 13 feet high, are just about all there is to see. Such cairns usually contain burials, and a small deposit of burnt wood and a few small pieces of cremated bone were found in a kist, or small, slab-lined compartment buried near the edge in the northwest quarter of the cairn. Only a few pieces of carbonized wood were recovered from a little kist within the smaller cairn.

Although the large cairn appears to be the most important component of Kintraw, Thom was more interested in what could be seen from the spot rather than in the cairn itself. A line of sight to the island of Jura, 27 miles to the southwest, puts a last gleam of the winter solstice sunset in a notch formed by two mountains—Beinn Shiantaidh and Beinn á Chaolais, also known as the Paps of Jura. A prehistoric astronomer could have used them to determine nearly the exact day of winter solstice, or at least he could have done so if an intervening ridge did not block the view. This was a complication that troubled Thom.

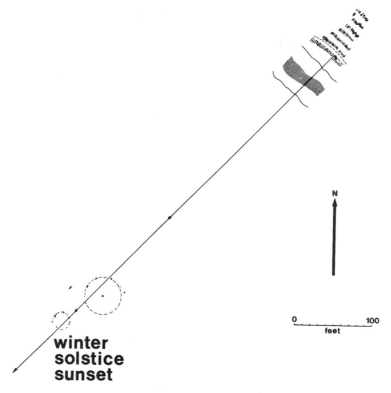

winter
solstice
sunset

Kintraw, a prehistoric site on Scotland's west coast, may provide an observational sightline to the winter solstice sunset. The backsight, or "place to stand," is on a narrow, flat ledge on a steep hillside that rises from a deep gorge. All of these features are in the plan's upper right. This line crosses two circular cairns as well as a large upright stone, and continues southwest. *(Griffith Observatory, after A. Thom)*

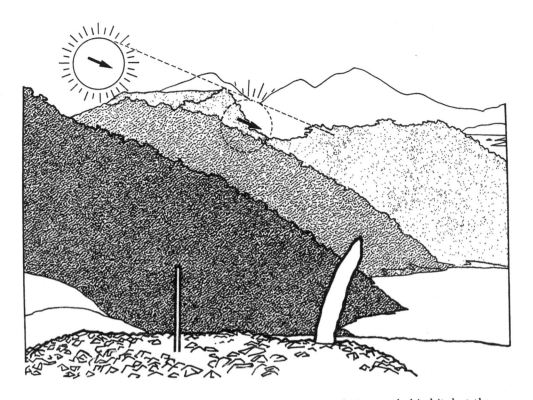

The profile of the island of Jura traps the setting winter solstice sun behind it, but the sun momentarily escapes when a flash from its upper limb appears, a short time later, in the notch between Beinn Shiantaidh and Beinn a' Chaolais, the "Paps of Jura." This is the type of observation that would be needed to establish, with precision, the date of the winter solstice. This drawing's viewpoint is from the boulder backsight on the steep hillside's narrow ledge. The top of the large cairn (shown with an upright pole to mark a socket discovered in its center) and the menhir, as it looked when it was still tilted, are in the foreground. *(Griffith Observatory, after Euan MacKie)*

At Kintraw, a very precise solstice alignment was possible. The Jura notch can just be seen from the top of the cairn, which was large enough to be used as a platform for a series of observations a few days before and after the solstice. But from the ground the notch cannot be seen. The question that intrigued Thom was: How did the builders know where to put the cairn?

While examining the site, Professor Thom discovered a ledge on the very steep hillside on the other side of a deep gorge, providing a good view of the Jura notch and ample room for several days' observations. A partially buried boulder on the edge of the platform turned out to mark the exact place one must stand to see the winter solstice sun set behind Beinn Shiantaidh and watch it reappear with a momentary burst of light a few minutes later when it cut through the notch.

Glare can make even the sunset hard to watch. At Kintraw, however, it is the momentary beam of sunlight—a beam that can be seen on the right day only from the right place—that tells the bronze age skywatcher it is the solstice. Kintraw's platform and boulder seem to say the winter solstice sunset was watched carefully there, but two other questions begged for an answer: Could the ledge be a natural formation? Could the buried boulder's position be a coincidence?

Dr. Euan MacKie, an archaeologist and Assistant Keeper at the University of Glasgow's Hunterian Museum, tried to answer these questions in 1970 and 1971 by excavating the platform for signs of human activity. None were found, but other clues emerged. J. S. Bibby, a soil scientist with the Macaulay Institute for Soil Research, analyzed the orientation and layering of the stones and pebbles that had collected on the ledge and may have given it a cobbled surface. He concluded that the stony layer resembled an artificial layer far more than it looked like a natural accumulation of fallen rock.

Critics of Thom's interpretation of Kintraw and of his general proposals about the precision and purpose of prehistoric astronomy have attacked the Kintraw study on several counts. Aubrey Burl believes less precise, symbolic alignments were built into some megalithic monuments but sees no evidence for "scientific" astronomy at Kintraw. The large standing stone sometimes associated with the winter solstice sunset alignment is close to but not exactly on the line. Burl thinks it works better with the smaller cairn as a symbolic indicator of summer solstice sunrise. After the Kintraw menhir fell in March, 1979, MacKie had a chance to see its original socket and so pinpoint its position. He confirmed that it agreed well with the alignment to Jura.

Sunwatchers at Kintraw might have stationed themselves for the sunsets a few days before and after the winter solstice at positions on a line perpendicular to the alignment with the Paps of Jura. As the solstice approached, the observer would have to shift a little to the northwest until, on the evening of the solstice itself, the sunwatcher would occupy the spot marked by the boulder V.

Even if Bibby is wrong about the ledge, no critic has dealt with the two boulders and the V notch they form. Their position on the solstice line could be coincidental, but their placement and orientation are consistent with their function as a backsight marker. They were found and excavated after—not before—the need for a marker at their position was recognized.

I am certain Kintraw could have been used to observe the winter solstice sunset with high precision. And there is additional evidence the site was built with the winter solstice sunset in mind. D.D.A. Simpson's excavations provide us with a cremation burial. While the large cairn could be a tomb for whomever was burned and buried there, the cremation deposit looks more like it was intended to consecrate the site. Kintraw may have been the stage for a winter solstice ritual. Although it differs considerably from Newgrange, a prehistoric winter solstice shrine in Ireland, some themes from Newgrange are repeated there. Simpson found a concentration of quartz

Excavation revealed that the backsight boulder on the Kintraw winter solstice line actually is a pair of large stones. Each ends in a wedge or "point," and together they create a *V* in which some cobbles have accumulated.   *(Robin Rector Krupp)*

The Jura profile visible from the Kintraw backsight is recorded by this photograph taken at eye level (about 5 feet 2 inches) from the boulder notch at the edge of the ledge. The listing menhir had fallen completely over a few years ago and was reerected upright in its socket.   *(E. C. Krupp)*

around the burial kist. So much more quartz was scattered in and around the cairn that Simpson concluded it originally wore an envelope of the gleaming white crystal. Ritual deposits of quartz, quartz-veined stones, and entire boulders of quartz are encountered at many stone rings and chambered cairns. The entire façade of Newgrange is wrapped in quartz, and the plugs in its winter solstice window box were quartz. Whatever the stone meant in detail to the megalith builders, it seems to have had something to do with death, ceremony, and the sky.

At Kintraw, even though we have no ethnographic evidence and precious little physical evidence to confirm it as a prehistoric observatory, there is no reason why it couldn't have been used annually to confirm the date of the solstice. It actually is more like what we should expect a genuine observatory to be: simple, easy to use, and more or less out of the way. It is not so different from the Hopi sunwatcher's rooftop station or the simple church bench of the Ossetian calendar keeper.

Reaching for the Moon

Professor Thom and his son, Dr. A. S. Thom, identify, in addition to precise prehistoric solar observatories, several dozen bronze age sites as megalithic lunar observatories. At even the best of these, the monuments are relatively simple. For example, nothing more than a very tall, very flat upright slab marks the lunar backsight at Ballinaby on Islay, an island in Scotland's Hebrides famous for its malt scotch whiskies. Near the iron age forts at Dun Skeig on Scotland's Kintyre peninsula, two low stones—one 2 feet high, the other, 4—provide another moon line. This site is known as Clach Leth Rathad, and, like the Ballinaby slab, it points toward a horizon feature that marks an extreme position of the moon with great precision. The level of precision the Thoms claim for these and the other lunar sites is what classifies them as observatories and is also what draws the fire of critics.

Thom's lunar observatories are not really instruments that measure the position of the moon. The stones that comprise them simply mark the correct places to stand. The action takes place on the horizon, but to see it, the moonwatcher must be in the right place.

According to the Thoms, megalithic moonwatchers kept track of the extreme positions of the moonrise and moonset throughout the 18.6-year cycle of variation. At the times when the moon reaches its greatest monthly limits, or oscillates, between its minimum monthly limits, the extreme moonrises and moonsets change very little from month to month. In a way, these periods of lunar extremes, or standstills, are like the situation at the solstices, when the positions of sunrise and sunset change very little from day to day.

Careful work with temporary markers—wooden stakes, perhaps—over more than one 18.6-year cycle must have preceded the erection of the megalithic lunar observatories. Aubrey Burl's symbolic lunar alignments in the recumbent stone circles of northeastern Scotland and the lunar alignments at Stonehenge certainly imply that prehistoric people in Britain were aware of the 18.6-year cycle and the lunar extremes. In addition to this knowledge, the Thoms claim that early bronze age "astronomers" found and then marked places to stand that would allow them to detect even subtler behavior of the moon.

At Ballinaby, the 16½-foot-high stone is about 3 feet wide but only 9 inches thick. It stands, then, like a single slat of fence, in the open, rolling pasture it shares with the farm's cattle, and points to a distinctive rocky peak that is isolated on the skyline about 1¼ miles to the northwest. When watched from the slab, the moon never sets farther north than this crag. According to the Thoms, details in the peak's silhouette even allow a careful observer to notice when the moon experiences a little extra gravitational tug from the sun and so advances slightly farther north than the average—just far enough to detect a momentary reappearance of the moon downslope.

An extremely thin slab, almost 17 feet tall, was oriented, according to Alexander Thom, toward the northernmost setting point ever reached by the moon, as seen from here at Ballinaby on the Hebridean island of Islay. *(Robin Rector Krupp)*

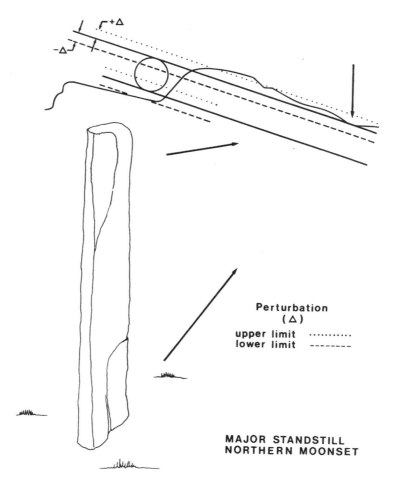

Perturbation
(△)
upper limit ··········
lower limit --------

MAJOR STANDSTILL
NORTHERN MOONSET

In Professor Thom's interpretation of Ballinaby, careful monitoring of the path of the setting moon permitted prehistoric astronomers to measure small changes in the northern extreme reached by the moon. Even the slight perturbation shift (△) that varies in time with the cycle of eclipse seasons could be detected by watching exactly where—if at all— the upper limb of the moon reappeared after slipping behind the rocky eminence to the northwest. *(Griffith Observatory)*

As the earth carries the moon around the sun, the moon constantly reacts to the sun. The effect is cyclical and takes about 173 days to complete. Because the change is so small, however, it is usually lost in the other motions of the moon. Prehistoric astronomers would have been able to detect it only at the lunar standstills, when it carried the moon a little bit outside its average maximum extremes or a bit inside its average minimum extremes.

This little shift in the moon's position is linked—by the length of its 173-day cycle—to the times when eclipses are possible, about every 173 days. The Thoms believe eclipses could have been predicted by reading the moon right. When they find more than just general alignments on the standstill positions of the moon—but, rather, precise alignments that detect the sun's small perturbation on the moon—they conclude eclipse prediction was part of the goal of the megalithic moonwatchers.

Critics doubt that successful eclipse prediction was possible with megalithic observatories and do not believe that precise observation of the moon was of interest in the Early Bronze Age. Although the debate at times has involved complicated and abstruse details, it boils down to a fairly simple issue. Could prehistoric astronomers obtain enough accurate observations of the moon—with sufficient precision—to allow them to see the subtler patterns in the moon's behavior and relate them to the occurrence of eclipses? The answer is yes, if they could determine when the moon reached the absolute limit of its excursion—including the extra little kick given by the sun's perturbation.

Unfortunately, there are too many variables and actual difficulties to permit such exactitude. Prehistoric lunar observatories pinpoint the moon only when it is on the horizon, and the time of moonrise or moonset might not coincide exactly with the moment the moon reached its limit. Even if the moments of the month's extremes occurred close to the time of moonrise or moonset, the moon itself might be anywhere in its monthly phases. The moon, therefore, might rise or set during the daytime. It would be difficult under those circumstances to see it low on the horizon. Partial phases also could be difficult to measure even at night, if the moon's dark edge happened to be at the critical spot behind the foresight. On top of these difficulties, bad weather or clouds, at least now and then, would mask the moon and make the megalithic astronomers miss a night.

Other factors would also complicate the moonwatcher's work. The air, for example, bends the moonbeams a little. This is refraction, and if it does not change from night to night, there is no problem. It does vary somewhat, however, and that introduces a little error into the observations. Also, the moon's orbit is not a perfect circle. Sometimes the moon is a bit closer to us, at other times a bit farther. These variations also can affect the exact position of the moon on the horizon. Finally, the moon is not a point of light, but a fairly large object in the sky. Somehow the prehistoric astronomers had to make sure they always measured the position of the same point on the moon's disk. Failure to do so would introduce a little more error.

All of these problems are real. To be useful, lunar observations have to be accurate to a few minutes of arc—about, say, a tenth the size of the moon's disk. Every source of error introduces a little more uncertainty, and when the uncertainty is larger than the effect itself, detection is very difficult. The Thoms' critics catalog all of these trouble spots and contend the solution was beyond the moonwatchers' reach. The only way the megalithic astronomers could have sidestepped these problems was by devising a technique that would let them figure out the true extreme position of the moon without actually seeing the moon reach it. Thom believed they could do this by working with observations of the moon made the day before and after the time the moon actually reached the extreme position of the standstill.

Professor Thom was well aware some kind of system for extrapolation had to exist if his ideas were correct. To overcome the numerous observational headaches, he devised schemes that made use of other ingredients sometimes found at sites with precise lunar alignments. For example, distances between various features near the Temple Wood stone circle in Argyll; "fans" of small stones in Caithness in the north of Scotland, and the vast and enigmatic arrays of standing stones at Carnac in Brittany can, with varying degrees of success, help locate where the observer would have had to stand to see the extreme position of the standstill moon had it actually reached it when the moon was visible on the horizon. Usually the moon would not have been seen there but a little short of it. Thom believed that the megalithic moonwatchers were aware of the absolute extreme position even though they might only rarely see the moon occupy it.

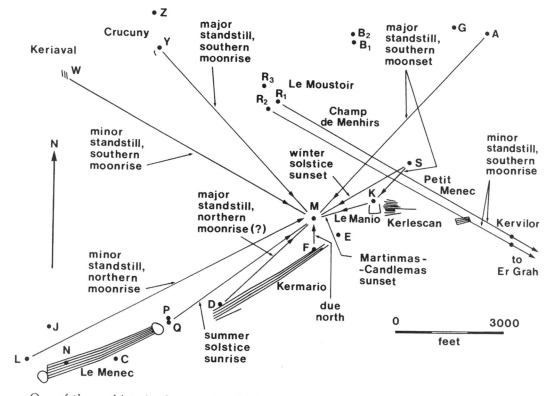

One of the prehistoric observatories Professor Thom believes he has found at Carnac is centered on a large, tall (16 feet high) stone known as Le Manio (*M*). If this interpretation is correct, Le Manio was the foresight for a number of the astronomical alignments shown on this plan. *(Griffith Observatory, after A. Thom and A. S. Thom)*

The Thoms think that one Carnac observatory was centered on a massive standing stone, 16 feet high, that occupies a hill known as Le Manio. Various stones within a radius of about 1½ miles mark backsights for lunar standstills as well as for the winter solstice and other dates in the solar calendar. An even more ambitious observatory at Carnac may have involved lunar sightlines between 2½ and 9½ miles long that converged on a genuinely monumental stone known as Er Grah ("the Stone of the Fairies") or Le Grand Menhir Brisé ("the Great Broken Menhir"). Le Grand Menhir now lies in four pieces. Together they probably weigh about 300 tons. Whether the stone, when whole, ever stood upright is in considerable dispute, but the way the fragments now lie favors the idea that the megalith builders of prehistoric France did once manage to erect this huge, 70-foot-long stone. An obvious use for its great height would be astronomical sightings from considerable distances away, but many of the Thoms' critics feel the remains they have selected in Brittany are too arbitrary, fragmentary, and ruined to satisfy the theory of megalithic lunar observatories. The Thom interpretation of the Carnac monuments is consistent, however, with their conclusions about the standing stones and stone circles in Britain and it is the most ambitious, developed, and detailed attempt to understand the mysterious Breton menhirs.

Another lunar observatory proposed for Carnac focused its alignments on a truly gigantic standing stone, Le Grand Menhir Brisé. Near the backsight position for this stone's minor standstill northern moonrise line, Thom noticed one of the odd fans or grids of stones he interprets as devices for establishing the exact position of a lunar backsight. This, then, is the "grid" at St. Pierre, and we are looking west along its stone rows. *(Robin Rector Krupp)*

Those who challenge the concept of precise lunar observatories doubt that the extrapolation methods devised by A. Thom really work. These methods still leave many features of the stone grids unexplained and require intricate and abstract reasoning to put into operation the features that do work. We are left somewhat mystified,

The uprights near the prehistoric Temple Wood stone circle form a line of possible moonwatching stations, according to Alexander Thom. More than one "place to stand" is needed in order to detect fine detail in the changing position of the northernmost and southernmost moonsets. The Thom interpretation has been challenged by some who cite a somewhat similar arrangement of stones at Barbreck, just 5½ miles north. That site does not seem to have lunar sightlines. *(Griffith Observatory, after A. Thom)*

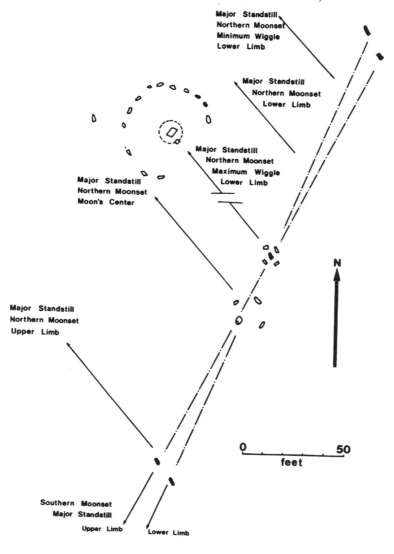

Major Standstill
Northern Moonset
Minimum Wiggle
Lower Limb

Major Standstill
Northern Moonset
Lower Limb

Major Standstill
Northern Moonset
Maximum Wiggle
Lower Limb

Major Standstill
Northern Moonset
Moon's Center

Major Standstill
Northern Moonset
Upper Limb

N

Southern Moonset
Major Standstill

Upper Limb   Lower Limb

0   50
feet

then, by alignments that indicate extreme moonrises and moonsets with high precision, higher perhaps than seems possible.

For example, 3 miles south of Kintraw, near Kilmartin and its extensive collection of neolithic and bronze age cairns, there is an unusual X-shaped arrangement of standing stones. Three of the four "stations" on the X are backsights for the extreme northwest positions of the upper limb, lower limb, and center of the major standstill setting moon—as seen in a notch about 1¼ miles away. These three settings are roughly oriented toward the notch, and the one pair of stones that doesn't point to the notch does combine with the pair at the southwest end of the X to align with another horizon feature and the major standstill *southern* moonset. The line from the tallest backsight also passes through the center of the Temple Wood stone circle. One of the Temple Wood alignments works for the center of the moon's disk. Because the exact center of the moon's disk is very hard to establish without modern

The southwest pair of standing stones at Temple Wood marks the correct place to stand in order to see the upper limb of the moon set in the indicated notch when the moon reaches its northern extreme. *(E. C. Krupp)*

instruments, this alignment is a bit unexpected. A rough, symbolic alignment with the moon's center would not be surprising, but this one is precise. Thom accepts the alignment and tries to explain how it might be determined from the positions of observers watching the moon's upper and lower limbs in the same notch.

It is possible, of course, to select sites that agree with one interpretation of their use and ignore other monuments that fail to fit the pattern. In their most recent analyses, the Thoms do try to deal with this problem. In any case, however, most investigators do not deny that at least some of the Thoms' sites are intentionally pointed toward the standstill moons. The issue is precision and purpose. Were these "observatories" designed to facilitate eclipse prediction? Eclipses, the critics argue, are easier to predict with calculations based on observed cycles of their recurrence. This requires written records, however, and there is no evidence for that in megalithic Britain.

There may be a way out of this dilemma, but it won't be easy. Both sides will have to give some ground. Skeptics may have to accept horizon notches and precise alignments. They may be able to stand their ground, however, on the meaning of that precision. For their part, the Thoms may have to concede that their case for "scientific astronomy" cannot be proven. What this means is that the megalithic alignments represent the moonwatchers in the midst of trying to deal with a vagrant moon. Already we have seen they were aware of the 18.6-year standstill cycle. The ritual alignments we encounter imply a more fully developed astronomical footing, just as the Stretching of the Cord ceremony and the Shang oracle bones hint at more systematic observation. The "megalithic lunar observatories" may be attempts to mark the moon's extremes, perhaps even to refine them over more than one cycle. But the simplest interpretation suggests they were each established through an actual observation. In this sense, they would be determined with some precision if the moonwatchers placed the stones in the positions from which they actually saw the standstill moon.

The accuracy of these alignments, however, would be a bit off. Each alignment would coincide with a particular and single moonrise actually seen from the site. From one site to the next, the moon's measured extreme would vary depending on luck and error. This could explain why some sites are precisely aligned and without a systematically precise astronomy.

Of course, if we abandon the notions of perturbation monitoring, extrapolation, and eclipse prediction, it is difficult for either side to see any reason for moderately precise observations. That is because both the partisans and the skeptics view the moonwatching process from the point of view of a modern, technical, secular society. Among the megalith builders, however, the arrival of the moon at a standstill might have been itself meaningful. The moon's rambles around a notch in the horizon must have been mystifying but probably were accepted as the way things are. Contained in every standstill was a confirmation of the expected as well as the surprise of unforeseen variations.

Shadows from the Sky?

An entirely different astronomical proposal also has been offered to explain the large standing stones. Instead of marking a spot from which the horizon would be viewed, it is imagined they cast shadows that would permit timekeeping and measurements of the passage of the solar year. A vertical, shadow-throwing pole is called a gnomon, and many traditional peoples—Borneo tribesmen, Babylonians, and Ionian Greeks among them—established the length of the year and the time of the solstice by measuring the length of the shadow at, say, noon. Some find it easy, therefore, to call any upright stone a gnomon, but no one has worked out the details that might permit us to conclude the shadows of stones like Le Grand Menhir Brisé and Le Manio were used in this way.

A famous antiquity from pre-Columbian Peru is also alleged to have something to do with shadows and the sun. It is the enigmatic *intihuatana*, or "hitching post of the sun," and it occupies the top of a terraced, rocky spur at Machu Picchu, the "Lost

Analysis of the shape and orientation of the *intihuatana* at Machu Picchu is unable to verify—at least so far—that this elegant and enigmatic sculptured stone was in any way used as a gnomon. *(E. C. Krupp)*

City" of the Inca. There actually was more than one intihuatana, but because their religious significance made them political liabilities, they were destroyed in the late sixteenth century by the Spanish authorities. The Viceroy Francisco de Toledo and the clergy battered and broke them wherever they were found. The upright stone column on the intihuatana at Pisac, an Inca town perched on the hillside above the valley of the Urubamba, was knocked off long ago, and only the mauled stump remains. Hiram Bingham, the American explorer who rediscovered Machu Picchu in 1911, found the stone intihuatana there still intact, however—a sign that the Spanish conquerors of Peru had never found the place.

The intihuatana at Machu Picchu is a mass of granite about the size and height of a large dining room table. Carved from the rocky summit of an 80-foot-high natural pyramid, this peculiar combination of oddly angled surfaces, corners, and projections has defied interpretation. Its name, along with incomplete descriptions in the chronicles, suggests that it had something to do with measurements of the position of the sun. Offhanded and cavalier astronomical interpretation has declared the intihuatana to be a gnomon, but assertions of this sort solve nothing. How did it work? Was there some kind of ruler or template or other piece of auxiliary equipment that was used in conjunction with the monument for measuring shadows? If so, how precise was the technique? Was the intihuatana used for systematic measurement or ritual observation? Superficial pronouncements about the intihuatana obviously have not been confirmed by evidence from the site, and the meaning of the "sun's hitching post" still eludes us.

Windows on the Sky

Another "observatory" at Machu Picchu may be within our grasp, however. In the summer of 1980, Dr. Ray White and Dr. David Dearborn, both astronomers at Steward Observatory in Tucson, Arizona, conducted an EARTHWATCH expedition to Machu Picchu and measured the orientations of windows in one of the site's most exquisitely masoned buildings, the Torreon.

Built upon a natural mass of rock, the Torreon, also known as the Caracol, or "snail," starts out as an ordinary-enough structure with straight walls and a rectangular floor plan. The east wall curves, however, into a semicircle that partitions the room into an outer hall and an inner chamber. It is the curved wall, reminiscent of a snail's convoluted shell, that prompts the local nickname. Bingham noticed the similarity between this semicircular temple and the Coricancha, or "Temple of the Sun," at Cuzco, and thought that the Torreon also might have been a Temple of the Sun. Trapezoidal wall niches—a standard feature of Inca architecture—and what seemed to be an altar carved from the protruding natural rock confer a formal, ceremonial character upon the Torreon.

Three apertures in the walls of the Torreon's inner chamber open onto the spectac-

An "altar" of natural rock was cut and carved by the Inca and then enclosed by the beautifully curved wall they built to make Machu Picchu's Torreon. The June solstice/Pleiades window, beyond the stone altar, overlooks the deep valley of the Urubamba River. Two wall niches appear to the right, and a corner of the southeast window can be just seen beyond the wall in the foreground. *(Robin Rector Krupp)*

ular landscape that surrounds Machu Picchu. Whatever the reasons that induced the Inca to situate this sanctuary 2,000 feet above the serpentine goosenecks of the Urubamba River, the view must have been one of them. Art historian George Kubler wrote of Machu Picchu:

> The mild climate, with its theatrical fogs, sunsets, and milky distances, affords one of the most picturesque settings in the world.

The vistas of Machu Picchu seem perfectly staged. Inca stonework is among the very best the ancient world has to offer, and at Machu Picchu it creates the impression of an important ceremonial center. The environment inside the Torreon seems to echo with ancient ritual, and one of its windows may have acted as an invitation to the sun.

White and Dearborn discovered that the Torreon's northeast window is centered on the June solstice sunrise, and the details imply that more than ritual observation may have been involved. At its west end the upper surface of the altar was cut and carved to create a straight, flat, vertical surface that divides the altar in half and is perpendicular to the window. The south half, then, is a few inches higher than the north, and the surface that separates them is like the sheer cliff of a miniature mesa. June solstice sunrise light just reaches this far end of the altar, and the shadow of a weighted strand suspended in the window could be used in conjunction with the cut surface of the altar to establish the date of the solstice. Although the technique is simple, none of the hypothetical auxiliary equipment has been discovered. Corner blocks in the outer face of the window have peculiar carved knobs that protrude from the otherwise uniform wall, however, and White and Dearborn propose that these stone "pegs"—which are known from other Inca sites as well—could support a removable frame from which the solstice plumb line could hang.

The Torreon's southeast window also is equipped with stone pegs, but its astronomical use is not so clearcut. White and Dearborn could see the sky through it only if they sat on the floor in front of it, backs against the altar. From this position the stars in the tail of Scorpius (the scorpion)—an Andean constellation sometimes known as *Collca*, the "storehouse"—could be seen rising through the window. In late Inca times these stars would be coming up when the June solstice sun went down. Anthropologist Gary Urton researched the importance of the constellation of the Collca among traditional Peruvian peoples today and discovered the Pleiades—even more often than the Scorpion's tail—were known by this name. This is interesting because the Pleiades are a kind of astronomical counterpart to the tail of Scorpius. The two groups of stars are on opposite sides of the sky, and, in Inca times, beginning about a month before the June solstice, the Pleiades appeared in the morning sky and rose in the winter (June) solstice window.

Opening to the northwest, the third aperture in the Torreon faces no obvious celestial event. Bingham called it "the problematical window" and thought a golden

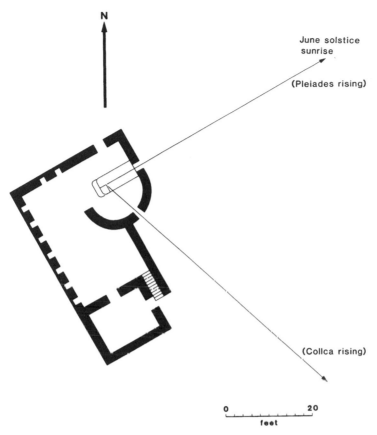

N

June solstice
sunrise

(Pleiades rising)

(Collca rising)

0   20
feet

Windows in the Torreon at Machu Picchu could have been oriented to permit viewing of stars known to be significant even today among the Quechua-speaking Indians near Cuzco. *Collca* is what we identify as the tail of Scorpius. Its setting point is opposite the rising point of the Pleiades, also known as *Collca*. The appearance of both groups of stars is seasonally and directionally significant to this Andean people. The "Pleiades window" also permits a precise assessment of the date of the June solstice with a minimum of hypothetical auxiliary equipment. *(Griffith Observatory)*

image of the sun might have been suspended in the portal as was done in the somewhat similar—but northeast—and June-solstice-oriented "tabernacle" in the Coricancha. This last window in the Torreon is still problematical, but we know the June solstice was observed and celebrated in ancient Peru in the important Inti Raymi festival. Architectural similarities between the Coricancha and the Torreon support and almost demand a June solstice alignment in the Torreon. A more intricate sequence of observations at this time of year may have involved the Pleiades and the tail of the Scorpion, although many other stars are visible through both windows, too. White and Dearborn reasonably argue that a genuine Inca observatory occupied sacred ground at Machu Picchu.

The Tower of Venus

Astronomically aligned windows seem to make an observatory out of another architectural "snail"—the Caracol of Chichén Itzá in Mexico's Yucatán. In this case, an association with the familiar gastropod is drawn because the building's cylindrical tower comes furnished with a winding passage and spiral staircase that provide access to the tower's upper rooms.

Yucatán was the focus of Maya civilization after the end of the Classic period and the abandonment of the great cities and ceremonial centers farther south in Mexico, Guatemala, and Honduras. When a group of Toltec or Toltec-influenced Putun Maya took control of Chichén Itzá, they helped create a hybrid style of architecture that betrays both Maya and Toltec influence. The Caracol's cylindrical tower resembles the round buildings of central Mexico, but its lower platform is closer to Classic Maya design.

Although Chichén Itzá was one of the most important post-Classic Maya centers, it was virtually unknown outside of Mexico when John Lloyd Stephens, an American lawyer, and Frederick Catherwood, an English artist, visited the site in 1842 and

Although the Caracol of Chichén Itzá looks domed like a modern observatory, its present shape is really the result of the ruin into which it has fallen. Originally, its shape was cylindrical, and the upper tower, some of whose windows are still visible, sat like a pillbox on top. *(E. C. Krupp)*

found among its ruins the unique Caracol. Its rather large tower sat upon a massive, two-terrace platform with grand staircases on the west side. Stephens noticed several doors and corridors in the lower tower and tried to clear the rubble from the inside passage in order to have a look at the upper tower. The spiral staircase was choked with stones, however, and the roof was tottering. Stephens was unable to describe much more of the upper tower, but he did note the presence of at least one large opening at the top, out of which an observer could look. Stephens, however, said nothing about the place being an astronomical observatory.

Augustus Le Plongeon, a cosmopolitan, pioneer explorer of the Maya realm noted for his unorthodox ideas, probably was the first, in 1875, to call the Caracol an observatory. Why he did so is not clear, but it seems likely he simply responded to the Caracol's resemblance to the great observatory domes of nineteenth-century telescopes. The Maya had no telescopes, of course, and if the Caracol looks as though it is domed, it is because much of the original tower has fallen away. Actually, with two cylinders—the smaller perched atop the larger and both supported by the double platform—the Caracol, in the words of the eminent Mayanist J. E. S. Thompson, stood like "... a two-decker wedding cake on the square carton in which it came."

There was no dome on the Caracol, but nearly a century after Le Plongeon, Erich von Däniken repeated the same sentiments in defense of his ancient-astronaut notions. In *Chariots of the Gods?* he writes that the Caracol "looks like an observatory" and tells us "... in the dome there are hatches and openings directed at the stars ..." Le Plongeon can plead ignorance of today's knowledge, but von Däniken has no such excuse.

Late in the nineteenth century, the English archaeologist Alfred P. Maudslay examined the Caracol's upper tower before its eastern half had fallen away. Maudslay mentioned six remaining doorways, or windows, but unfortunately made no sketch of their placement or measurements of their orientation. Although Le Plongeon was the first to say the Caracol was an observatory, he did not explain how it might have been used. Oliver Ricketson, an American archaeologist sponsored by the Carnegie Institution of Washington, proposed an astronomical interpretation for the Caracol's upper tower windows in 1925. More than half of the upper chamber had fallen away by then, but three windows still survived 41 feet above the pavement of the platform's upper terrace. Window 1, on the west, is about 3 feet high and large enough to allow one to crawl through it nearly 8 feet to the precipitous outer face of the upper tower. By contrast, windows 2 and 3 are both quite small, roughly 9 inches square. Window 2 opens to the southwest, and window 3 faces nearly due south.

Ricketson measured the orientations of the windows' center lines as well as—and this was a novel approach—the directions of the diagonals from each inner jamb to the opposite outer jamb. He thought this would make a precise sighting possible even though the window were wide. One of window 1's diagonals turned out to point due west, and Ricketson concluded its intended target was the equinox sunset. Ricketson

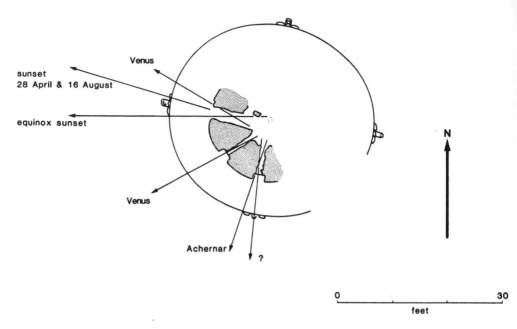

sunset
28 April & 16 August

equinox sunset

Venus

Venus

Achernar

?

N

0   30

feet

Some of the windows in the ruined upper tower of the Caracol may have been targeted on Venus. The planet's extreme northern and southern setting points do coincide with diagonals across two of the three remaining ports. *(Griffith Observatory, after A. F. Aveni, S. Gibbs, and H. Hartung)*

thought the other diagonal, which pointed northwest, was sighted on the northernmost moonset. Similarly, one of the diagonals across the window was thought to be aimed at the moon's most southern setting. Finally, Ricketson claimed that one of window 3's diagonals pointed south.

The definitive archaeological survey of the Caracol was performed by Karl Ruppert, also under the auspices of Carnegie, which published his report in 1935. Ruppert reconstructed the outer recesses of the tower's windows in 1930, but he felt the repairs he had made did not modify the alignments intended by the original Maya builders. He carefully repeated Ricketson's measurements and obtained slightly different results. The moonset lines were thrown in some doubt, for Ruppert found orientations about 1 degree different (two times the size of the full moon). According to Ruppert, the due south line was 3 degrees off. Ruppert didn't feel the moonset deviations were great enough to dismiss Ricketson's interpretation, but Ruppert did speculate that the due south line really may have been oriented on magnetic south.

Ricketson had been very enthusiastic about his astronomical interpretation of the Caracol and even went so far as to say it was the only use to which the building could have been put. Ruppert was more cautious. He supported the idea of an astronomical use but also thought the upper levels may have served as a military watchtower.

As innovative and valuable as Ricketson's and Ruppert's reports were, we no longer have to rely on them to assess the Caracol's astronomical potential. Anthony F. Aveni and Horst Hartung have resurveyed and reanalyzed the Caracol in collaboration with Sharon Gibbs. Their report, published in *Science* in 1975, is the most comprehensive study to date. Aveni, Gibbs, and Hartung affirm both practical and symbolic astronomical use of the Caracol, but their results differ from those of their predecessors. Their measurements of the "moonset alignments" showed they missed the moon by more than 2 degrees, and, in fact, the northernmost moonset is not even visible through window 1! Instead, they suggest, Venus may have been intended. Venus is the third brightest object in the sky, and its settings reach extremes on the horizon analogous to the sun's. The exact extremes of the planet are slightly different, however, because its orbit is tilted 3.4 degrees with respect to the path of the sun. Also, Venus reaches its extremes in a more complex fashion and arrives at them on varying dates. The Caracol's "moonset alignments" fit the Venus extremes the Maya

Looking along the diagonal to the northwest across the jambs of the largest window in the Caracol's upper tower directs a line of sight through a narrow "slit" and on to the place where Venus set farthest north. *(E. C. Krupp)*

would have seen from Chichén Itzá around A.D. 1000 better than the moon's although even the Venus fit is not perfect.

Ricketson's equinox sunset diagonal was also rechecked, and the result turned out to be somewhat puzzling. The window jambs do produce an accurate east-west line, but the inner right jamb is not vertical. The wall looks like it has slipped out and down about 3 inches to the right of where it should be had the wall once been vertical. A check of the "restored" corner—reconstructed with a piece of suspended tape—showed that it misses the equinox by 2 degrees. This is a large error and corresponds to the position of the sun six to eight days after the vernal equinox. It may be just a coincidence that the present positions of the jambs of window 1 provide an alignment with the equinox sunset. Furthermore, this event can only be observed by lying awkwardly on the floor of the upper tower. The line seems doubtful.

The ambiguous situation with window 1 led the trio to explore other possibilities for the upper tower windows. The center line of window 1 agreed with the sunset on April 28 and August 16 in A.D. 1000. Also the Pleiades pass through the window. Vincent H. Malmstrom, a professor of geography at Dartmouth College, feels the August 16 date is sufficiently close to August 13, the starting date, in our calendar, of the Maya Long Count system. The Maya associated the date with the Creation, which in our chronology would fall in 3113 B.C., according to the most widely accepted correlation of Old World and New World history.

Several other astronomical alignments were identified in various features of the Caracol, and some of these also link the building with Venus and with the god Kukulcán. In central Mexico the same deity was known as Quetzalcóatl, and he was clearly identified with Venus. Venus, we know, had tremendous significance in Mesoamerica, and its various appearances prompted strong reactions in Maya communities. Floyd Lounsbury's analysis of the date inscriptions that accompany the famed murals at Bonampak that depict Maya battle scenes and prisoner sacrifice shows that ritual raids against other neighboring centers took place at what the Maya considered celestially appropriate times. For Bonampak, an inferior conjunction of Venus coincided with the date of the battle.

The few surviving Maya codices originate from northern Yucatán—and probably not far from Chichén Itzá. The Dresden Codex contains an extensive schematic tabulation of the intervals in the cycles of Venus for divination. Aveni's study of the movement of Venus in the tenth century shows that the planet's extreme positions on the horizon varied noticeably in amount and date. Maya astronomers interested in its appearances and disappearances would have been strongly motivated to build an observatory to keep track of the planet. Alignments on Venus at the Caracol therefore should not surprise us. They had to observe the planet from somewhere.

Alignments in the double platform also seem convincing and emphasize the celestial intent of the building. Although the platform's corners do not form right angles and the sides are not symmetric, the northeast-southwest diagonal is nicely oriented with summer solstice sunrise and the winter solstice sunset. The flat Yucateco horizon allows the alignment to work in both directions.

Other towers are known, too—at, for example, Mayapán and Paalmul (on the coast, in Quintana Roo). They all lie close to the same latitude and face somewhat north of west, but the Mayapán and Paalmul towers are now too ruined to tell much more about them.

As an observatory, the Caracol does have some things to offer. Not only does it incorporate some reasonable alignments, but it is mounted on a raised platform. It is hard to realize how important this is until you climb above the platform into the upper tower. Together, they put an observer above the trees and scrub that cover Yucatán. From the ground level, it seems impossible that anyone could do practical astronomy here. The horizon is completely blocked by relatively dense growth. Once above the vegetation, however, the incredibly flat and featureless horizon of Yucatán and the vault of sky overhead are spectacular. Suddenly, astronomy becomes possible.

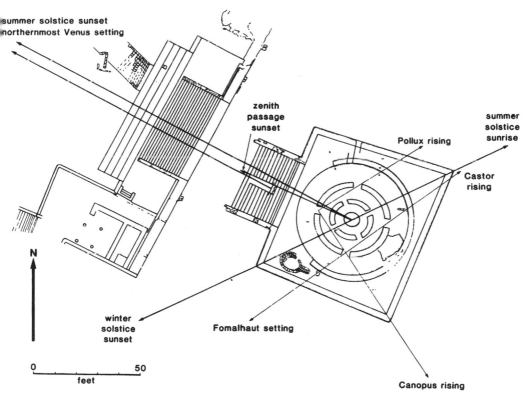

A number of astronomical alignments have been proposed for the Caracol at Chichén Itzá. Among the most convincing are the solstice diagonal of the upper platform, the zenith passage sunset perpendicular, and the Venus line from the niche that splits the upper platform stairway. *(Griffith Observatory, after A. F. Aveni, S. Gibbs, and H. Hartung)*

The number of steps up each platform is symbolic: 13 on the Upper Platform and 18 on the Lower Platform. Both of these are important numbers in the Maya calendar.

It is also important to keep in mind that a building like the Caracol, if an observatory, doubtless incorporated a great deal of impermanent auxiliary equipment that has long since disappeared. Interpreting the Caracol may be a lot like going to Palomar Mountain and finding the 200-inch telescope dome empty save for the ruined fragments of the great equatorial mount.

Actually, the situation is not so nebulous, for there is another clue. As one travels through the ruins of ancient Mesoamerica, it is impossible to miss the undeniably impractical and dangerously steep stairways that seem to adorn every pyramid, palace, and temple. The risers are high; the steps are shallow. At the Caracol, however, the stairs meet the needs of a sane climb. Each step is low and wide. It is very clear that the stairs here were intended to be functional. When all others embrace a fashion that departs from common sense, we may depend on the astronomer to insist that the daily climb to the observatory be accomplished without hazard or strain.

The Quest for Precision

One of the most important things about the astronomical tradition of ancient China is that so much of it is written down. In content and form Chinese astronomy was part of a virtually continuous cultural tradition that extends back to the Bronze Age and 1800 B.C. To be sure, it changed through the millennia and was influenced by outsiders, but the basic pattern remained the same. Written records in China, therefore, amplify, illuminate, and explain what in other traditions has been lost.

For example, the idea of the use of the gnomon in Incaic Peru or megalithic Europe may remain unresolved, but in ancient China the situation was much less ambiguous. On bronze age Shang dynasty oracle bones the early Chinese ideogram for "post" or "pillar" looks like a wand held by a hand and topped by a sun disk. Certainly this suggests the idea that the gnomon was closely associated with the word for post. An explicit reference to the observation of a shadow from a special tower designed for that purpose and careful measurement of the shadow's length shows up in Zhou dynasty writings in the seventh century B.C. From then on, more detailed accounts of Chinese gnomons allow us to trace the use and development of this simple but useful astronomical instrument through the centuries of Chinese civilization. In north central China the high summer solstice sun produced a shadow close to 1½ feet long with a gnomon 8 feet high. Accounts from several centuries before and after the birth of Christ imply the existence of an official gnomon installation at a town known as Yang-chhêng. Today, known as Gao cheng zhen, it is about 50 miles southeast of Luoyang, the chief capital in Eastern Han times (A.D. 25–220) and the home of one of China's great astronomers, Chang Heng.

During the Han era the town of Yang-chhêng had cosmographical significance; it

A T'ang Dynasty (A.D. 618–906) gnomon is preserved at Gao cheng zhen, a place where astronomical observations were carried out for at least 1,500 years. *(Robin Rector Krupp)*

was regarded as the center of the world. It acquired this reputation because either the sun's shadow was measured there, or the gnomon was installed there because it was considered the center. In either case, the association between the order of time and the order of the world is expressed in the gnomon at the "world's center."

By A.D. 725 a series of field stations—each equipped with an 8-foot gnomon and all strung out like widely separated beads on a single meridian longitude over a span of nearly 2,200 miles—was set up by a Chinese Buddhist monk, I Hsing. He, too, was one of the great astronomers in China's antiquity. (One of his surviving writings is called *Mnemonic Rhyme of the Seven Stars of the Big Dipper and Their Tracks.*) His gnomon installations were intended to determine the dates of summer and winter solstice. Of course, one of the stations was set up at Yang-chhêng. Several stations were built at widely separated locations because the difference in the length of the solstice shadow from one gnomon to another could be used to estimate geographical distances with greater accuracy.

Eight-foot gnomons were adequate for normal calendar calibrations, but Chinese astronomers working at the frontier of knowledge built them even larger. In 1279, during the Yuan dynasty, the astronomer Guo Shou jing built a platform out of brick 40 feet high. This truncated pyramid was itself the gnomon, for the shadow of a horizontal bar supported by the upper level fell upon the top surface of a low wall that extended perpendicularly from the pyramid's north face for 120 feet. This is contradicted by the most authoritative Western accounts, which say that a high pole stood in the vertical slot on the pyramid's north face and that this was what cast the shadow. But inspection of the monument reveals no sign of any structure or pit that might have held and supported such a high pole, and in any case the pyramid itself obviates the need for such a pole. Although this may be news outside of China, the little museum at the site shows how the observatory was used without a pole, and auxiliary equipment, based on descriptions in the historical records, has been refabricated to permit duplication of the measuring techniques of Guo Shou jing.

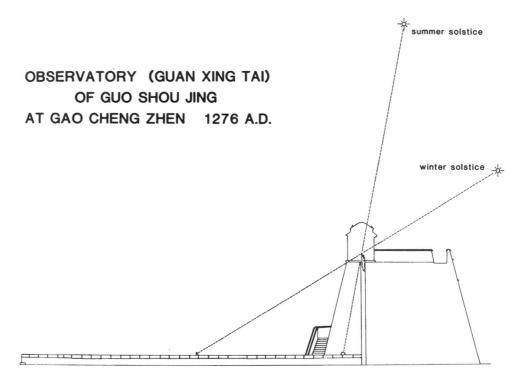

OBSERVATORY (GUAN XING TAI)
OF GUO SHOU JING
AT GAO CHENG ZHEN   1276 A.D.

summer solstice

winter solstice

Among the observations carried out at the tower of the thirteenth-century Chinese astronomer Guo Shou jing were measurements of the shadows cast upon the long low horizontal wall by the sun at noon on the summer and winter solstices. Because the tower is about 40 feet high, the difference in length between the two shadows is great, and this permitted extremely precise determination of the length of the tropical year. *(Griffith Observatory)*

Looking due south along the shadow measuring wall (or "Sky Measuring Scale") toward the brick tower of Guo Shou jing, we can see a horizontal bar (reconstructed) in the center gap at the top. It was the position of the shadow cast by such a bar that was observed in order to establish the exact dates of the solstices here at Gao cheng zhen, near Dengfeng in Henan province. *(Robin Rector Krupp)*

The long, low wall was known as the Sky Measuring Scale, and it had to be very long to accommodate the shadow of the low winter sun. Graduated for precise measurements and fitted with water troughs to test its levelness, the Sky Measuring Scale permitted Guo Shou jing to estimate the length of the year with high precision.

The tower was a real observatory. Accounts by early historians tell us that side chambers on top of the platform housed a water clock and perhaps an armillary sphere for measuring the positions of celestial objects. This "Tower for the Measurement of the Sun's Shadow"—one of the most important observatories of its day—is located, as one might guess, at Yang-chhêng—the traditional "center of the world" until the Ming emperor Yong Le moved the "world's center" to Beijing (Peking). This structure at Yang-chhêng tells about more than ancient Chinese astronomy, however. We know it was an observatory, and we know it was precise. Chinese astronomy was official imperial business and not exactly true science, but here we can see that high precision was still desired. Exactness was the handmaiden of authority and stability, and that accounts, at least in part, for the Chinese quest for precision.

3

The Gods We Worship

A perusal of nearly any ancient pantheon reveals the obvious: At least some of the gods, often the most important ones, are objects in the sky. The metaphoric reasons are not difficult to understand. The regular motions of celestial objects made them agents of order that helped give meaning to the world below; endless repetition of their appearances and disappearances suggested immortality; their light commanded attention and connoted power. And being in the sky, with such a perspective on earth below, it was only natural to assume that the gods must know all because they could see all: To see the world, one's eyes must be in heaven. In fact, the sun and moon were described by the Egyptians as the eyes of Horus, the falcon god of the sky.

Although particular gods may differ in terms of the resources they are believed to control, control is the attribute they share. What they control, and how they do it, determines exactly what sorts of gods they are. Celestial gods control the passage of time by marking it and measuring it. They control direction and space through the locations of their comings and goings. As masters of time and space, they move the world. They make it change. Day changes into night. Winter melts into spring. Rivers flood and fall. Grain sprouts, grows, and ripens. In these cycles of the world and in our daily lives we see patterned change, and it is driven by the sky.

The importance of the sky is best revealed in the things we take for granted, the little things that mean so much. The sky permeates our lives so completely that the days of the week, one of the most basic units of organized time, bear the names of celestial objects. In English, the origins of three of the names are fairly obvious. Sunday is, of course, the sun's day; Monday is the moon's day; and Saturday is Saturn's day. We have inherited this nomenclature from the ancient Romans, whose names for these days were *Dies Solis, Dies Lunae,* and *Dies Saturni.* The names of the other four days of the week come from Norse gods:

Tuesday	=	Tyr
Wednesday	=	Odin
Thursday	=	Thor
Friday	=	Frey

In English, the names of the first two days sound more like *Tiw* and *Woden,* the Anglo-Saxon counterparts of Tyr and Odin. All four were chosen, however, because they were equivalent to the Roman gods who governed those days of the week, and, again, the celestial connections are clear:

Tuesday	=	*Dies Martis*	=	the day of Mars
Wednesday	=	*Dies Mercurii*	=	the day of Mercury
Thursday	=	*Dies Iovis*	=	the day of Jupiter
Friday	=	*Dies Veneris*	=	the day of Venus

So, the names of the days of the week refer to the seven moving celestial objects: the sun, moon, and the five "wanderers"—the planets. And although Latin is no longer a living language, this issue is not dead. Equivalent names have been carried into all of the Romance Languages, which derive from Latin. By the power of their names these old celestial gods still live and drive the week through time.

Immortality and Divinity

If we are seeking immortality, the sky is a good place to start. We see endless repetition there. Although we know that we ourselves will die, we see the sun, moon, and stars survive night after night, month after month, year after year. They may disappear, but their absences are only temporary.

The sky in Egypt was Nut, a goddess whose outstretched body canopied the earth. Although she swallowed the sun each night, it was reborn from her loins each dawn. As the sun rose, she consumed the stars and gave birth to them again when at sunset the sun crossed her lips once more. The Egyptians also described the predawn reappearances of the *decans*—the stars that heralded the hours of the night—as rebirths from the "death" each experienced when, at its own time during the year, it vanished like Sirius for 70 days from the nighttime sky. Because circumpolar stars never went below the horizon, and so never left the sky, the Egyptians called them "imperishable" and "undying."

In another myth, Re, the Egyptian sun god, passed each night into a dangerous realm, the kingdom of death. He sailed through 12 perilous territories—12 hours of the night—and steered a course through the frightening creatures that inhabited them. Gigantic snakes writhing in their lairs, multiheaded serpents that walked on human legs, fire-spitting vipers armed with knives, a scorpion goddess: all these awaited Re. Despite confrontation and mortal danger, however, the sun survived to rise again each day.

The sky is one of the few things that provides concrete images upon which our conception of immortality might condense. The sky is itself eternal, and its occupants are continuously resurrected. There, in the celestial passages and returns, is the contrast between what is mortal and what is divine.

The power of the celestial gods was revealed by their light. Anyone standing in sunlight senses its energy. Its warmth is unmistakable. Though obviously weaker, the moon and planets also commanded respect. They shine not only in the blackened vault of night but, on occasion, in the brighter twilight sky, and some can even be seen in broad daylight. Again and again, gods were associated with light.

For example, Any, or An, was the greatest of the Sumerian gods. His name was the word for "sky" and "high," and the written symbol for his name was shared with the word *Diugir*: "Shining." And in the celebrated tomb of the Egyptian pharaoh Tutankhamun, one of the four nested boxes, or shrines, that enveloped his sarcophagus portrays the sky goddess Nut. An accompanying inscription reads:

Nut, the Great, the Brilliant One

Among the old Indo-Europeans, from whom most modern Europeans are descended, the sky's ruler had the name Djevos, or something very close to it. From his name, other forms evolved: Dyaus (Sanskrit), Zeus (Greek), and Jovis (Latin). The name Jupiter derives, in turn, from Dyauspitar, or Zeuspater, "Father Zeus." In terms of its original Indo-European root, the name Zeus means "resplendent" or "shining." The collective Indo-European name for the celestial gods was *daevos,* "the Bright Ones." Even today, the shamanistic religion of the Ostiaks of northwestern Siberia includes an intimate relationship with the sky, in which their chief god resides. His name derives from a word that means "luminous, shining light."

Equal reverence for the softer, indirect light of the moon is evident from a text from Ur, in the Mesopotamia of the third millennium B.C.

> Nanna, great lord
> light shining in the clear skies,
> wearing on his head a prince's headdress
> right god for bringing forth day and night,
> establishing the month
> bringing the year to completion.

Another Sumerian prayer invokes the brilliance of Inanna, the goddess Venus, in the evening twilight:

> The pure torch that flares in the sky,
> the heavenly light shining bright like the day,

the great queen of heaven, Inanna, I will hail . . .
Of her majesty, of her greatness, of her exceeding dignity
of her brilliant coming forth in the evening sky
of her flaring in the sky—a pure torch—
of her standing in the sky like the sun and the moon,
known by all lands from south to north
of the greatness of the holy one of heaven
to the lady I will sing.

The Gods in Heaven: Sun and Moon

The particular appearance and behavior of certain celestial objects have often led different peoples in different places at different times to assign the same symbolic values to them. The sun, for example, is both powerful and dependable, as it pursues its orderly course through the seasons, and these characteristics have inspired many peoples to see in it the source of all authority, law, and social order.

In Egypt the sun, Re, was ruler of the day. His steady course ordered the world. The Egyptian environment was self-contained and stable. Government was ordered. Society was ordered. All this was a gift from the sun. The Nile, too, was well-behaved, but its gift was life. The real source of order was the sun. Each day, each season, each year the sun emerged from the darkness with its offering: *Maat.* Maat was sunlight and the rightness of things—congruence and the natural order. Personified as a goddess, Maat sailed the sky in the company of Re.

In ancient Babylon, the sun was Shamash. His watchful eye noted all things and judged everyone. Justice resided in him. Hammurabi, the great codifier of Babylonian law, is shown standing before Shamash on the stone column, or stela, inscribed with this king's famous Code. Through law the sun's order was transferred to earth.

In Vedic India—the period of Indo-European, or Aryan, invasion—Varuna was the guardian of cosmic law and order. To do this he calipered earth, air, and sky and pegged the four quarters, or cardinal directions. His "calipers" were Surya, the sun.

Peoples of the New World developed similar metaphors. The Inca of Peru claimed Inti, the sun, as their ancestor and legitimized their sovereignty in terms of the sun's divine authority. Religious, social, and political organization of Inca life was focused upon Inti's shrine at the heart of Cuzco, the Inca capital. In Mexico, the tribal god of the Aztec, Huitzilopochtli, was elevated to preeminence as Aztec fortunes rose. Their god eventually assumed a solar dimension as part of the Aztec's legitimization of their right to rule. For the Aztec, the actual sun was Tonatiuh, and his name means "He Who Goes Forth Shining." The continued existence of the entire world was wrapped up in his well-being. The present age was named for him. Without him time would end.

Compared to the sun, the moon's rapid changes make it seem practically vagrant,

(Above left) Varuna, a Vedic sky god, was the source of cosmic order. The sun shined, the moon moved, the winds blew, the rivers flowed, and the stars appeared all according to his regulations and by his power. *(W. J. Wilkins,* Hindu Mythology*)*

(Above right) Surya, the Vedic sun god of India, travels—like Phoebus Apollo—by chariot. His driver is the Dawn, and in later tradition his seven horses were replaced by one with seven heads. *(W. J. Wilkins,* Hindu Mythology*)*

Tonatiuh was one incarnation of the sun among the ancient Aztec. Behind his head is a distinctive rayed disk. *(T. A. Joyce,* Mexican Archaeology*)*

but it is useful as a timekeeper, and many peoples have accorded it divine status as such.

To the Egyptians the moon was Khonsu, the "runner." The earliest Egyptian calendar was based on the moon, and the gods associated with each lunar month appear often in the paintings and reliefs in Egyptian temples and tombs. Thoth, another form of the moon, in combination with his identity of scribe, was the regulator or measurer of time and seasons. His records and computations set the days for the festivals.

Babylonia's moon god was Sin, the "lord of knowledge." He presided over the calendar and astrological divination. In accord with the approximate number of days in a month, 30 was his sacred number.

The Inca considered the moon to be the sun's wife. She was the calendar-keeper and scheduler of feasts.

Through its monthly cycle of waxing and waning, the moon was often linked with life and fertility. The myth of Osiris provided a clear example, and in ancient India similar connections were felt. The Desana, in southern Colombia, call the moon the "night sun," and they credit it with reinforcing the world's fertility by providing the dew.

The Wandering Gods

In the tradition of associating the ancient gods with the planets, new planets discovered with the telescope also were given names of gods: Uranus, Neptune, and Pluto are now in the celestial roster. After the English astronomer William Herschel first observed Uranus in 1781, and before he proved it was a new planet, he called it *Georgium Sidus,* or "George's Star," after King George III. This name satisfied the English, but no one else, and Uranus was the name that prevailed. In addition, numerous minor planets, or asteroids, have been named after other Classical gods, goddesses, and demi-deities; Ceres is the largest of these. When we ran out of Greek and Roman names, we borrowed from other pantheons. Today Lilith, Hel, Quetzalcóatl, Frigga, Aten, Gilgamesh, and many more orbit the sun.

In very earliest times the Greeks and the Romans do not seem to have differentiated the planets. Writing in the fourth century B.C., the Greek philosopher Plato described the five "wanderers" as gods and mentioned that the practice of associating them with specific Olympian gods was introduced by foreigners. The foreigners probably came from either Egypt or Mesopotamia. The latter is the more likely source since the attributes and characteristics of Babylonian planetary gods parallel those of the Greek gods, while the early Egyptian representations of planets do not.

In ancient Babylon, Marduk was honored as king of the gods and quite specifically associated with the planet Jupiter. In Greece, Zeus was chief of the Olympians, with dominion over the planet Jupiter. In that sense he was the counterpart of Marduk. By contrast, the Egyptians portrayed Jupiter—and Mars and Saturn as well—with the falcon head of the sky god Horus. Among Jupiter's Egyptian titles were "Horus Who Illuminates the Two Lands" and "the Star of the South."

The role of Jupiter-Marduk was preeminent in Babylon, for he was credited with the world's creation, bringing order out of chaos. Texts of the Babylonian creation myth are preserved on cuneiform tablets, some from the library of Ashurbanipal, king of Assyria in the seventh century B.C., but the tale itself is much older, apparently deriving from the Old Babylonian empire, about 1800 B.C. In the myth, Marduk establishes order by killing Tiamat, the dragon of primordial chaos. From the monster's body he fashions the sky and the sea. Then he prepares to take advantage of his victory. His price for his service is the right to fashion an ordered cosmos. First, he organizes the sky, apportioning it among the other gods, symbolized in the constellations overhead. The year is next. Marduk decides how long it will be and subdivides it into months, their passage regulated by the stars he chooses. More celestial references, contrived by Marduk, put the world in order. He also marks the horizon, the zenith, and the points where the sun might emerge and depart. He puts up the moon and assigns it to light the night and count the days of the month. Clearly Marduk was the ruler of the sky.

Jupiter's course through the sky, Marduk decides, will guide the stars and planets.

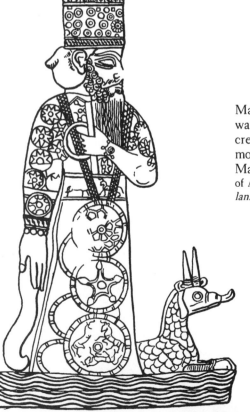

Marduk, the Babylonian celestial sovereign, was associated with the planet Jupiter and credited with the destruction of Tiamat, the monster of primordial chaos, shown here at Marduk's feet. *(J. A. MacCulloch,* The Mythology of All Races—Semitic. *Used by permission of Macmillan.)*

This may seem like an odd choice to make. The constant sun, perhaps, would define things better. But Jupiter's path through the sky follows the ecliptic, the annual path of the sun, more closely than the other planets known to the ancients. Also, Jupiter's configurations in the stars repeat themselves almost exactly every 12 years. For example, Jupiter will come into opposition (that is, be opposite the sun in the sky) 12 times in a span of time just five days longer than 12 years, and the last opposition will occur among the same stars as the first.

These aspects of Jupiter's movement, combined with its brilliance among the stars of the nighttime sky, probably influenced early astronomers to use the planet as a reference, a function reflected, it seems, in the myth. There are uncertainties, however. The actual name for the planet used in the text is Nebiru. Although this did mean Jupiter, it meant other things as well, and sometimes it meant pole, or pivot. The north celestial pole is a key reference for the sky's rotation, so either or both meanings may have been intended in the creation epic.

The other planets also played important, often similar, roles in the pantheons of ancient cultures. And so, the Babylonians associated Ishtar, their goddess of love and fertility, with the planet Venus, another parallel—and perhaps direct antecedent—to Greek and, ultimately, Roman tradition.

Apart from its brightness, the most distinctive feature of Venus is its cycle as a morning star and evening star. Accordingly, the Egyptians sometimes called it "the crosser" and in a later time portrayed it with two falcon heads. More often they symbolized Venus as the *Bennu,* a heron-like bird commonly equated with the phoenix. The Bennu belonged to Osiris, probably because the Egyptians associated death and resurrection with the planet's evening and morning appearances, or perhaps with its conjunctions behind the sun and its periods of visibility. Something similar may be behind the Mesopotamian myth of Ishtar's descent into the Underworld.

Mercury's Egyptian names refer to Set, the adversary of Osiris. Usually the planet is shown with the head of Set's mysterious, unidentified totem animal and it is called *Sebeg.* Among the Babylonians, Mercury was Nebo, the record keeper and messenger of the gods. Its status as messenger may be related to the quickness of the planet in its circuit from west of the sun to east of the sun and back to the west again. Mercury's swiftness also made him the gods' messenger in Greece and Rome, as well as the escort of the souls to the realm of the dead.

Compared to the other planets, Mercury is not often seen, which may account for its assignment to the troublesome Set. In the course of its 116-day cycle, Mercury is invisible one day for every two days it can be seen. Its conjunctions last about five days and 35 days respectively, and its morning and evening appearances are each about 38 days long. This, of course, is the theoretical situation. Because of its nearness to the sun, Mercury generally is difficult to see, and some appearances are more favorable than others. We say erratic and fickle behavior is "mercurial," and the planet's antics in part explain why.

It is easy to pick out Mars in the nighttime sky. Its red color sets it apart from the other planets and from most of the stars. The color—the same as blood—also explains its association with gods of war: Nergal in Babylonia, Ares in Greece, and, of course, the Roman Mars.

The celestial path of Mars is similar to that of Jupiter and Saturn, but it is considerably closer to the earth and so appears to move more quickly. The effect is especially striking when the earth in its orbit swings past Mars, and the latter appears to move backward for a little over two months—a much more dramatic retrograde motion than those of the other outer planets. Egyptian texts refer to Mars as "Horus the Red" and "Shining Horus, the Star of the East in Heaven that Runs Its Course Backward." Apart from the "East," these names are easily understood. Both Mars and Saturn were called the Eastern Star and the Western Star in various inscriptions, however,

Finally, Saturn, the last of ancient "wandering stars," was known as Ninib to the Babylonians. After an initial career as a sun god and patron of the ancient city of Nippur, Ninib became affiliated with springtime and planting. His closest analog in Greece and Rome was Kronos, or Saturn, the youngest Titan, father of the Olympian gods, and also a god of agriculture. The Egyptian Saturn was "Horus, Bull of Heaven," and its falcon head was often shown topped with a set of bull's horns.

The Stellar Gods

In most, if not all, ancient cultures the stars also were made divine. In Egypt, as noted, the stellar timekeepers—the decans—were depicted as gods in Egyptian temples and tombs: Sirius and Orion, as Isis and Osiris, occupied an important place in Egyptian religion.

New World civilizations also connected the stars with gods. Tezcatlipoca was a warrior god of the ancient Aztec and their Nahuatl-speaking predecessors in central Mexico. His name meant "Smoking Mirror," a reference both to obsidian and to the black, nighttime sky. His color was black, and he was, appropriately, a lord of the night. Tezcatlipoca's stellar associations are even more specific. His realm was the north and his stars were those we visualize as the Big Dipper. Usually he is shown without his left foot, the bone protruding from the severed ankle. The foot was ripped away by the Earth, and the story is a metaphor for the Dipper's loss of stars as its handle swings below the northern horizon. Tezcatlipoca's animal identity is the jaguar, a nocturnal creature whose spotted coat mimics, in reverse, the starry sky. Tezcatlipoca gazed into his dark mirrors to glimpse the future and watch the world. This is an activity of shamans, specialists in ritual magic and mystic quests to the sky.

The steadiness of the celestial pole and the constancy in the circuits of the northern stars may make them seem to be more benevolent gods than the sun, whose hot summers and cold winters keep life in a precarious balance. California's Chumash

In this detail from an ancient Mexican codex, or picture writing manuscript, Tezcatlipoca, a lord of the night sky, limps along with his black obsidian "smoking" mirror in his hand. His foot is severed at the ankle, and bone protrudes. *(T. A. Joyce,* Mexican Archaeology*)*

Indians thought of the sky gods this way. They saw a balance of nature and the world order in terms of a nightly gambling game played between two teams. Sun was the captain of one team, while the pole star, Polaris, led the other. Polaris was known as Sky Coyote, and its pivotal position among the stars made it a symbol of the night. The two teams would play until dawn, the Evening Star on Sun's team, the Morning Star on the team of the night. We can understand the assignment of team members in a figurative way: Evening Star is on Sun's team because it lingers for awhile in the evening twilight and then follows Sun below the horizon. Night prevails. In the dawn, Morning Star is up before the Sun, but as Sun floods the sky with light, Morning Star fades with the rest of the night's team. Day prevails.

The game's stakes kept interest in the game high: The winners lived, their opponents died. For a full year they played, and at the winter solstice the score was tallied. Moon, that expert at counting out the days, kept score. If, at the winter solstice, Sun were the winner, it would go bad for people on earth. Rather than return upon his yearly journey back to the north, he might just continue on south and leave the earth in the dead of winter, with the cosmos out of balance. Sky Coyote was a benefactor, a benevolent influence. If his team won, the order of things would be restored.

There is a suggestive image at Painted Rock, in the territory of another California tribe, the Yokuts. An animal with all fours—and, possibly, a tail—stretched out, and

Sky Coyote seems to be balancing Sun on the tip of his nose in this Yokuts Indian rock art. These two cosmic gamblers are painted on the ceiling of the cave known as Painted Rock in California's central San Joaquin Valley. *(Griffith Observatory)*

what looks like a sun disk poised on his snout, is painted on the ceiling, oriented toward the shelter's opening on the southeast. The figure could well represent the exchange between Polaris and Sun.

The Chumash were unusual for regarding the stars, in the person of Sky Coyote, and not the sun, as protectors of their world order. Still, it was the sky that provided orientation and established the sacred order. Clearly the sky game can be played any number of ways. Its point, however, remains the same.

The Purpose of Celestial Power

Our scientific viewpoint distinguishes the meteorological activity in the sky from the astronomical behavior of objects far beyond the confines of the earth's sheer slip of atmosphere. To our ancestors, however, the distinctions we make may not have been clear. In their minds, the sky was a place. Its geography was not our geography of outer space. The gods that lived there were not just astronomical objects that shine down upon earth, but also rain, clouds, thunder, lightning, rainbows, halos, and many more.

Our ancestors' vulnerability to the storm and their dependence upon rain prompted them to imagine numerous gods of thunder, weather, and wind. Because the themes of their lives were intricately entwined—just as our own are—it is often difficult to tell where one metaphor leaves off and another begins. We see this rich fabric of allusion and symbol contrived from threads of association between life and the sky. Rain, for example, falls from the sky and fertilizes the land. The process is paralleled when the man impregnates the woman. And in agriculture, the plow is the male. It cuts the female earth, and the furrow is where the seed is planted. Through such a network of images and associations, the sky, the rain, and semen are interchangeably symbolic.

In the Vedic period of India, for instance, in the second millennium B.C., Indra was a warrior among gods. He was the shatterer of mountains, the bringer of rains, and the hurler of lightning. Varuna was the first sovereign of the sky and the proprietor of cosmic order, but Indra was the force in the sky. His association with the storm linked him with the rain, and the life-giving rain linked him, in turn, with fertility. A storm god like Indra naturally acquires a reputation for sexual excess.

Unlike most ancient peoples, the Egyptians personified the sky as a goddess. But in Egypt this makes sense: It rarely rains. The source of life-giving, fertilizing water is the Nile, not the sky. The river, as Osiris, is male, then, and the sky, as Nut, is female.

Other examples of nonastronomical sky gods are numerous, and their roles, like Indra's, involved the weather. Baal was the Canaanite warrior god of fertility and rain. Teshub was the god of weather and rain among the Hurrians in northern Mesopotamia and, later, the Hittites of Asia Minor. These Hindu and Near Eastern gods are part of the Indo-European tradition. We can, therefore, see some of the elements of later European belief in them. Among the Indo-European peoples, the sky, in all of its aspects, was an obvious reservoir of power. The way that power was conceived— what it was thought to do—is visible in the organization of Indo European society.

What is the business of the gods? To what use is their power put? According to mythologist Georges Dumézil, in *The Destiny of the Warrior,* we can divide Indo-European society into three groups, each with its own function. These components of the social structure were ruler-priests, warriors, and food producers, and each specialist was represented in Indo-European mythology. In the broad sense, the functions of these three groups were actually powers visible in nature: sovereignty, force, and fecundity

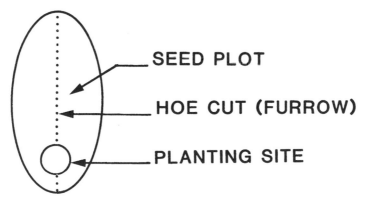

When the Dogon in west Africa's Republic of Mali draw this design as part of one of their ceremonies, they produce a sexual mnemonic that equates agriculture with human procreation. The oval and circle represent the female genitalia. The enclosed area, then, is also the seed plot. The line down the center is the furrow the hoe or plow cuts into the earth, and the circular aperture is the place where the seed is planted.   *(Griffith Observatory)*

Because myths are a means by which a people explains its identity to itself, the main actors in the myths, the gods, represent and endorse the principles by which a people lives. Thus, Indo-European gods adopt the functions of importance in Indo-European life, with the warrior gods of weather and storm identifying with the idea of force. Through life-giving, fertilizing rain, fecundity, also, was linked to the sky. Force animates the world, and fecundity fills it with living things.

Creation, by contrast, was the province of the sky gods because the sky orients the landscape and orchestrates the rhythm of time. We see a fundamental pattern in the celestial realm and frame from it what seems to be the cycle of cosmic order and the way of the world: creation-growth-death-rebirth. We seek our own past, present, and future in that cycle. Our interaction with the sky is a reflection of our own identity. It is so important, we involve the gods.

Creation involves more than just promoting activity or procreating life, however. True creation is the establishment of order. This is why the sky is required, and this is where sovereignty comes in. By sovereignty, Dumézil meant preservation of sacred order. This responsibility was split between two regents—one to sustain cosmic order, the other to preside over social order and its laws.

The dual nature of creation—coordinating as well as originating—is evident in many myths. For Marduk, for instance, fashioning the world from the body of the dead monster Tiamat was only the beginning: He went on from this to establish a continuing order by regulating the motions of celestial objects and creating a calendar.

The Norse creation myth follows the same pattern. In this case, by the time the myth was written down, Odin had assumed the role of chief of the gods and the sky. He was supposed to have created the world as we know it. Before his time, there had been a void, a yawning gulf of nothingness, a boiling fountain of waters congealed from the chaos, cold, and fire. From this source all rivers flowed, and in one of them ice hardened out and spread throughout the void. Ymir, a giant, condensed and quickened into life there. Later, after the gods were born, Ymir fought them and lost. From his corpse they crafted the world. His skull was thrown up to make a sky, and four parasites were picked from Ymir's carcass, fashioned into dwarfs, and set to stand at the cardinal points of the compass: north, south, east, and west. The landscape was organized. Sparks from Muspelheim, the realm of fire, thrown into the sky by Odin and his brothers, became the sun, moon, and the stars. Their places and paths in the sky were next set, and from this effort it was possible to count out the days and nights and to measure the years. The ruling gods, through this act, established Odin as sovereign of the sacred order. As with Marduk, it was not enough to throw the world together from a giant's carcass. Order must be established formally once chaos is defeated.

Even if we depart from the Indo-European roots of our own civilization and look at a completely independent tradition, we find the same themes repeated. The Fon, for example, are a West African people living in the nation of Benin, once known as Dahomey. In their mythology a giant snake created the world by first setting up four posts. Each was placed at one of cardinal points in order to hold up the sky. Caparisoned in black, white, and red robes, the serpent wrapped itself around the posts to brace them upright. As the great snake's coils slowly rolled, the proper colors would light the sky: black at night, white at day, and red at evening twilight and dawn. This myth makes sense. The serpent's movements are in its coils, and its coils are the eternal turns of the sky. They send all the celestial objects along their paths. It is the change, the daily movement of the sky, that puts the cardinal directions in their proper places. By organizing the landscape, a world becomes possible. It is no coincidence that this concern for orientation is expressed so early in the myth.

Heaven's Mandate: The Sky Reaches Earth

Celestial order generally was transfused into human society in ancient times through the sovereignty of the ruler. The mandate of heaven sanctified kingship. By invoking the sky, kings and their institutions gained special authority and meaning. Like the sky, they could not be carelessly compromised. To do so would jeopardize the people's way of life and their sense of their own identity. Mircea Eliade, a specialist in comparative religion, emphasizes that orientation—the order of space and time—is what makes the world sacred for traditional, nonsecularized peoples. Once the world is founded and ordered, it acquires meaning. We can then have a sense of place in it.

In Egypt, an intricate network of language and ceremony reinforced a resonance between the sun and the pharaoh. The name of the sun was incorporated often in the cartouche of the king. The king wore the *uraeus,* the divine solar cobra, on his crown. The king's public appearance on the throne was equated with the sun's arrival on the horizon. Both concepts shared the same verb, a hieroglyph of the sun emerging from the primordial mound of creation. Enemies of the king were supposed to suffer the same defeat as Apep, the serpentine nemesis that faced Re in the last hour of his journey through the night. A festival of royal rejuvenation, the *Sed,* when held, was scheduled for the first day of the first month of the first season. Timed, then, at the New Year, the *Sed* invoked the notions of re-creation and reestablishment of the world order. Just as the sun was renewed, the pharaoh was reenergized, and so, too, was the land. Like the sun, the pharaoh was divine. He was the son of the Sun.

The mandate of heaven could invoke other imagery, however. The Maya of ancient Mesoamerica made use of some of these alternatives. Numerous dynastic monuments of the Classic (ca. A.D. 300–900) Maya rulers incorporate celestial symbols into the badges of office and the signs of authority. Portraits of high-ranking persons appear on stelae—carved, upright stones frequently found at Maya ceremonial centers. The person usually is shown standing, and often he carries a manikin scepter. This ceremonial object acquires its name from the imp on the top of the stick. It looks, in fact, a lot like a stick puppet from a Punch and Judy show.

We know who the dwarf on the wand really is. His flamboyant nose and upturned upper lip mark him as God K, or Ah Bocon Dzacab. Mayanist Michael Coe emphasizes his association with royal lineage and sees parallels with the central Mexican sky

The face that glares from the darkness of the Temple of the Sun at Palenque, Mexico, is usually thought to belong to the Maya sun god in one of his aspects, the Jaguar God of the Underworld. Recently, however, Floyd Lounsbury of Yale University has shown that dated inscriptions on the temple may force us to recognize here the face of another Maya celestial deity and a kind of "night sun": the planet Jupiter. In either case, the presence of a celestial god in this temple is linked with the concept of the sovereignty of the Maya king.   (Robin Rector Krupp)

god Tezcatlipoca—"Smoking Mirror." Ah Bocon Dzacab sometimes is shown wearing a mirror on his forehead, and smoke issues from it. This is the same imagery seen in representations of Tezcatlipoca. Also, God K, like Tezcatlipoca, is missing a foot. One of his legs ends in a serpentine dragon. God K's connections with the sky are strengthened by identification of the "dragon" with Itzam Na, the benign lord of creation, whose reptilian body encompasses the universe: sky and earth. The manikin scepter is a sign of cosmological authority.

Astronomers John Carlson and Linda C. Landis have analyzed another symbol of Maya dynastic power—the "skyband." A skyband is a chain of symbols, each in its own rectangular frame. Carlson and Landis have shown that these individual symbols relate to the sun, the moon, Venus, day, night, and the sky itself. The ribbons of symbols are equated with the body of a serpentine dragon, and they appear on monuments in a variety of ways. The decorated belt worn by a chief or the hem of a cloak may actually be such a skyband. Sometimes a ruler is shown holding a ceremonial bar of authority. Dragon heads terminate both ends of the bar, and the bar itself—the two-headed dragon's body—is a skyband. Individual symbols are the dragon's "scales." Skybands also decorate the platforms and thrones of the elite, and they appear in the elaborate headdresses of the Maya chiefs.

Correspondences between the skyband symbols and the patron deities of the "months" (20 days long) in the *Haab,* or 365-day year, imply some connection with the sun's annual cycle through the sky. Agriculture was, of course, structured in terms of the solar year. Production and distribution of food was organized and administered by the elite. Their tool was the calendar. Its foundation is the sky. No wonder the Maya rulers costumed themselves with the sky. They presided with a nod from heaven.

The notion that the organization of human affairs is closely related to the order of the cosmos is by no means restricted to ancient Egyptians and the Maya, of course. Consider the Gospel of St. Matthew, in which the birth of Christ is heralded by the sky. A brilliant, miraculous star drew "wise men" from the east, presumably Persian or Babylonian astrologers, to Bethlehem. The historical circumstances are in dispute, and the star may have been a detail contrived by St. Matthew. None of the other gospels mentions the star. We know, however, that several impressive conjunctions of planets and stars occurred during the most likely period of Christ's birth, and any of them could have been the Star of the Magi. But real or not, the Christmas Star has impact. It means new life and new order, and it appeared in the sky.

December 25, the traditional date of the birth of Christ, also signals the world's rebirth. The date owes more to celestial cycles and seasonal symbolism than to any historical record. It was borrowed from an early competitor of Christianity, Mithraism. This religion originated in Persia and passed to Rome, where it absorbed elements of the old Roman paganism.

Mithra was an avatar of the sun. He drove a solar chariot, judged the world, and preserved the cosmic order. His birthday was December 25, and, in the old calendars,

The Maya ruler portrayed on Stela 10 at Seibal, in Guatemala, carries a ceremonial bar of authority decorated as a skyband to acknowledge the mandate of heaven. The glyph on the left that looks like a *w* or a Greek *omega* is a Venus symbol. His belt is also a skyband, and the four-petaled "flower" symbol on the left is the *kin* glyph, which can mean "time," "day," or "sun." Hanging over his spotted jaguar-skin skirt, his breechcloth bears a portrait of the cross-eyed sun.   *(W. G. Turner,* Maya Design Coloring Book*)*

Seated in a niche on the back side of Stela I at Quirigua, in Guatemala, the Maya figure is surrounded by a skyband to create almost a "halo" of celestial authority around him. *(E. C. Krupp)*

it coincided with the winter solstice, when the sun turned back from its southernmost path. As the annual rebirth of the sun's light, the winter solstice was important in most parts of the world. In fact, the Romans already had an ancient winter festival whose seven days bracketed the solstice. Derived from very old fertility rituals of the harvest, the Saturnalia—the festival of the god-planet Saturn—marked a period of license and intoxication. Choosing the birth of Christ as December 25 successfully integrated long-standing popular traditions with the imagery of a new religion, and the theme of renewal is still part of Christmas. Its story of the star can still excite wonder.

(Left) The Star of Bethlehem is the celestial sign of a new order in the Christmas story. In this Gustave Doré engraving, the three Wise Men and their retinue are following the brilliant star to the Christ Child's manger.

4

The Tales We Tell

Myths are not simple tales, but tales told simply. And these tales are not just idle chatter. The stories are important. They reflect, in symbols, the deepest concerns of our minds. For this reason myths are worth analyzing. The richest myths have multiple levels of meaning. They are networks of thought. Because they link a variety of perceptions and incorporate more than one theme, they endure. Our goal is to understand the use we make of the sky, and here mythology can help. We are hunting descriptions of the world's structure, explanations of natural phenomena, and accounts of the passage of time. We find them all in myth.

We must be thoughtful, however, about what we do with myths. We can identify the celestial connotations in a story, but then to declare the tale to be a "sky myth" shortchanges the metaphor. The sky symbolism does not necessarily explain the myth. The myth, on the other hand, may help explain what the sky means to us. Celestial metaphors still speak to us. To know what they say and why it matters, we must understand, in the words of anthropologist Claude Levi-Strauss, "not how men think in myths, but how myths operate in men's minds without their being aware of the fact."

What emerges in myth is a connection between the sky and cosmic order. The theme may appear in the imagery of the cycle of the seasons or the orientation of the world. These myths link the sky to society's institutions and so reflect the role of cosmic order in human affairs. We find in these tales recitals of the creation, maintenance, destruction, and restoration of order. The sky is both the source of this order and the place where it is challenged.

Greek mythology provides a clear narrative of one such challenge: the insistence of

Phaëthon, the mortal son of Helios, god of the sun, on driving his father's golden chariot. A journey to the east brought Phaëthon to the shining palace of the sun. On its silver doors the twelve signs of the zodiac symbolized the sun's proper path through the stars. Inside, Helios was enthroned, attended by the Day, the Month, the Year, the Four Seasons, and the Hours—incarnations of the order of time. The god welcomed his son and swore, as evidence to all of the young man's divine lineage, to grant whatever favor he asked. When Phaëthon made his injudicious request, Helios tried to convince him of his folly, but failing in this, reluctantly instructed his son in the perils of the ride and led him to the sun chariot.

As the fiery horses climbed, they sensed a novice at the reins and left the proper course. Phaëthon, panicked by their reckless run and by the chariot's wild sway, was not able to regain control. The mad path they followed scorched the earth and set fire to the sky. Mother Earth, cracked and ablaze, appealed to Jupiter in heaven for aid. She warned that if the earth's poles, smoking now from the heat of the vagrant chariot, were to burn through, even the celestial palace of Jupiter would tumble.

Calling, then, on all the gods to witness that the safety of the entire universe was at stake, Jupiter hurled a thunderbolt at Phaëthon. The chariot was shattered, and into the Eridanus River Phaëthon fell in flames.

The myth of Phaëthon could represent any of the several celestial phenomena. Perhaps Venus, whose path crosses to either side of the sun and is in some ways erratic, is intended by the allegory. Or perhaps some dramatic, unexpected, and unwelcome cosmic visitor inspired the myth. A comet may seem wayward and unsettling. Some evidence in the myth implies that the story concerns the order of the year and the fear the sun might abandon its normal path, an anxiety akin to the notion that in December the sun threatens just to continue its way south and maroon the world in winter. Although the chariot's departure with Phaëthon was styled as a sunrise in the sun's daily course, the sun's annual motion on the ecliptic may be the story's real core. Helios advised Phaëthon the road would pass by the Bull, the Archer, the Lion, the Scorpion, and the Crab. These constellations define the better part of the zodiac, territory the sun would never encounter in but one day's travels.

There is a cosmic theme here: the challenge to world order. Mother Earth, when she calls out to Jupiter, reminds him that if "the sea perish, and the land, and all the realms of the sky, then we are back again in the chaos of the beginning!"

Primordial Chaos threatens a return to the cosmos. Because the sky is the visible framework of world order, the sky suffers if Chaos reemerges. In the version of Phaëthon by the Greek writer Nonnos, "There was tumult in the sky shaking the joints of the immovable universe; the very axle bent which runs through the middle of the revolving heavens." Despite threat of storm or eclipse, the sun—holding to its steady courses, its predictable cycle—measured out the world's order in time and space. No more vivid image of the peril of chaos could be contrived than the chariot of the sun out of control.

The Order of the Seasons

When the sky is linked with a story, it may tell us something about the sky's importance in our perception of the world. But it is not always easy to see the link. To understand myth, we have to understand the language of metaphor—and we have to know the cultural context from which the particular myth comes. A seemingly simple, ribald tale may have much deeper meanings than meet the uninitiated eye. For example, take the tale of Yawira, told by the Barasana Indians, an Amazonian tribe that inhabits the region of the Río Píra-Paraná in southeastern Colombia, near the border with Brazil.

Yawira, a woman who figures in several Barasana myths, agreed to an amorous appointment with Tinamou Chief. (A tinamou is a ground bird that looks like, but is unrelated to, the partridge.) Another participant in Barasana myth, Opossum, overheard the plans for the tryst and tricked Yawira into meeting him instead. Yawira was to take a path marked by the feather of a japú bird and not the fork posted with the feather of a blue macaw, but Opossum ran ahead and switched the feathers. The plume from the foul-smelling japú led instead to the home of the foul-smelling Opossum.

Yawira lived for some time with Opossum, but had no desire to stay. While bathing in the river she disobeyed Opossum's instructions and looked downstream, where she saw Tinamou Chief. Yawira swam down to Tinamou, but he rejected her because she had such a bad odor. Sleeping with Opossum had given her a stench. Despite Tinamou's initial rejection, Yawira stayed with him, and eventually Opossum came looking for her. Tinamou and Opossum quarreled, and Opossum was killed. At that moment, a downpour began, an omen Opossum had predicted would occur upon his death.

What does this story have to do with the sky? The connections become apparent only when the symbolic associations of the main characters are known. Tribes in this part of the Amazon associate the tinamou with the sun on account of its yellow plumage. In *The Palm and the Pleiades*, Dr. Stephen Hugh-Jones, the British anthropologist who collected this Barasana myth, goes on to describe an even more intricate web of relationships. The woman, Yawira, is equivalent to another character in Barasana mythology, Romi Kumu—and she is the mother of the sky, or the sky itself. The sky, in turn, is described as a gourd, but the gourd is special. It is the same as the gourd in which the shaman keeps beeswax, a substance burned on certain ritual occasions. For the Barasana the gourd's own shape and the distinctive aroma of the wax smoke have connotations of female sexuality.

Now Romi Kumu ages as the day passes, and by nightfall, she has grown old. Her youth and beauty are restored, however, each morning. The gourd, a source of fertility and renewal, gets the credit for this transformation. By equivalence, the sky is also a source of cyclic renewal. As we shall see, these associations can get even more specific. The Pleiades (a well-known and unique cluster of stars) are also equated with the gourd and its fecundity.

More sky symbolism must be collected to make sense of the myth, however. It is Opossum who provides additional bonds between the story and the sky. In certain Indian names Opossum and the Pleiades are linked, and these associations are significant because Opossum and the dry season are related. The seasons and the stars are the hidden parts of the tale.

In the tropical rain forest of the Barasana, there are four well-defined seasons. A long dry period, starting in December, is followed by a long wet period, beginning in March. A second, short dry season begins in August and ends in September, when another rainy season—but a short one—intrudes, until December's drought.

Another Amazonian myth, about a contest of hibernation and fasting between the tortoise and the opossum, provides additional clues. Opossum buried himself in the ground and was due to emerge at the end of the dry season, but died instead. As in the previous Barasana myth, Opossum's death is followed by heavy rains. And significantly, it is in March—when the drought ends, when Opossum dies, when the rains come—that the Pleiades are just east of the sun. They set as the sky darkens. They have "fallen" from high in the sky to the western horizon—swum "downstream" to the sun—to die as the dry season ends.

And the myth-metaphor continues. The Pleiades disappear from the nighttime sky in May and remain invisible until late June, when they first become visible in the east, before sunrise. Each day shifts the Pleiades a little to the west. They rise a bit earlier each dawn. By December they appear to rise in the east, as the sun sets in the west, and December marks, once again, the beginning of Opossum's long dry season. For the Barasana the sequence of drought and rain orders the year and is punctuated by the Pleiades, with the long dry season coinciding with the months that the Pleiades rule the nighttime sky.

| March | The Pleiades set in the west after sunset. | Long dry season ends. Long rainy season begins. Opossum dies. |
| May | The Pleiades set in the west with the sun and disappear from the night sky. | |

| June | The Pleiades rise in the east before the sun and reappear in the predawn sky. | |
| December | The Pleiades rise in the east at sunset and are visible throughout the night. | Short rainy season ends. Long dry season begins. Opossum lives. |

Nothing in the sky is quite like the Pleiades. They are a conspicuous collection of stars, six of which can be seen with the unaided eye under normal viewing conditions. Now there are plenty of individual stars in the sky, and people throughout the world have associated various stars together into constellations. These may differ from culture to culture, but the Pleiades are recognized by nearly everyone as something special. Worldwide, they are seasonal heralds. Their comings and goings coincide, more or less, with one change of weather or another, and so their risings and settings have been used to regulate calendars, festivals, and rituals.

Fortuitously, the Pleiades are located near the ecliptic, the sun's path through the sky. This position is what makes them useful, for the round of seasons is reflected in

The stars of the Pleiades are grouped closely, but they remain bright and distinct. The unique visual appearance of the Pleiades and their position—fairly close to the path of the sun—have made a special seasonal signal out of the cluster for many peoples. *(Curtis Leseman)*

In the network of associations that binds Barasana myth, the shaman's beeswax gourd is the same thing as the Pleiades. This Barasana drawing of a gourd shows some other relationships between the gourd and the sky. For example, the fringe around the edge is also the rays of the sun. This gourd is also a womb and is the source of renewal of life, just as the Pleiades bring the rains and the seasonal renewal of the year. *(S. Hugh-Jones,* The Palm and the Pleiades, *Cambridge University Press.)*

the sun's annual motion. Any spot on this path automatically splits the year in some way. Any star would do, but the Pleiades are compact, conspicuous, and unmistakable.

There are no simple equations in the Barasana myth. All of the players have multiple and overlapping associations. For example, Yawira and Romi Kumu can also represent the Pleiades because they are women. The year is cyclically renewed through the Pleiades, and the rains are symbolically identified with the flow of menstrual blood. The first rains therefore are Romi Kumu's menstrual blood. Also, in Barasana thought, the Pleiades can be the wax gourd, and the wax can be the blood. The Pleiades become the source of the rain as the gourd is the source of the wax and the vagina is the source of the blood. From the "fall" of the Pleiades and the death of Opossum comes renewal.

A single principle unites all these ideas, and it is periodicity. The repeated cycle defines the world. It creates order. The Barasana confirm this in their most important ritual, known as *He* House. This ceremony can take place only at the junction between the long dry season and the long rainy season, when Opossum and the Pleiades die. Wax is burned, and the wax gourd is used in the ceremony along with special musical instruments that women are forbidden to see. Male and female emblems are brought into conjunction as the two seasons are joined. Thus, hot dry sun (male) and rainy sky (female) are invoked. Their interaction creates the year's cycle. They are opposite but complementary. They make the world complete.

The purpose of *He* House is reinstatement or reconsecration of natural order, and it does this through agents linked with the sky. The seasonal cycle is the organizing principle of the world, and the Pleiades signal the seasons.

The Structure of the World

In the *He* House ritual of the Barasana, wax is ceremonially burned. The smoke is fanned by the presiding shaman toward the four cardinal directions, a recognition of the structure and orientation of the earth. Cardinal directions—north, south, east, and west—are what we use to orient the landscape, and these references acquire their meaning from the sky. In the northern hemisphere considerable evidence, and common sense, support the idea that this system of directions evolves first from recognition of the north celestial pole. This is the fixed point that defines all other directions—the cosmic axis, sometimes conceived of as a sacred mountain, pole, or tree, around which the sky turns. It, too, is the subject of many myths.

On the Great Plains of Nebraska, the Omaha Indians preserved a myth of a sacred cedar tree upon which the stability of their society was founded. During a time of tribal disruption, brought on by rivalries between the Omaha chiefs, one of their sons encountered a burning cedar tree. Although enveloped in flames, the tree did not burn up. The chief's son noticed that the forest animals had worn four trails to the tree, one leading to each cardinal direction, and in the tree's branches roosted thunderbirds, which are emblems of the sky and the talismans of warriors.

Upon learning of the burning cedar, the Omaha warriors dressed for battle and attacked it. They brought it down and transported it to their village. There the tree was reerected, and responsibility for it was assigned to one family. Through the tree, the once-threatened social order of the Omaha was strengthened. Problems, disagreements, and troubles were all to be brought—with presents and prayers—to the sacred pole. Leadership of the tribe was invested in the keepers of the pole, and only through them could authority over the tribe be transferred to others.

As with the "simple" Barasana story of Opossum, the Omaha myth operates on many levels of meaning. For example, it is no coincidence that the burning tree is a cedar. The cedar is coniferous—an evergreen—and so represents enduring life. Meanwhile, its position at the center of the four cardinal directions identifies it as the axis of the world. In this way, the symbol of the tree fits with the analogy of the celestial axis renewing life through the day-night cycle.

These associations are by no means unique, and mythologist Mircea Eliade has, in

At the age of nine, Black Elk, an Oglala Sioux shaman, experienced his "Great Vision." In it he was transported to the Center of the World. The Six Grandfathers—the powers of east, west, north, south, earth, and sky—presented him with gifts. One of these was a living branch, a Tree of Life, "the living center of the nation." Then, when Black Elk suddenly found himself looking down upon the world from a great height, he saw the sacred stick blooming in the center of the world. Through it crossed two roads: one from east to west, the other from north to south. This drawing, based upon a watercolor by Standing Bear, shows Black Elk at the Tree at the Center of the World receiving the gifts from the Grandfathers of the world's four quarters. Like their linguistic and cultural neighbors, the Omaha, the Sioux associated the cosmic axis with the upright Tree of Life and combined it with the four world directions into a system of cosmic order on which the well-being of the people depended. *(Griffith Observatory, after a watercolor in* Black Elk Speaks *by John G. Neihardt)*

Shamanism—Archaic Techniques of Ecstasy, explained the link between such trees and shamanism. As the world's axis, the tree is a route to the sky, where the shaman seeks to ascend. Because the sky is in everlasting cyclical motion about the celestial pole, the tree equates, in turn, with the principle of continuous renewal. This renewal marks the tree, or pole, as a limitless source of life. It is the Tree of Life.

Because the sacred tree or pole is associated with renewal of life and the world's order, it is a reservoir of sacred power. Our Omaha myth tells us quite explicitly that this power restored order in the tribe. As a mechanism for authorizing leadership, the miraculous cedar tree secured the social order and the prosperity of the tribe. Its

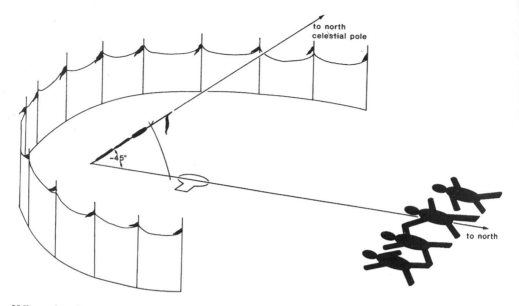

When the Omaha Indians set up their Sacred Pole for an annual ceremony of reannoint-ment, they arranged it to lean in a crotched stick at an angle of about 45 degrees toward the center of the encampment. Because the ceremonial location for the Sacred Pole was in the south quarter of the circular arrangement of the camp, the pole pointed to the north as well as to the "center" of the people. The latitude of the Omaha territory extended be-tween 41 and 43 degrees north, and the Sacred Pole, therefore, was very nearly oriented on the north celestial pole. *(Griffith Observatory)*

power to do this is clearly tied to the sky, also a reservoir of sacred order. Cardinal directions and the celestial pole are products of the sky's most obvious cycles. We draw upon this natural framework and orient ourselves. The celestial pole is a beacon of stability. It steadies the world. It gives it structure. It is, therefore, sacred.

The Maintenance of Order

People in organized societies trace the lineage of their institutions to sources with enough power to justify the way things are. We find it important to demonstrate that our ways of governing ourselves, of organizing ourselves, are part of the natural order. This is done in terms of myth, because myth is the realm of the sacred.

By common consent, reverence and respect are signs of our acknowledgment of what is sacred, which, by definition, transcends the ordinary. It is immune to chal-lenge and interference. It is like the sky. That is why the god-kings of ancient times allied themselves with heaven, and why the divine right of kings was very much an issue in European history until comparatively recent times. In seventeenth-century France, Louis XIV, "the Sun King," enveloped himself and his monarchy in solar imagery. This was but an echo of ancient tradition, but the meaning was clear: From the Sun King all power radiated, by this authority all order was sustained. The sacred solar framework was presented as the model for society's order.

Once order is created, it must be maintained. Recurrent events—the sunrise, the moon's phases, the departure of the Pleiades from the nighttime sky, the circuit of the stars around the pole—are both the source of cosmic order and the signs of its preservation. The ancient kingships guarded the social order by invoking heaven in myths. The myths authorized their power. The Aztec myth of the birth of Huitzilopochtli illustrates how the cosmic order is maintained.

The Aztec, or the Mexica as they called themselves, were the last of a series of northern tribes of barbarians who forced their way into the Valley of Mexico. They were hostile and predatory, and Huitzilopochtli, their tribal god—the source of power—was a warrior. He was associated with the high-flying, aggressive eagle, and the high-flying, day-ruling sun. After several conflicts with the tribes already settled in the valley, the Aztec retreated to a small and disagreeable island in Lake Texcoco. There they saw the sign Huitzilopochtli had said would signal the end of their wandering: Perched upon a cactus an eagle consumed the fruit. (There are several variants of this legend, and in some the eagle also consumes a serpent. This is the scene depicted on Mexico's flag.) Upon that spot they built the shrine of Huitzilopochtli and founded, in A.D. 1325, what would become Tenochtitlán, the capital of the mighty Aztec state.

Huitzilopochtli had numerous solar attributes. His name means "hummingbird on the left," and in Eva Hunt's intriguing analysis, *The Transformation of the Hummingbird*, these associations are cleverly explained. Hummingbirds, like the sun, do not fly at night. In the rainy season (spring and summer), the sun is high, and hummingbirds are flying. In the dry season, the sun is low, food is scarce, and the birds are less active. The bird's feathers are bright, like the sun, and shine most brilliantly in sunlight. Able to hover and fly backward the hummingbird mimics the sun's hesitation at the solstices and the annual back-and-forth shift of the sunrise upon the horizon.

Huitzilopochtli also was known as the Hummingbird of the South, and this identity, too, can be understood in terms of the sun's motion. Throughout Mexico, the sun's daily path falls south of the zenith during most of the year. By facing west, the direction of the sun's daily motion, the sun is placed on the left and across the south. In Nahuatl, the language of the Aztec, another name for the sun was, in fact, "the left-handed one."

Contradictions and unresolved details in the mythology of Huitzilopochtli make his identity complex and his origin problematic. Some traditions suggest that the first Huitzilopochtli may have been a man, an actual warrior chief, who, at some point in Aztec history, was transformed into a god. We know for certain that as the Aztec forged their empire they refashioned their history to conform with their imperial needs. In the fifteenth century, Tlacaelel, the chief counselor to three Aztec kings,

Huitzilopochtli, the tribal god of the Aztec, was closely associated with the sun. He was known as the Hummingbird of the South, and seated here in his regal hummingbird regalia he carries a sunbeam, or fire-serpent, as a weapon. *(after Fray Diego Durán,* Book of the Gods and Rites, *Griffith Observatory)*

engineered an image of the Aztec as a chosen people dedicated to a sacred mission. Under Tlacaelel's direction, old codices were burned, and a new mythology, more serviceable to Aztec ambitions, was contrived. Expressed in ritual and hymn, but not codified into an official version, Huitzilopochtli was protean, a versatile god whose origin became a drama set upon a cosmic stage and whose undiluted power was as public as the sun. His story begins with Coatlicue, his mother. Her name means "serpent skirt," and she is Mother Earth. Visitors to the Aztec Room in Mexico City's National Museum of Anthropology usually are unsettled by a monumental sculpture of her. She is portrayed symbolically, her "face" formed by two serpent heads, in profile, fang to fang. Dressed in a skirt of writhing snakes and a necklace of severed hands and human hearts, she is decorated with emblems of life and death. Snakes crawl, of course, upon the surface of the earth, and many peoples have associated the serpent's shedding skin with the idea of birth and renewal. The skull upon Coatlicue's breast, as well as her necklace, suggest sacrifice and death. The earth is the mother of all life, but she consumes living things as well. All that lives also dies upon her. Even the stars are her children. The *Centzon Huitzinahua,* or "Four Hundred Southerners," are the stars, brothers who are born from her body each night. They emerge from the eastern horizon and parade to the west until dawn.

The story of Huitzilopochtli's birth is told in the *Florentine Codex,* one of several sources of the myth. This manuscript is really an ethnographic account of Aztec life, collected by Fray Bernardino de Sahagún from native informants shortly after the Spanish conquest of Mexico. Sahagún called his book *General History of the Things of New Spain,* and in its third section, "The Origin of the Gods," we are told that Coatlicue was doing penance at Coatepec, or "Serpent Hill," the mountain of the world's beginning. While she was sweeping, a ball of feathers fell from the sky, and she tucked them under the coils of her skirt for safekeeping. When she later reached for them and could not find them, she realized she had conceived another child.

The other children of Mother Earth, the stars of the southern sky, were outraged by their mother's scandalous condition. They convinced their sister, Coyolxauhqui, who some scholars believe is the moon, to join them in a march to the mountain to

The skirt of snakes worn by this monumental figure identifies her as Coatlícue, the mother of Huitzilopochtli. Fabricated from a variety of symbols of life and death, she represents the earth, which creates life and consumes it.   *(E. C. Krupp)*

As the Moon and Stars marched to Coatepec, the home of Coatlícue, she gave birth to Huitzilopochtli. He emerged from her womb fully grown and fully armed. This portrayal of Huitzilopochtli's miraculous birth comes from Book 3 ("The Origin of the Gods") of the *Florentine Codex* (or *General History of the Things of New Spain*), by Fray Bernardino de Sahagún. *(Griffith Observatory)*

slay their mother. Coyolxauhqui, too, was angered by the mystery and shame of her mother's pregnancy.

In the Aztec language Coyolxauhqui means "adorned with bells." An army of stars, led then by the moon with her bells jingling at her ankles and wrists, approached the sanctuary of Mother Earth. Coatlicue trembled, fearful for herself and her unborn child, but from within her womb the child called, "Have no fear. Already I know what I shall do."

As the starry children of Coatlicue approached their mother's door, Huitzilopochtli was born. Suddenly and fully grown, he burst from her womb—a savage warrior armed with serpents of fire and light. First, he struck Coyolxauhqui. A fire serpent sliced her neck and severed her head from her body, which rolled, tumbled, and fell to pieces at the foot of the mountain. Huitzilopochtli then set upon his older brothers, the southern stars. He chased them from the mountain top and pursued them around its base. They were slain or scattered and dispersed from the sky.

The tale is straightforward enough. Huitzilopochtli is the sun. His light devastates the night. He kills the stars. His sunbeams, or fire serpents, slice up the moon. Coyolxauhqui falls, as does the old waning moon, a thin crescent fading into the dawn. To rise each day and rule it, Huitzilopochtli must be fed. Human hearts, obtained from prisoners captured in the "Flowery Wars," were offered to him. The Flowery Wars were contrived battles, instituted by Tlacaelel and intended, as he said, to be "a convenient market ... where our god may go with his army to buy victims and people to eat as if he were to go to a nearby place to buy tortillas."

In the rituals of sacrifice these warrior victims were equated with the Four Hundred Southern Stars. The myth of Huitzilopochtli's birth was reenacted each year in a great public festival, which culminated in the sacrifices at Templo Mayor, the great twin pyramid of Tenochtitlán.

An ambitious program of excavation on that site, close to the main square, or Zócalo, of Mexico City and its Cathedral, is revealing much new information about the Aztec and their capital. In February, 1978, a spectacular discovery was made in the course of digging for new electric cable. There, 10 feet below the street called Calle Republica de Guatemala, Coyolxauhqui had lain imprisoned. Workmen found a huge stone disk, 11 feet in diameter and nearly 20 tons in weight. The dismembered goddess is carved upon it. This sculpture is as heroically proportioned as the so-called Aztec Calendar Stone, so extensively promoted as *the* emblem of the Aztec.

Another large stone sculpture representing the decapitated head of the Coyolxauhqui is displayed in the National Museum of Anthropology. On the goddess's cheeks are the emblems of bells and the hieroglyphs for gold. Fire-serpent darts penetrate her ears and nose, and her eyes are half closed in death.

A huge carved disk bearing the image of the Aztec goddess Coyolxauhqui was uncovered during excavations near Templo Mayor in Mexico City in 1978.   *(Robin Rector Krupp)*

The newly discovered portrait of Coyolxauhqui has the same symbols on the face, while also depicting the body with floral lacerations at the shoulders, neck, and thighs. Bones protrude from her severed legs. Fire-serpent tourniquets throttle her limbs, and more sunbeams, in the form of talons and claws, nip her heels, elbows, and knees. Fragmented and frozen, she floats on the stone, at the base of the pyramid, a victim of sunrise. Although the myth makes sense with Coyolxauhqui as the moon, some experts on Aztec tradition doubt that interpretation. But in any case the story clearly concerns the sun's victory over the nighttime sky, and Coyolxauhqui belongs to the vanquished night. She was ravaged by Huitzilopochtli, and it is her doom to circle through this death month after month. By such sacrifices the days are counted. In hearts and blood the world order was sustained.

Despite the obvious solar dimension of Huitzilopochtli's character, it would be imprecise to say he was simply the sun. As the seasoned warrior talisman of the Aztec, he was born in violence. Conquest, not pedigree, established his titles. He was a latecomer to the company of central Mexican gods, just as the Aztec were the last to arrive in the Valley of Mexico. The Aztec had no standing, no lineage, no traditional right to territory in the valley. They seized power through force. Huitzilopochtli has no clear lineage either. His celestial brothers and sister were shamed and angered by what they construed as their mother's illegitimate pregnancy. Like his chosen people, Huitzilopochtli traded in fear and illegitimacy. Yet through him the world order was maintained. His myth may have been as fabricated as Louis XIV's Sun King, but Tlacaelel sensed correctly the power in the sacred. Huitzilopochtli's link to the sky legitimized Aztec sovereignty and powered a ritual that sustained it.

The Intrusion of Chaos and the Restoration of Order

We don't perceive the world as chaos. We don't scramble space and time. Instead, we watch the sky and decide that there is order in the world. But we never quite forget the possibility of chaos. Certainly, the Aztec had a sense that the whole cosmic structure might collapse, and ritual sacrifice was their way of preventing this.

In many creation myths, *creation* means the defeat of chaos and the introduction of order. Chaos, however down, is not out. It lurks somewhere, submerged below the world's structure of space and time. We allow it to intrude in myth, and permit ourselves to banish it again. Set kills Osiris, but Osiris is resurrected and order is restored. Our myths of the intrusion of chaos and the restoration of order have a celestial dimension, for the sky and the seasons mirror this conflict and its end. We gaze upward and harvest the metaphors we find there.

A clear account of the retreat of cosmic order is contained in the Japanese tale of the sun goddess, Amaterasu Omikami. It is one of the Shinto myths, first written down in the eighth century after Christ and collected in two compilations, the *Kojiki,* or "Record of Ancient Matters," and the *Nihongi,* or "Chronicles of Japan."

Amaterasu ruled the Plain of Heaven and resided in a palace there. Her name means Great and August Heaven Shining Spirit. She was created with Tsukiyomi, the moon, and Susano, a spirit of force and storms.

As the sun, Amaterasu was the gentle, radiant source of celestial life. Susano was certain to clash with her. He was as arrogant and hostile as the clouds and rain that wall out the sun. His name means Swift, Impetuous Male and emphasizes rude, even violent, sexuality. Amaterasu also was associated with fertility and procreation, but her character was benign, her behavior measured.

Susano's advances upon the sun goddess were tempestuous, and, in character, he broke down her rice field boundary ridges and blocked up her irrigation ditches. (Establishment of field boundaries and control of irrigation are fundamental expressions of ordered life in an agrarian society; it is significant that both of these visible signs of order are associated with the sun.) In addition to this symbolic sacrilege, Susano desecrated the Temple of First Fruits, the shrine of Amaterasu, fouling the place with excrement. This vivid detail is not actually needed to propel the plot of the myth forward or to clarify the identity of a chief character. Instead, it emphasizes how serious is the threat to the natural order. The inviolable is violated; the sacred is profaned.

But the challenge does not end here. Next Susano stripped the skin from a spotted horse and threw the corpse into the "sacred weaving hall" where Amaterasu's handmaidens were weaving the "god's garments." Looms went flying, and some of the women, struck in their genitals by shuttles, died. The piebald horse is a suggestive element in the myth. Such a horse, or at least its skin, had ascended to the sky in a myth connected with the raising of silkworms and production of silk, but the connection here is unclear. There is no ambiguity about Susano's impact, however. The rainstorm is virile and violent. The attendants of the sun goddess are overwhelmed by the procreative power of the tempest. Confronted by such prodigal force, Amaterasu withdrew in anger and closeted herself in the cave of the sky.

Amaterasu's departure meant more than darkness and night. Chaos consumed the world. Demons and evil spirits spread calamity and doom, and the 8,000,000 spirits of the cosmos saw the end coming. Without the golden light of Amaterasu, the universe must fold.

Assembled on the dry bed of the celestial river, the Milky Way, the nearly numberless gods played their last card. One among them, "Thought Hoarder," was asked to devise a plan to coax the sun from her cave. Thought Hoarder is styled as a god of wisdom and memory, but his name also suggests the idea of consciousness.

At the celestial cave's barricaded entrance the spirits set up a tree and decked it out with streamers of white cloth, strings of jewels, and a large bronze mirror. Roosters were kept crowing by lanterns that mimicked the dawn. Then, in the crackling light of bonfires, Ama no uzume, a young, exuberant goddess, began to dance. She stamped the ground, removed her belt, and loosened her kimono. Using an upturned

Ama no uzume

Ama-no-uzume dances upon an upturned bucket and discards her clothes to rouse the laughter of the gods and the curiosity of Amaterasu. *(J. Hackin,* Asiatic Mythology*)*

bucket for a stage, she continued her striptease, revealing first her breasts, then showing off her belly. The assembled spirits were amused and delighted as her dance grew more provocative and lewd. When her kimono fell completely open, she let it drop, lowered her skirts, and let them all see the door to *her* celestial cavern.

A spontaneous shout of laughter rose from the entire company and prompted Amaterasu to peek outside and see what was causing such a fuss. "How," she asked, "could the multitude laugh in the gloom she had left behind her?" Ama no uzume answered, "We laugh and celebrate because we now have a new goddess, more stunning than the sun." At this moment, according to Thought Hoarder's scheme, several gods pushed the mirror up to Amaterasu to show her the new goddess. Intrigued, she stepped from her cave, sunbeams spilled back into the world, and a straw rope, or *shimenawa,* drawn across the entrance, precluded another retreat. As order returned to the world, the evil spirits were dispersed.

We may not say for certain what Amaterasu's seclusion and return symbolize, but clear references to the seasonal cycle, the sunrise, and the world's fertility are present in the myth. Susano's unbridled vigor is too much for the world. After the fall harvest of first fruits, the sun's attendants are killed by his antics. Fertility is slain by blows to their wombs. Amaterasu's departure from the world echoes the winter sun's retreat,

Amaterasu is tempted to peek out the door of her sanctuary by the laughter of the Eight Million Gods, and as she does, sunbeams spray back into the world. The sacred sakaki tree by the cave is decorated with jewels, cloth tassels tied as *shimenawas,* and the sacred mirror in which she saw her own reflection. *(Yeitaku, from John C. Ferguson and Masaharu Anesaki,* Mythology of All Races—Chinese/Japanese. *Used by permission of Macmillan.)*

and her reemergence is the sun's return, accompanied in spring by the laughter of all the spirits of nature who recognize in Ama no uzume's lascivious dance the reawakening of life. Other allusions abound: The tree is a world axis and, by the offerings upon it, a tree of life. The mirror and jewels are morning paraphernalia; the crowing roosters are a salute to the dawn.

Susano is an agent of death and confusion, but his seasonal storms make his tantrums unavoidable. He assaults the ordered, agrarian world, but he is part of the natural cycle. The real threat to world order is Amaterasu's withdrawal. When she withholds her light from the world, true chaos intrudes. Winter holds death for the world, and death, though not the same as chaos, is like it. It makes sense then to use the winter's gloom as a metaphor for the chaos we feel is out there. But winter, too, retreats. The *shimenawa* that keeps the sun goddess from abandoning the world also hangs above the temple threshold. It is a symbol of renewal and decorates the streets at the New Year. Life and the sun return to the earth. Our winter of discontent is over.

For awhile.

5

The Dead We Bury

Each person's life echoes the cycle of cosmic order we see played out in the sky. Our ancient ancestors saw a parallel between cosmic creation and each individual birth. Each life meant growth and the preservation of order. Each death was an intrusion of chaos. And just as our ancestors saw rebirth, renewal, and restoration of order in the sky, they also believed in an afterlife for the soul, a resurrection. Death, the key transformation in each soul's share of the cycle of cosmic order, initiated a transcendental journey, like the mystical ascent of the shaman, to the realm of the gods and immortality. Very often the destination of the dead was the domain of sacred order—the sky itself.

For this reason, in culture after culture, we find celestial metaphor in the ceremonies of the funeral, celestial imagery in the paraphernalia of the dead, and astronomical significance in the architecture of the tomb. By recognizing the relation between death and the sky, we can understand what death meant to our ancestors and what they judged was the role of the dead.

Great Pyramid Astronomy

Death meant no rest for the pharaohs of Old Kingdom Egypt. Their destiny was overhead. Prayers carved into the stone walls of chambers within Fifth and Sixth Dynasty (*ca.* 2494–2181 B.C.) pyramids describe the pharaoh's "ascent to the sky among the stars," where he "regulates the night" and "sends the hours on the way." In some of these *Pyramid Texts*, in the monuments at Saqqara (about 40 miles up the Nile from Cairo), he joins the circumpolar stars and governs them. They never rise and never set, and so they never die. By joining these eternal stars, the pharaoh becomes eternal.

The pharaoh also makes another celestial journey: to Orion. This constellation is a symbol of his soul's rebirth because Orion stood for Osiris and the great cycle of birth, life, death, and resurrection.

In the company of Osiris and guided by Sirius (in roughly the same part of the sky), the dead king maintains the calendar and administers the seasons. The pharaoh is busy in heaven. According to the texts, he takes "possession of the sky, its pillars and its stars" after death. On earth he maintained order and vitalized the system; in the sky, among the spirits, he activates the great cosmic cycles. His soul, or *ba*, has become a "living star at the head of his brethren." The tomb in which he was buried was the point of his departure for the sky.

Later Egyptian tombs are filled with astronomical imagery of the journey of the sun, the body of the sky, and the destinations of the stars: like the pyramids that contain hieroglyphic texts, they confirm that the Egyptians sensed a link between death and the sky. The builders of the earliest pyramids, however, and the builders of the greatest—those Fourth Dynasty (*ca.* 2613–2494 B.C.) mountains of stone blocks at Giza—left no written records of their intentions. Later commentators filled this void with fanciful speculation, downright misinformation, and plain silliness.

About 80 pyramids are known, all built on the Nile's west bank—the realm of the dying sun—but the Great Pyramid outstrips the others in fame as well as in size. Erected about 2600 B.C. it was recognized by Classical writers as one of the seven wonders of the ancient world, and it is the only one of the seven still standing. With amazing success, it has resisted the assaults of the desert, earthquake, quarrymen, and tourists. Herodotus, the Greek historian, marveled at its antiquity—and he visited Egypt in the fifth century B.C. Now with more justification than ever, an old Arab proverb counsels, "Man fears time, but time fears the pyramids."

Various astronomical interpretations of the Great Pyramid at Giza have been proposed, involving alignments of its internal corridors and shadows cast by its profile. Some people have even suggested that it was used as an observatory. This pyramid's dimensions also have been said to encode practically everything from the number of days in the year to prophecies of the world's history. In our supposedly rational era, the Great Pyramid—or at least its shape—has been touted as a focuser of an as-yet-unidentified energy that can do everything from sharpening razor blades to mummifying milk.

It has been estimated that the volume of the Great Pyramid is as large as that of St. Peter's in Rome, St. Paul's and Westminster Abbey in London, and the cathedrals of Milan and Florence combined. Nearly 2,300,000 blocks of stone, each weighing on the average 2½ tons (the heaviest being 15 tons with none less than a ton and a half) were piled into an artificial mountain whose four sloping sides originally merged at a point 481.4 feet above the limestone plateau of Giza. Today, the top, including the gold-covered capstone, or pyramidion, is missing, and the Great Pyramid is 31 feet short of its original height.

The sides of the Great Pyramid once competed in brightness with the golden cap, for the rough limestone blocks that comprise the bulk of the pyramid were covered with a fine surface of shiny, white tura limestone. Only a little of the original tura now remains, however, in the bottom course on the north side.

Sheer size qualifies the Great Pyramid as "great," but its architectural details make it unique. First, whatever else may be hypothesized about the pyramid, there is no doubt that it is aligned astronomically, and with extreme precision, the four sides of its immense base (covering more than 13 acres) running north, south, east, and west. The worst agreement of any side with exact cardinal orientation is on the east, and even there the misalignment from a true north-south line is but 5½ arcminutes. The other three sides agree even better. Accuracy of this sort is possible, even with very simple techniques, provided that care is taken in setting out the lines. Preserving this accuracy on the monumental scale of the Great Pyramid, however, means not "twisting" the sides at higher levels, and the Egyptians' success is impressive. The corners of the base are very nearly right angles, and the entire base is almost a perfect square. The greatest difference in length between any two sides is 7.8 inches. This accuracy emphasizes the concern the builders had for the four cardinal directions, what the *Pyramid Texts* of the later Fifth and Sixth Dynasties call the "pillars" of the sky.

The internal details of the Great Pyramid are equally remarkable. A cross-section reveals a confusing collection of corridors, chambers, and shafts. Numerous claims for

From his pyramid, the pharaoh Khufu departed for the sky. His two celestial destinies were symbolized in the "air shafts" that emerge from the King's Chamber. One pointed to Thuban, the "north star" of the Old Kingdom, and the other was oriented toward the belt of Orion.   *(Griffith Observatory)*

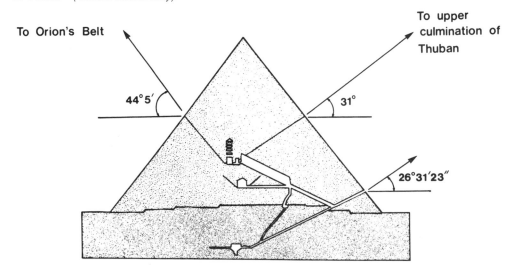

astronomical alignment of these vents and passages have been made, and most are wrong. There is astronomy among them, however. To appreciate it we must enter the pyramid.

One main passage, the Descending Corridor, opens on the north face, 55 feet above the ground, and extends down into the pyramid and the bedrock for 345 feet. At this point the passage, which is just high enough for crouching, levels out. Twenty-nine feet beyond this bend it ends, at an unfinished subterranean chamber. A blind passage, rough and unfinished, continues a short distance into the bedrock of the chamber's south wall. This and other unfinished internal details suggest that plans were modified several times during construction of the pyramid in favor of progressively grander designs.

The Ascending Corridor breaks off from the Descending Corridor, continues upward through the immense stone blocks, and then opens into one of the features that really makes the Great Pyramid great, the Grand Gallery. This is one of the strangest architectural spaces in the world. It is a monumental corbelled vault, its ceiling disappearing in the darkness and peaking 28 feet overhead. The proportions are peculiar: 153 feet long but just under 7 feet wide, with a steep, but not unclimbable, floor. Neither photographs nor drawings really capture the feeling of the Grand Gallery. Despite the fact that you are deep within the pyramid, not even a third of the way up through its tons of limestone blocks, there is no sense of claustrophobia.

One last step at the end of the Grand Gallery allows access to the King's Chamber, a spacious room oriented north-south, east-west. The high ceiling and all of the stone blocks and slabs that form the chamber's walls and floor are hard granite from the Aswan quarries 500 miles to the south. The room must have meant something special to justify transporting that special granite such a great distance.

Apart from a large, lidless granite sarcophagus, no ancient artifacts were found in the King's Chamber. Probably a wooden coffin, containing the body of the interred pharaoh, once filled the empty sarcophagus, but all traces of it are gone. We know, however, whose pyramid it was. Even in the time of Herodotus, our most ancient source, Khufu, the second pharaoh of the Fourth Dynasty, was remembered by the Egyptians as the builder of the Great Pyramid. In the Greek of Herodotus, Khufu's name was Cheops.

Despite the emptiness of the King's Chamber and the absence of symbolic paintings and hieroglyphic texts, the room contains two peculiar features that confirm the Great Pyramid as a tomb with specific celestial symbolism built into its architecture. Two openings, one on the north wall, the other on the south, both just a few feet above the floor, are actually the open ends of shafts that extend all the way from the King's Chamber to the outer faces of the pyramid.

Only 9 inches high and 9 inches wide, these shafts are sometimes called "air shafts," but their real significance was revealed in 1964 through a joint analysis by Professor Alexander Badawy, an Egyptologist, and Dr. Virginia Trimble, an astronomer. Badawy thoroughly punctured the "air shaft" thesis by explaining that the Egyptians knew a lot more about air conditioning than these shafts would imply. Certainly, driving two ducts, which open no wider than two Kleenex boxes standing side by side, does not seem to be the easiest way of freshening the air. Besides, Badawy noted, the Egyptians are not known to have ventilated any other tombs.

Trimble, meanwhile, confirmed the good alignment of the northern shaft with the uppermost transit, or arc, of the star Thuban. She also demonstrated that the southern shaft pointed toward the part of the sky occupied by the conspicuous belt of Orion during the period between 2700 and 2600 B.C. Of the three stars in the belt, Alnilam, the middle one, seemed to be best aligned for transiting the right patch of sky. Alnilam then had a declination (an angle of celestial position analogous to latitude on the earth) of $-15\frac{1}{2}$ degrees. Because the Great Pyramid is located just $1\frac{1}{3}$ miles south of 30 degrees north latitude, the celestial equator, or 0 degrees, crosses the sky at an angle of 60 degrees. By subtracting the declination of Alnilam ($-15\frac{1}{2}$ degrees) from this, one gets the angle of its passage across the southern meridian, $44\frac{1}{2}$ degrees above the horizon. This is very close to the actual angle of the southern shaft. Uncertainty in the exact date of the pyramid's construction makes the other two stars in Orion's belt, Alnitak and Mintaka, candidates for the alignment as well. No bright

For the ancient Egyptians, the circumpolar zone was the realm of immortality, the home of the Imperishable Stars. In following their circular paths around the north celestial pole, those that never rose and never set became synonymous with eternal life. Through a time exposure, the circular arcs that mark the trails of stars in this photograph record a part of the never-ending march of the circumpolar stars.   *(Curtis Leseman)*

stars, other than those in Orion's belt, came within 1½ degrees—three times the size of the full moon—during the era considered.

Both shafts, then, are oriented toward celestial zones that are mentioned in the *Pyramid Texts*. Thuban, as the North Star, was in a sense first among the circumpolar stars or "Imperishable Stars," as the Egyptians phrased it to acknowledge their ability to circuit eternally without ever setting. Thuban, then, made an appropriate target for the north shaft. Meanwhile, the stars in Orion's belt, the objective of the south shaft, were among the decans, the stars whose risings or transits marked the hours of the night throughout the Egyptian year. Orion, of course, was Osiris, who presided over the resurrection of souls.

These celestial alignments don't make an observatory out of the Great Pyramid, for neither Thuban nor Orion's belt could actually be seen through the shafts' openings inside the King's Chamber. Both shafts bend horizontally for a short distance before they reach the King's Chamber and also at their other ends, prior to opening on the north and south faces of the pyramid. The shafts are symbolic references to the pharaoh's celestial destinies and to the starry realms that ordered heaven and earth. They mark the Great Pyramid—the tomb of Khufu—as the site of his celestial transformation.

Starry Tombs in the Valley of the Kings

Allusions to the same stellar destinies of the pharaoh appeared in the royal tombs of the New Kingdom, 1,500 years after the building of the pyramids. Most of these tombs are cut into the cliffs in the Valley of the Kings, near the old capital of Thebes on the Nile's west bank, across from modern-day Luxor.

Numerous representations of the "Northern Group," a figurative portrayal of the circumpolar constellations, were painted on the corridor ceilings in the tomb of Ramesses VI, a Twentieth Dynasty pharaoh. Perhaps the finest version of this theme appears overhead in the burial chamber of the tomb of Seti I, a member of the Nineteenth Dynasty, who ruled to about 1292 B.C. Against a rich blue field, we see the tawny forms of the constellations of the "northern Heaven." One of these, an upright hippopotamus with a crocodile on her back, leans upon an odd instrument, most often translated from references in the texts as a "mooring post." It is the north celestial pole, and the golden lines attached to it lead through the hands of an unidentified human figure to a bull. Sir Norman Lockyer, the British astronomer who discovered helium in the sun, devoted time and energy to astronomical interpretation of Egyptian myths and monuments, and he equated the hippopotamus with at least some of the stars of Draco the Dragon, the site of the north celestial pole in the early period of Egyptian civilization. It is actually broader than that and encompasses most of the northern heavens.

Just as the orientation of the Fourth Dynasty pyramids to the four cardinal direc-

Astronomical images that represented the realm around the north celestial pole are marked with dots that stand for stars. This painted ceiling from the burial chamber of the tomb of Seti I includes a bull, a portrayal of the Big Dipper. Just as the Big Dipper must pivot around the north celestial pole, the bull is tethered to the post held by the hippopotamus. *(E. C. Krupp)*

tions stressed the cosmic order, so the burial chamber of Tutankhamun (*ca.* 1345 B.C.) incorporated north-south and east-west alignments. References to the stellar destiny of this pharoah may now be seen in Cairo's Egyptian Museum. The inscriptions on the four gold-covered, boxlike shrines that formerly encased the sarcophagus of the deceased identify the king with Osiris, and Isis is described as following behind him, just as Sirius follows Orion across the sky. On the door of one of the shrines, Isis is portrayed as a bird with extended wings; on another she says: "I am thy protection, I am behind thee."

The tomb of Seti I, whose burial chamber ceiling illustrates the Northern Group, the stars around the north celestial pole, on one half, includes Sirius and Orion (as boat-borne figures of Isis and Osiris) on the other. Both stellar destinations of the pharaoh, portrayed here, repeat the traditions expressed in the Great Pyramid's "air shafts." On the ceiling of another burial chamber, that of Ramesses VI, *The Book of the Day* and *The Book of the Night* frame, each on its own side, the double body of the sky goddess. These texts chronicled the sun's journey, hour by hour, through the day and the night. They imply yet another celestial destiny for the pharaoh: the path of the sun.

Accompanying the images of Sirius and Orion in the tomb of Seti I, and on the ceilings of many other temples and tombs as well, are lists of the decans. Diagonal "star clocks" listed the names of the timekeeping stars inside the lids of coffins from the Ninth through Twelfth Dynasties (2160–1786 B.C.). These clocks are really grid diagrams of the decans in the sequence of their appearance. A single star shifts its position in the grid, up one square and over one, to mark an earlier hour of the night every ten days. Each name, in hieroglyphs, traces a diagonal path through the grid and gives the design its name. Additional decorations usually include images of the Bull's Leg (that is, Set, the Big Dipper), Isis, and Osiris. Another type of star clock was painted in the Twentieth Dynasty tombs of the Ramesside pharaohs, from about 1300 to 1100 B.C. These, too, appeared together with the Northern Group, Sirius, and Orion. There is no reason to think Egyptian astronomers ever entered tombs to consult the star clocks. Their purpose was funereal. They belonged to the dead.

The decans themselves were selected because they imitated the behavior of Sirius and disappeared from the nighttime sky for 70 days. During this time they were up in

Epagomenal Days																					DECADES																			
40	39	38	37	36	35	34	33	32	31	30	29	28	27	26	25	24	23	22	21	20	19		18	17	16	15	14	13	12	11	10	9	8	7	6	5	4	3	2	1

																						hours of the night
A	25	13	1																			19						1	I
B	26	14	2	A																		12						2	II
C	27	15	3		A																	12						3	III
D	28	16	4			A																	12					4	IV
E	29	17	5				A																12					5	V
F	30	18	6					A																12				6	VI
G	31	19	7						A															12				7	VII
H	32	20	8							A															12			8	VIII
J	33	21	9								A														12		9	IX
K	34	22	10									A														12	10	X
L	35	23	11										A													12	11	XI
M	36	24	12	L	K	J	H	G	F	E	D	C	B	A	36	35	34	33	32	31	30	29 28 27 26 25 24 23 22 21 20 19 18 17 16 15 14 13 12	12	XII

1—36   **Regular Decans**   A—M   **Epagomenal Decans**

Coffin lid star clocks are really arrangements of the names of the stars used to monitor the hours of the night. Any particular star—say that represented by the number 12—shifts up and left to a new position after each *decade*, or ten-day interval, and marks an earlier hour of the night. Because of this, the names of any star form a diagonal line through the grid. The epagomenal decans were used similarly during the five nights that finished out the year. *(Griffith Observatory)* the daytime, with the sun, and remained, therefore, invisible. In Egyptian texts, a decan's departure from the night sky was called its death. It resided in the dangerous realm of Tuat, where the sun traveled each night. An astronomical treatise from the second century after Christ, the *Papyrus Carlsberg I*, quotes inscriptions from the cenotaph, or symbolic tomb, of Seti I at Abydos. Explicit references liken the decanal cycle to death and resurrection:

> Their burials take place like those of men. . . .
> The one which goes to earth dies and enters the Tuat. It stops in the House of Geb the earth 70 days.
> It is in the Embalming House . . . It sheds its impurity to the earth. It is pure and it comes into existence in the horizon like Sirius.

Another type of grid or star clock appeared on the ceilings of the tombs of the Ramesside pharaohs during Egypt's New Kingdom. This one comes from the tomb of Ramesses VI, and the Northern Group of constellations is painted above the hieroglyphic inscriptions and checkerboard of the star clock. *(E. C. Krupp)*

Here the language of the decans is the language of the mortuary. In a direct turn-about the mortuary customs mirrored the stars. The deceased was said to be a star, and mummification, also a process of purification, took 70 days. Then the dead pharaoh was reborn, like the decans.

Buried Under China's Skies

Ching k'o's poisoned dagger sliced through the air and struck one of the bronze pillars in the imperial audience chamber. It narrowly missed the target of the assassination attempt, the emperor himself, Qin shi Huang di. A sleeve from the royal robe lay where it had fallen after the emperor wrenched himself free of Ching k'o's grip. Sleeve torn, the emperor tried to reach his sword and pull it from the scabbard. He could not, however, keep dodging his attacker and draw his weapon at the same time. He would have died had the royal physician not struck Ching k'o' with a medicine bag. Ching k'o, in a last desperate move, threw the knife at the emperor. The miss was fatal, for by then China's first emperor had unsheathed his sword and he struck the assassin dead.

This attempt on the life of Qin (pronounced *chin*) took place in 227 B.C., and it was not the last. But Qin survived them all and ruled a united China for 18 more years, until he died of illness in 210 B.C.

When Qin died, he was buried within a huge pyramid of earth—an artificial "mountain" at the foot of Mount Li shan, about 25 miles east of the present city of Xi an in northwest China's Shan xi Province. The monument has been recognized as Qin's tomb for the 22 centuries it has stood there, but archaeologists have never breached its inner chambers. These were described, however, about a century after Qin's death, in the historical records compiled by Si ma Qian. He wrote that the emperor's sarcophagus floated in a river of mercury, surrounded by a miniature landscape of palaces, pavilions, office buildings, waterways, fields, and hills. It symbolized the world Qin ruled, and on the ceiling overhead, "all the heavenly constellations" were depicted. No one has seen that starry ceiling since 206 B.C., when the Han general Hsiang Yu reportedly desecrated the tomb. Many tombs from later periods are known to have domed ceilings that imitate the night sky, however.

The world ignored Qin and his tomb until 1974, but then a remarkable discovery was made. Members of the Hsiyang Village People's Commune were digging a well, and when they reached a depth of 13 feet, one of them came upon a lifesize human head, molded in terracotta. Soon the rest of the statue was uncovered. It was a soldier. Reasoning that with Qin's tomb not more than 3 miles to the west, the sculpture might have something to do with the first emperor's burial, the peasants informed archaeologists of their find. What they uncovered next was unprecedented. One vast pit, 689 feet long and 197 feet wide, held an estimated 6,000 terracotta soldiers arranged in 11 orderly ranks and accompanied by six life-size chariots, each drawn by

The burial site of Qin shi Huang di, the first emperor of China, is a massive pyramid of earth, located near the city of Xi an. *(Robin Rector Krupp)*

four terracotta horses. Test excavations in two other pits brought the total number of soldiers to perhaps 7,500. An entire army, then, of infantrymen, charioteers, cavalrymen, archers, and officers guarded Qin's east flank, even in death. The buried army of Emperor Qin is one of the most significant and spectacular archaeological finds of this century. But what does it have to do with the sky?

Celestial and cosmological considerations affect the shape and orientation of Qin's tomb. The mound is nearly square, about 1,200 feet on a side, and each side faces one of the cardinal directions. The two enclosures that surround the tomb are rectangular. The inner is 2,247 feet by 1,896 feet, and the outer is 7,129 feet by 3,196 feet. The sides of the enclosures also face the cardinal points, and the main axis of each is north-south.

Cardinal directions derive their meaning from the daily rotation of the sky. The north pole of the sky—around which everything else seems to turn—establishes the primary direction; the other three follow quite naturally. Discovery of the buried army on the east side of Qin's tomb certainly prompts speculation that other impressive finds remain to be discovered in the other three directions. In fact, two four-horse chariots with their drivers, cast in *bronze,* were found in test excavations on the west side of the tumulus early in 1981.

To understand the celestial symbolism of Qin's tomb, we have to appreciate who he was and what he did. In his lifetime Qin unified the feudal states of China into a single nation. His impact on Chinese society was enormous. The power that was once so dispersed among many feudal lords was centralized by him and wielded by a single government. At its center stood Qin shi Huang di.

An empire is held together by the standards that operate from one end of its territory to the other. Qin put cohesion into China by establishing a uniform system of writing and a universal code of laws. We may imagine that he authorized the calendar each year. He standardized the system of weights and measures, so that even the axles of carts and wagons were built to the same established length. This maximized their usefulness on the nationwide system of roads Qin built. The roads complemented the most ambitious network of navigable canals in antiquity, and an even greater wonder—the Great Wall of China—was pulled together by Emperor Qin. He built his capital on the Wei River near what is now Xi an, and his palace incorporated cosmic symbolism. It stood for the "Apex of Heaven," that is, the north pole of the sky. The heavens revolved about the Apex of Heaven and the world revolved around the emperor. He was the steady spot at the center that provided stability and order to the world.

Qin's tomb expressed the same principles of cosmic order. The mound has three levels, or tiers, which stand for the earth, human beings, and the sky. The mound's summit is the Apex of Heaven. As a "spirit city," the entire complex was a sacred precinct oriented by the sky.

Other tombs from other periods of China's long history also incorporate astronomi-

The tomb of Emperor Qin, like the pyramids at Giza, was aligned with the cardinal directions. To the east of his tomb, Qin buried an army of more than 7,000 terracotta soldiers—each practically life-size—to guard the eastern flank.   *(Griffith Observatory)*

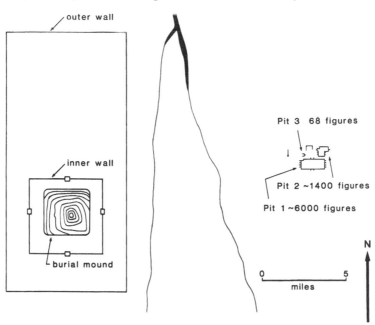

cal imagery and celestial orientation. The Qian ling, about 42 miles northwest of Xi an, is the tomb of Gao zong, the third emperor of the T'ang Dynasty (A.D. 618–906), and his wife and successor, the Empress Wu Ze tian. He died in 684 and she was buried in 705. Like the "spirit city" of Qin, the Qian ling is organized upon a north-south axis, well defined by an avenue of statuary, obelisks, and towers. Although this cardinal orientation is obvious again here, not all Chinese tombs—or even all imperial tombs—followed this rule. The rules for tomb orientation were more complex and involved a geomantic consideration of many aspects of the local topography. The sky, of course, was part of this environment, too, and so astronomically aligned tombs occur.

Not far from the Qian ling is a much smaller but still very impressive tomb. It belongs to Princess Yung Tai, the granddaughter of Emperor Gao zong, and the shaft that leads to its chambers is oriented north-south. A mural painting of the Green Dragon of the East guards the east wall of the tomb's entrance corridor, and on the opposite wall the White Tiger of the West fulfills a similar function. This dragon and tiger are two of the four cosmic animals of Chinese tradition. They can be traced back at least to the Han Dynasty (206 B.C.–A.D. 220) and probably originated even earlier. The other two are the Red Bird of the South and the Black Tortoise (or Dark Warrior) of the North. These same symbolic animals were associated by the Chinese with four zones, or "palaces," along the celestial equator and were equated with the four seasons.

The White Tiger of the West, one of the four cardinal and cosmological emblems of ancient China, growls upon the west wall of the entrance corridor into the tomb of the T'ang Dynasty princess Yung T'ai. *(E. C. Krupp)*

Farther into the tomb of Yung Tai are two domed chambers. The second is the burial chamber itself. The resemblance to a planetarium theater is enhanced by the ceiling, painted to mimic the starry sky. A red sun is in the east, a white moon in the west. And from one side to the other the River of Heaven, or Milky Way, bridges the vault.

Domed ceilings that mimic the vault of the sky are known in several other Chinese tombs. One of the earliest astronomical ceilings is in a tomb of the Han Dynasty, near Luoyang, in north central China. The Milky Way, along with several identifiable constellations, the Big Dipper among them, stretches across the top of the burial chamber. This painted map is oriented correctly with the cardinal directions, and the sun is shown in the east, symbolically penetrating the tomb entrance.

A similar starry ceiling survives in a Wei Kingdom tomb, about fifteen centuries old. Here, too, the Milky Way crosses the dome, and the circumpolar stars are emphasized. These include the Purple Forbidden Palace and the Great Bear. The Bear, of course, is the Big Dipper, and the Palace frames the sky's pole, the seat of celestial government.

Like a petrified planetarium, the hemispherical ceiling that domes the burial chamber of Princess Yung T'ai portrays the night sky. Through the many stars painted overhead, the Milky Way runs from one horizon to the other.   *(E. C. Krupp)*

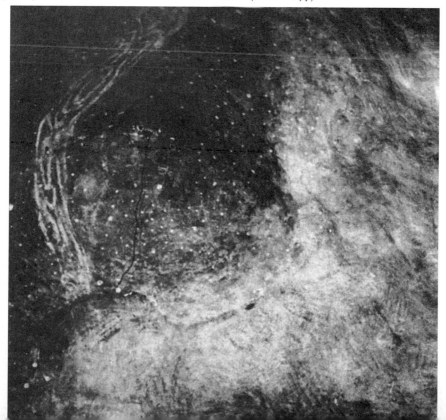

Astronomical manuscripts have been found in some ancient Chinese tombs, and celestially symbolic artifacts have been recovered from others. When, in 1973, the tomb of a Han dynasty noblewoman was opened at Mawangdui, close to Changsha in central China, archaeologists uncovered an unprecedented find. After nearly 2,200 years underground, a perfectly preserved 12-foot-long silk funerary banner was exposed again to human eyes. Painted with a complex composition of symbols, figures, and scenes, it seems to depict the journey of the deceased to the celestial paradise and includes the emblems of the sun and the moon. The former is a red disk, in it a black bird. A white crescent, accompanied by a toad and a hare, stands for the moon. The toad is Heng O, the moon goddess, who stole the pill of immortality from the Queen Mother of the West. Fleeing to the moon, Heng O encountered the white rabbit who lives there. (Many people see the shape of a rabbit in the light and dark areas on the face of the moon.) According to tradition, the hare pounds and grinds his drugs into the elixir of immortality at the foot of a cassia tree. This evergreen produces cinnamon and is also associated with immortality. As Heng O drank the white rabbit's potion, she was transformed into a toad and enjoyed immortality in her lunar home.

It makes sense that the moon—which first appears as a crescent, grows full, and then wanes—is linked with the afterlife and immortality. In that celestial cycle of birth and death there is always another new moon. Through the cycle of cosmic order the moon guarantees its own immortality, and it is a fitting image for the bunting of the dead.

In China, then, as in many traditional cultures, the dead were associated with the sky. This may seem strange at first, but we can understand the ideas behind it. The souls of the dead journey to a celestial paradise. Paradise is a realm of immortality and divinity, and that is what people see in the sky. Celestial objects, through their cycles of rising and setting, coming and going, waxing and waning, seem to die but are always reborn. They are the immortal gods, and the patterns of their movements establish order in time and space. The dead have access to their kingdom. Tombs are points of the soul's departure for the celestial realm. The astronomical components of these tombs are intimations of immortality. Such celestial references also charge the earth with the cosmic order they represent, and the burials create hallowed ground.

Cahokia's Mystery of Mound 72

Burials of the honored dead incorporate metaphors of cosmic order in recognition of the soul's journey, but these same tombs also assist the living. With celestial symbolism expressed in their placement, orientation, or design, they consecrate the ground where they sit. The dead provide, in effect, access to the sacred for the living. Burials, bones, or ashes may be incorporated into temples, ceremonies, or even city plans, and by their presence, the dead bring cosmic order to earth. This tradition is by no means

The silk funerary banner found in the tomb of a Han Dynasty noblewoman at Mawangdui depicts in part, the journey of the soul to the celestial paradise. The crescent moon, accompanied by a toad and a hare, occupies the left end of the bar of the *T*, and the sun disk, with a bird ensconced, is on the right. *(J. Hay,* Ancient China. *Reproduced by permission of The Bodley Head.)*

limited to the highly developed civilizations of the Old World. An unusual burial at Cahokia—a huge prehistoric settlement of mound-building Indians in southern Illinois—was also part of a mechanism that organized everyday life by the order in the sky.

More than a hundred earthen mounds were built at Cahokia. Many had flat tops and supported buildings. Others were conical and probably enclosed burials. The most impressive monument of all is Monks Mound, a four-terrace earthwork 100 feet high.

Most of the mounds and plazas were rectangular and oriented with cardinal directions. Cardinal orientation seems, in fact, to be preserved in the entire "urban plan" of the site. This was defined by "ridgetop" mounds, special constructions whose sides slope up from a rectangular base and form a long ridge at the top. Archaeologist Dr. Melvin L. Fowler of the University of Wisconsin, Milwaukee, believes three of Cahokia's six ridgetop mounds were placed to establish the southern, eastern, and western "city limits." The other ridgetops may be special markers as well, and one of them, Mound 72, contains a remarkable burial.

Fowler's study revealed that Cahokia's main east-west axis was marked by the ridgetops at the east and west limits of the site. And the primary axis is a north-south line that passes through the southwest corner of Monks Mound, a ridgetop mound (number 49) south of the main plaza, and Mound 72—which is skewed about 30 degrees north of west—and ends at Rattlesnake Mound, the ridgetop at the south "end of town."

Already convinced that the positions of ridgetop mounds were important, Fowler decided to excavate the peculiarly oriented Mound 72. In 1973, Fowler dug into the mound where the prime axis of Cahokia cut through. Fairly soon his team found signs of a pit that penetrated 8 feet into the earth. At the bottom were timber stains left by a long-disintegrated post, nearly 3 feet in diameter. Smaller logs also had been arranged as a support crib. Evidently a high and massive post had stood, perhaps as a marker, in exactly this spot on the north-south line. Radiocarbon analysis of log samples dated construction at A.D. 950.

Surprises at Mound 72 did not stop with the discovery of the marker post. A series of extraordinarily rich burials lay beneath Mound 72. In less than a century two prominent individuals had been buried there, along with nearly 300 others of lower status. Most of these others were young women, and they probably were sacrificed at the interments of the two individuals whose grave goods dominate the finds in the mound. Caches of arrowheads—more than 300 points in one and more than 400 in the other—had been deposited near the bodies of the two principal people. In addition, Fowler found nearly two bushels of uncut mica sheets, a roll of unworked copper sheet, more than 20,000 marine shell beads, and other items. Most of these materials were exotic trade goods and not local to Cahokia. Sources for the items were as distant as Oklahoma, North Carolina, Lake Superior, and the Gulf Coast. All of them represent considerable wealth and influence.

In its day, Cahokia politically dominated the broad alluvial valley where the Missouri and Illinois rivers meet the Mississippi and influenced more than a million square miles of North America. Perhaps 40,000 people lived there at its height. Mound 72's burials confirm Cahokia's role as an administrative hub and mercantile center for long-distance trade.

We can see something else in the high-status burials of Mound 72. Everything about them marks their location as a special place. That spot, in turn, was part of the cosmically oriented plan for the entire site. Even without knowing the beliefs of the ancient Cahokians, we can detect their concern for celestial order in the burials of Mound 72.

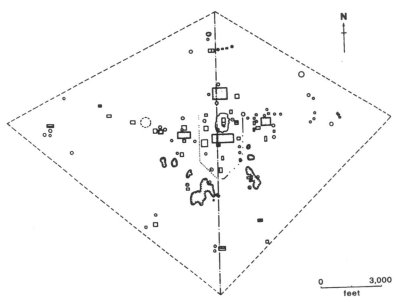

Cahokia, a Mississippian Indian metropolis whose ruins are located 7 miles east of St. Louis, was laid out according to a deliberate cosmological plan. Ridgetop mounds marked the 'city limits' at the corners of Cahokia's diamond-shaped territory, and these mounds defined the primary north-south and east-west axes of the site. The large feature near the center is Monks Mound, and the shaded rectangles represent plazas. A dotted circle to the west marks the location of the "Sun Circle." Mound 72 crosses the main north-south axis at an angle, near a "leg" of one of the irregularly shaped pits from which the Cahokians dug the earth to build their mounds. *(Griffith Observatory)*

Mound 72 is not a prominent feature in the Cahokia landscape, but it was discovered to harbor two important and rich burials. The mound is skewed to the rest of the site, about 30 degrees south of east, and it was penetrated by a high post that marked the mound's point of intersection with Cahokia's primary north-south axis. *(E. C. Krupp)*

At Palenque, in Chiapas, Mexico, the winter solstice sun sets behind the Temple of the Inscriptions along a path that mimics the descent of the first leg of a stairway, inside this Maya pyramid, that leads down to a chamber housing the sarcophagus and remains of Pacal, Lord of Palenque. *(Robin Rector Krupp)*

The Sun's Death at Palenque

Discoveries of important burials like Melvin Fowler's find in Mound 72 or Howard Carter's opening of the tomb of Tutankhamun command our interest with the romance of the search. No less exciting was the entry of Alberto Ruz, a Mexican archaeologist, down an undiscovered stairway, hidden within a temple-topped pyramid at a major ceremonial center of the Maya. The site was Palenque, in Chiapas in southern Mexico, and the pyramid is known as the Temple of the Inscriptions. In 1949, Dr. Ruz lifted a large slab in the temple floor, beneath which was a rubble-filled stairway. It took him four seasons to proceed down two flights of steps—sixty-six altogether, with a U-turn after forty-five of them—but the effort was definitely worthwhile. At the bottom was a large, vaulted crypt, almost completely filled by an elaborately carved sarcophagus. When the huge, 5-ton lid was lifted, the skeleton and grave goods of what must have been a high-ranking member of Palenque society were uncovered. The find was like no other. Until Ruz's opening of the tomb, Maya pyramids were thought only to support temples, not to harbor burials.

Since the discovery, some wild speculations have been made about the image on the lid of the sarcophagus. Erich von Däniken, author of *Chariots of the Gods?*, and other purveyors of the notion that the earth was visited by alien astronauts in ancient times, say the lid shows a rocketship blasting off. This sort of flummery sells books. The hieroglyphics on the tomb have been deciphered, and we know that the man buried there was Lord Pacal (or "Shield"), that he ruled Palenque, and that he died in A.D. 683. The Temple of the Inscriptions is his funerary monument, and it bonds him with the dying sun.

The famous "Palenque astronaut" on the stone lid is actually Shield-Pacal, and he is falling into the maw of the underworld, not flying into outer space. Dr. Linda Schele, an artist and professor of art history, has interpreted the symbolism of the sarcophagus carving and provided a convincing explanation for some of it.

Below Pacal, a frightening face is also framed by the reptilian jaws of the underworld. Across its forehead is a hat or headband, and it bears the *kin* glyph, a Maya word that can mean "day," "time," and "sun." A badge on top of the head includes three symbols: the cross section of a sea shell, a sting-ray spine or other bloodletting device, and a variant of the "crossed bands" glyph. According to Schele, these symbolize the realm of the dead, or underworld; the earth, or middle world; and the sky, or celestial world. The face is fleshed above the jaw, but skeletal below. To Schele this signifies a critical transition, from the middle world to the underworld, from life to death. By the glyph on the forehead, Schele identifies Pacal's partner, preceding him to the grave, as the sun. A similar portrait of the sun appears on the wall of the Temple of the Cross, nearby, and there its face is poised, half above and half below a band of glyphs. Those on the left are signs of the night, and in his temple left is west. Signs of the day are on the east.

Winter solstice and sunset mark the death of the sun. On its southernmost path the sun is weakest, and it enters the underworld by passing below the horizon. The moment of transition logged by the sarcophagus is winter solstice sunset.

From nearly anywhere on the west side of the platform of the Palenque Palace, a good view of the Temple of the Inscriptions is possible. Schele observed that the winter solstice sun sets behind the high ridge beyond Pacal's pyramid, in line with the center of the temple on top. The sun's path, as it heads for that horizon, follows the same angle as the first stairway that descends into Pacal's tomb. Although this angle gradually changes with the passing millennia because the angle of the earth's axis changes, the difference is too small to notice over the 12 centuries elapsed since Pacal's death. Pacal's death and entry into the underworld are still equated by the architecture of his tomb with the sun's death and entry into the earth.

The link we have seen so often between kingship and the sun is present at Palenque as well. Pacal's son, Chan Bahlum, succeeded Pacal on the Palenque throne, and the Temple of the Cross commemorates the transfer of power from father to son, from the dead to the living. The wall relief portrays Pacal, on the left and west side, visibly smaller than Chan Bahlum, on the right. Pacal holds a small head, the sun's, before him. On outer walls of the entrance Chan Bahlum stands, now dressed in the royal regalia, on the left, and offers a scepter topped with the inverted head of the sun

MERLE GREENE '75

(Left) The cover of Pacal's sarcophagus does not depict an astronaut blasting off in his rocket, as claimed by ancient astronaut enthusiasts, but catches Pacal for a moment as he falls from the celestial realm to the underworld and dies, just as the winter solstice sun falls into the maw of the earth and "dies." The odd face beneath Pacal is the sun, its jaw already skeletal. The *kin* glyph—looking like a four-petaled flower or a butterfly—dominates the sun's forehead. Both Pacal and the sun are enclosed in the jaws of the Earth Monster and are about to be snapped up. The right and left borders of the sarcophagus are skybands. In the upper left box is a Venus symbol. On the right, or east, side the center glyph in the band is a *kin* glyph for the sun and day. On the opposite side, a crescent glyph symbolixes the moon, night, and the west. Although the bird at the top of this scene serves as an admirable hood ornament for an ancient spacecraft, it is actually associated with night and the underworld and may symbolize the Big Dipper. (*Drawing by Merle Greene Robinson*)

Winter solstice sunset is best witnessed from the vicinity of the Tower, in the palace structure at Palenque. The sun disappears in the southwest, behind the hill in back of the Temple of the Inscriptions, at about 2:30 in the afternoon, but its last light falls at the feet of God L, carved in relief by a doorway to the Temple of the Cross, about three hours later. (*Griffith Observatory*)

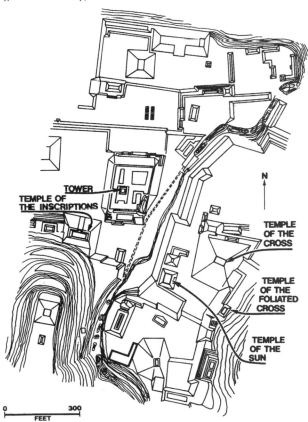

to God L, to the right of the entrance. God L is one of the principal lords of the underworld. What do these scenes mean? In the process of assuming the kingship, Chan Bahlum must offer something—the dying sun—to the rulers of the night. Pacal himself is that offering, for in death he goes below. It is his death, in fact, that makes it possible for Chan Bahlum to rule. A fitting symbolic climax to these transfers and offerings takes place as the last light of the dying sun falls into a pool at the feet of God L, who carries it off into the winter solstice night.

Sunshine in an Irish Tomb

Thirteen and a half hours before the last light of the winter solstice sun falls at the feet of a god of the night in Palenque, the first light penetrates a window of Newgrange, a prehistoric passage grave in Ireland, 26 miles north of Dublin. The light shines down 62 feet of megalithic corridor and illuminates a vaulted chamber, consecrated by deposits of human bone.

Newgrange was built by neolithic farmers, about 3300 B.C., roughly 500 years before the first stage of Stonehenge was erected in southern England. A few years ago the Newgrange ruin looked like a small hill in the valley of the Boyne River. Archaeologist Michael J. O'Kelly's recent restoration reveals, however, that the builders had a well-developed sense of design. Newgrange is heart-shaped and surrounded by the remains of a stone ring, also heart-shaped. With vertical walls, faced with white quartz, and a flat top, Newgrange now looks like a giant white pillbox in the Irish farm country. Many of the big stones in the curb around its perimeter—and inside as

The white quartz façade of Newgrange, a chambered passage grave in the Bend of the Boyne, gleams in the Irish sunlight, and the sharp geometric lines of the monument make it stand out in the landscape of soft, green, rolling farms. *(Robin Rector Krupp)*

The spectacular carved entrance stone of Newgrange partly obstructs the passage door-way, and the recently discovered window, or "roof box," is situated just above the entrance. *(Robin Rector Krupp)*

well—are elaborately incised and pocked with spirals, lozenges, and many more enigmatic designs, all made with stone tools.

According to an old local tradition, a triple spiral on one of the stones near the back of the vault was lighted by the rising sun on a special day of the year. Until O'Kelly's excavation this folklore seemed to be nonsense, for no light entering the door of the passage could reach the inner chamber. The uphill grade of the floor and other stones in the passage intercepted all light rays. Above the door, however, O'Kelly discovered a kind of window, a "roof-box" he called it. Two blocks of quartz—one still in place and the other found nearby—exactly filled the opening. Both showed scratches that indicated they had been removed and reinstalled numerous times.

On a hunch, Professor O'Kelly installed himself in the back of the tomb on the winter solstice, December 21, 1969, and waited for the sun to come up. Four minutes after the local sunrise a shaft of golden sunlight shot through the roof-box to the small chamber behind the main vault. For 17 minutes the inner stones and their carvings were touched lightly by the sun, until it passed out of the window again.

There is little doubt the astronomical alignment and the sunlight event were intentional. The entire monument reflects care in engineering and design. Grooves cut in the large roof slabs are well adapted to carry off rainwater and keep the chambers of the tomb dry. Even after 53 centuries the corbel vault, 20 feet high, remains intact. With its main axis nearly in line with the direction of the winter solstice sunrise, even the geometry of the entire structure emphasizes the astronomical orientation. Details like the roof-box, the quartz blocks, and a vertical "axis line" carved down the middle of the entrance stone confirm the intent. Symbolic elements—perhaps the quartz façade itself—suggest more connections with sun.

The people who built Newgrange left us no written accounts of their ceremonies. We cannot reconstruct the imagery of their beliefs. Even so, the themes of Newgrange are clear. Death, the sun, and the winter solstice are all interrelated there.

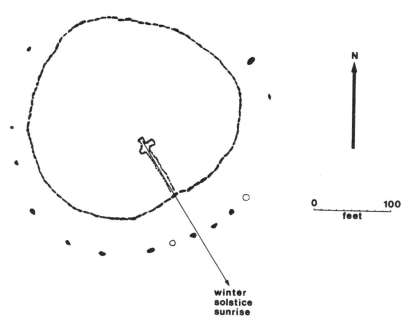

Both the tumulus of Newgrange and the stone ring that surrounds it share an axis of symmetry directed to the southeast and winter solstice sunrise. *(Griffith Observatory)*

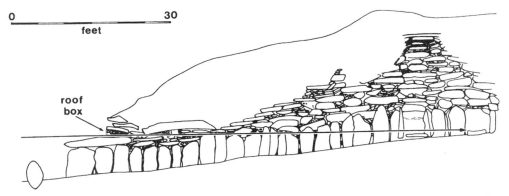

0   30

feet

roof box

With a floor that slopes uphill, the passage at Newgrange does not permit sunlight that enters the doorway to reach the inner chamber. The roof box incorporated above the entrance, however, does allow the winter solstice sunrise to splash into the far end of the tomb. *(Griffith Observatory)*

Winter solstice can be the death of the year and the birth of the year. Combining sunrise with winter solstice suggests both death and rebirth. Death, for certain, is part of Newgrange. It was, after all, a tomb. But the cremated bones cocooned within it seem too few to represent a community cemetery. The deposits of ash are small and perhaps special. They may be symbolic components of a structure designed more or less for ritual than for disposing of the dead. Penetration of a darkened chamber by revivifying sunlight makes the event seem like a metaphor modeled on fertility, sanctified by the dead, and mobilized by the sun. Newgrange as a communal enterprise served a communal purpose. It embodied a celestial rhythm that organizes the world and energizes life. It reflected back to the people who built it, the theme that prompted them to do so. Whatever it meant to them in detail does not matter. Their sense of community and their sense of place in the world were enhanced by a monument that brought the sky in touch with the earth. As in Egypt and China, in Cahokia and Palenque, their dead had joined the cosmic circuit. Their mortal remains, carefully consecrated in the symbolic landscape, preserved the world's order for the living left behind.

6

The Vigils We Keep

Whether the dead sink into the earth like the setting sun or ascend to heaven like the rising stars, their destiny is linked to the sky. Others besides the dead, however, can enter the supernatural underworld or the celestial realm above, but the journey is dangerous for the living soul and requires unusual experience and skill. It is a calling, the business of specialists.

Such specialists are shamans, and their mystical trances and vision quests carry them to the sky. There, shamans have access to the sacred. Their spiritual travels provide them with knowledge and power. For example, their knowledge of transcendental journeys of the soul allows them to treat illness with magic and ritual. Disease, among shamanic peoples, is perceived as a malady of the soul.

Shamans are community leaders who negotiate with the gods and spirits and enjoy intimacy with the sacred. In a sense, they are like priests, but they are not exactly the same thing. Priests represent institutionalized religion. They manipulate symbols of the sacred in ceremonies, sacrifices, and prayer. Shamans may do some of these things, but they are independent agents who encounter the supernatural directly through their mystic vigils.

Shamanism is most closely associated with the hunters and herders of central and northern Asia, but we encounter the essentials of the same tradition among most Indians of the Americas. These spiritual techniques traveled across the Bering Strait with the migrating peoples who emerged from Siberia and fanned out into the entire western hemisphere to become the ancestors of the Indians. Not surprisingly, elements of shamanism also show up in the religions of Oceania, Australia, southeast Asia, Tibet, China, and Japan—wherever shamanism was carried.

Mircea Eliade sees a link to the primordial religion of Old Stone Age hunters and

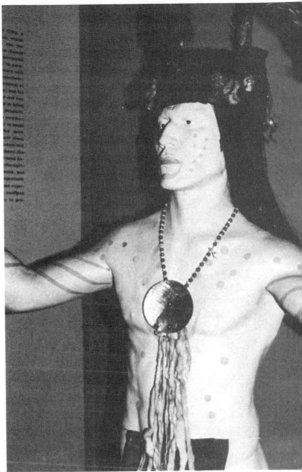

A Blackfoot Indian shaman, Bull Child, portrayed by this mannequin attired with his original regalia, is on display in the American Museum of Natural History. His body, painted yellow, was adorned with symbols, some of them astronomical. Of course, the crescent on his chest, below the shell pendant, represents the moon. The blue dots are stars, and the pattern on his cheek looks like the Pleiades. (E. C. Krupp)

gatherers in shamanism. Even the Neanderthals, 60 thousand years ago, may have had their shamans and buried one of them with his medicinal herbs in Shanidar Cave, Iraq.

Because shamanism is a system—and a well developed system at that—of symbol and ritual, it would be wrong to say shamanism was itself our first religion. Some of its aspects, however, are universal, and this tells us it may preserve some of our prehistoric ancestors' first perceptions of the sacred. In this sense, shamanism does have something to do with the origin of religion, and we can see this in many of its fundamental themes. The mystical ascent to the sky is one of these major themes, and this is, of course, understandable. The sky is the theater for the pageant of cosmic order and is, therefore, the source of the shaman's sacred power.

The shaman's metaphor is the cycle of cosmic order. It appears in his initiation into shamanic mysteries, for this ordeal of deprivation, illness, pain, and psychological crisis is treated as a death. In the process, the initiate may be guided to visualize his or her own death and reduction to nothing but bones. From these the shaman's body is reconstituted, usually with the help of substances charged with symbolic power, like

The Chukchi, a people of eastern Siberia, place the pole star, or the 'Nail Star,' as they call it, at the center of the sky. In this Chukchi drawing, four lines intersect at the pole, and the Milky Way descends diagonally to the left. The world in the lower left is the realm of the dawn; the world of dusk is in the upper right. Both are linked to the pole by the intersecting "skylines," as is the darkness of night represented by the world on the lower right. The sun is on the far left, and the crescent moon is on the right. The recognizable pattern of the Pleiades is in the top of the picture, near the center. *(Larousse World Mythology)*

quartz. The vision obtained from the mystical quest is transcendental and creates, in effect, a new personality and a new perspective. Initiation, then, like the sunrise or the first crescent moon, is a rebirth.

Each mystic experience is also an echo of the cosmic order. The shaman's trance is a death. His spiritual pilgrimage parallels the cycle of the natural world. To achieve a direct and personal revelation of the sacred, the shaman immerses himself in the sky. It is the reservoir and mirror of cosmic order, and he emerges from it renewed and filled with a vision of reality. This return to normal consciousness—transformed by contact with the sacred—is a rebirth.

The shaman's transaction with the sky helps us understand the sky's role in the human mind, but it also helps us recognize and understand why certain ancient monuments are astronomically oriented. This is particularly true in the New World, where evidence of such sites has accumulated rapidly in recent years. The purpose of some of these astronomical sites becomes clearer when we recall the shamanic traditions of the people who used them.

The Shaman and the Shrine

At Burro Flats in the Simi Hills, just beyond the northwest corner of the San Fernando Valley (and so within the official limits of Los Angeles), there is a panel of Chumash Indian rock paintings, the shrine of a tribal shaman. There are many such relatively small shrines. Their form varies. Some are oriented toward the winter solstice sun; others involve the summer solstice. They register the sun's rising or setting at its northern and southern extremes, with lighting effects that are always evocative and sometimes spectacular. An individual shaman or a small group might gather at one of them to participate in the rhythms of the cosmos at special times of the year and to make mystical journeys, often under the influence of hallucinogenic drugs.

The panel of paintings at Burro Flats consists of a complicated collection of images, including a variety of creatures, some with paws like rakes; winged figures with headdresses; animals—centipedes, perhaps—with segmented bodies; chains; handprints with abstract designs on them, and other odd patterns of dots, lines, crosses, circles, and concentric rings. It all appears to be the result of overpainting on several separate occasions, perhaps by different artists.

The rock paintings at Burro Flats are unusually well protected from the elements and—perhaps more important—from people. Rolling sandstone reefs, carved by water and wind into natural overhangs and smooth, rounded piles of rock make the place look like an ideal set for a cowboy movie. The Burro Flats art is canopied by one of these sheltering havens of rock. Rain cannot reach it and, most of the year, the sandstone canopy shades all the pictographs.

Nor is the site vulnerable, as are so many others in the American West, to mindless vandals, careless visitors, and graffiti scribblers who seek their own immortality by defacing the sacred symbols of those who occupied the land before them. The paintings are on land owned by Rockwell International, which protects the area with its own security force as part of the Rocketdyne Santa Susana Test Facility. The paintings, which record the involvement of the Chumash with the sky, are separated by just a ridge from the stands on which the huge moon-rocket and Space Shuttle engines were test-fired.

An astronomical element in the paintings at Burro Flats was first noticed in early 1979 by John Romani, a graduate student in archaeology at California State University, Northridge. He thought a natural cut—a kind of bottomless window—in the overhang above the western end of the panel of paintings looked like it might let sunlight pass through and strike a part of the otherwise shaded panel—at about the time of the winter solstice.

To test Romani's idea, a party of Northridge archaeologists and other interested observers, including myself, arrived at the Rocketdyne security post well before sunrise on the morning of December 22, the day of winter solstice in 1979. We drove by engine test sites and explosive storage depots, and continued to the end of a bumpy dirt road. Then we picked our way over a rocky streambed, flowing with runoff from winter rains, walked across the level ground below the sandstone ridge, and hiked a short way up it to the main shelter.

As our vigil began, the dawn grew brighter. The upper ledges caught the sun, and a golden sheet of light gradually edged down the ridge. At about 7:35 A.M., Pacific Standard Time, the first direct sunlight fell upon the "window" and produced a momentary image of a bright white triangle of light. It cut across a set of five concentric rings, painted in white, and pointed toward their shared center. The tip of the triangle rested on the second ring from the center. This brief impression of a triangle was followed by the full silhouette of the opening. It looked like a spike, its sharp tip cocked to the left and toward the center of the rings. Gradually, as the sun rose higher, the image shrank back from the rings to the base of the panel. For the rest of the day sunlight remained below this prepared rock surface and all of its most prominent paintings. Later observation confirmed that the same effect can be seen each morning for about a week before and a week after the winter solstice. During the rest of the year, however, the sun rises too far north and passes too high to illuminate any of the pictographs.

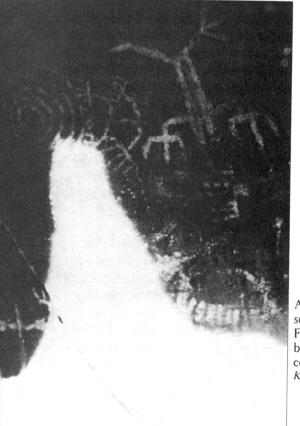

A few moments after the first light of winter solstice sunrise falls upon the rock art at Burro Flats, the small luminous triangle fills out to become this finger of sunlight pointing to the center of a set of five concentric rings. *(E. C. Krupp)*

The passage of sunlight through the "window" at Burro Flats is reminiscent of the play of winter solstice sunlight in the inner chamber at Newgrange. The two sites also are similar in that neither could be used as an observatory; they are not arranged precisely enough to pinpoint the actual day of the solstice to closer than a week. But unlike the Irish structure, Burro Flats definitely was not a tomb.

With no real evidence at Burro Flats other than the paintings themselves, it is necessary to rely on Chumash traditions to understand the site. Fortunately (despite centuries of contact with whites and the ensuing erosion of Chumash culture) tribal traditions persisted—albeit in fragmentary and sometimes distorted form—until the early decades of this century, when they were recorded by ethnographers, notably John Peabody Harrington.

The Chumash were hunter-gatherers, but their society was populous and complex. As many as 15,000 individuals inhabited numerous towns and villages along the Santa Barbara Channel portion of the southern California coast. Their settlements extended a hundred miles inland and out to the Channel Islands as well. They were active traders with a money economy and a well-deserved reputation for exquisite basket-work. Their religious life was organized by the 'antap, an important cult whose officers comprised an elite component of Chumash society.

Among the 'antap were astronomer-priests, astrologers of a sort, each known as an 'alchuklash, who were responsible for maintaining the calendar and determining the proper times for various ceremonies. The 'alchuklash also provided newborn children with names based upon the celestial circumstances of their births and administered Datura (jimson weed), a dangerous hallucinogen, in the ritual initiations of boys into manhood.

The 'alchuklash themselves consumed Datura at the time of the winter solstice in order to foresee the outcome of the gambling game between Sun and Sky Coyote (Polaris). The solstice was a critical time in the Chumash calendar. All resources had to be marshaled then to preserve the cosmic balance. An 'antap leader (who was also an assistant chief) assumed the role of sun priest and, with 12 helpers, conducted a ceremony, the Kakunupmawa, in which they erected "sunsticks" to coax the southern sun back north again. During this time, some of the 'antap priests also executed rock art at special shrines. In a fundamental survey on California Indian astronomy, "Solstice Observers and Observatories in Native California," three specialists in California Indians and rock art, Dr. Travis Hudson, Ken Hedges, and Georgia Lee, have assembled evidence from throughout the state showing that observation of celestial objects, winter solstice rituals, rock art, and shamanism are all interrelated.

Systematic study of California Indian astronomy was actually pioneered as recently as 1978, when Travis Hudson and Ernest Underhay published *Crystals in the Sky*. They assembled a considerable amount of ethnographic information, information that before had gone unnoticed and unappreciated.

It must seem that the whole idea of California Indian astronomy is a rather specialized and obscure field of academic inquiry, but the Native Californians' sky traditions have far-reaching implications. They tell us—at least in part—who some of the ancient skywatchers were and what motivated their vigils.

For the *'alchuklash,* or astronomer-shaman, winter solstice sunrise was not something watched with scientific dispassion. It was a religious experience, a revelation. This vigil, like the shaman's mystical visits to the kingdom of gods and spirits, helped the shaman acquire and exercise sacred knowledge and supernatural power. In the play of winter solstice sunlight upon symbolic rock art, the shaman saw the sign of cosmic order. He experienced a *hierophany*—a "showing of the sacred." The shrines themselves were sacred because they were the places where cosmic order was revealed.

The Chumash shaman must have waited alone—or with at most a few attendants—through the night at Burro Flats for the first flash of morning sunlight. It and other California solstice shrines are not suitable for large public gatherings. Our own observations of sunrises and sunsets on different sites demonstrate how difficult it is for a crowd to witness the event. Instead of great assemblies at these events, we must imagine a lone shaman striking out into the spirit lands of the soul from the privacy of a sacred shrine decorated by his own hand.

Houses of the Sun

Chumash sun shrines may well have been the departure points for the shaman's soul on its journey to the sacred kingdom. To reach a sacred zone, one must leave from consecrated ground. The rock art and solstice events sanctified these places and made them suitable for shamanic ritual.

Condor Cave, another Chumash rock art shelter in the Los Padres National Forest, provides more details about the shaman's vocabulary of symbols and incorporates them with observation of the winter solstice sun. The thin wall on the southeast side of the shelter's entrance is perforated by a small hole, about 2 inches in diameter. It probably was drilled through purposely. Around the time of winter solstice, the rising sun sends a beam through this opening and dramatically illuminates the cave with an intense shaft of light. Travis Hudson and Ernest Underhay speculate that a ritual device may have been placed where sunlight strikes the floor. Perhaps one of the sunsticks erected by the *'antap* in the winter solstice rites caught the sunrise beam.

Condor Cave also contains many rock paintings that suggest connections with the winter solstice. Among the pictographs are animals identified as a frog, a water strider, and a newt. All three are associated with water, the winter solstice, or the rains that follow it. Another figure, human perhaps, is black with red and white dots. This is similar to the body painting worn among other southern California Indians for the winter solstice rituals. Above the figure is a quartered circle, itself surrounded by 20 dots, and the whole, in turn, surrounded by an eight-pointed star. This may be a sun symbol, and above it is a problematic shape that may represent the three feather-topped poles reported at sun shrines. At a nearby rock shelter a possible link between the winter solstice and Chumash rock art was observed in December, 1978, by Steve Junak: Near solstice sunset a sharp line of shadow edges across six concentric rings painted on the inside surface of a kind of "pillar" that forms the light-admitting window of Edwards Cave.

Did Chumash shamans set up ritual paraphernalia to intercept beams of sunlight? We have no real evidence for this at Condor Cave. To prove it we would have to find the remains of a pit that might have held a sunstick, say, and the thick layer of debris on the cave floor has not been excavated. Still, the painted pillar inside Edwards Cave suggests the same idea, and a five-sided hole, about 3 inches deep, has been found on the floor of another cave, the higher of two at a site known as Painted Rock, on the high grasslands of the Sierra Madre. There is a sunset display of winter solstice light on the rock art, but the cave also has a roof aperture that admits light on the summer solstice.

The Painted Rock cave has a "sunburst," 18 inches in diameter, painted in red on the ceiling. This detail has helped Georgia Lee and archaeologist Stephen Horne to identify this site with a place described by a Chumash informant as *Sapaksi*, or "the House of the Sun." Dr. C. T. Hoskinson and Robert M. Cooper have observed that light transmitted through the opening in the cave's roof reaches the peculiar hole in the floor in the middle of the afternoon on the summer solstice. Experimenting with a natural quartz crystal, they found they could send rainbows all around the shelter by letting sunlight fall upon it. There is no proof the five-sided hole in the cave floor

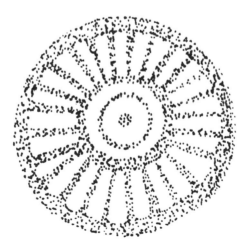

This radial design and other California Indian rock art remembered from Chumash tradition verified that its location, Painted Rock, was a sacred Indian shrine known as "the House of the Sun." *(Griffith Observatory, after Georgia Lee)*

actually was used to hold such a device, but the notion may not be so farfetched. In Chumash tradition, Sun was said to live in a quartz crystal house, and such crystals were part of the ritual paraphernalia of Chumash shamans. Quartz seems, in fact, to have sun and sky associations among shamans from Australia to the Amazon, and rock crystal sometimes is called "solidified light." The Chumash 'antap valued the power in quartz crystal to disperse light and associated the spectra produced by it with the celestial rainbow that bridges the sky.

Signs of shamanic interaction with the sky are still evident at other Chumash shrines. One of these is at Vandenberg Air Force Base, where rockets and satellites are launched into space. Here archaeologist Laurence Spanne has discovered what seems to be a winter solstice sanctuary and—appropriately—a place, perhaps, where the shaman began his mystical ascent to the sky.

Up a long slope from Honda Canyon, in a remote southern quarter of the base, a natural rock shelter provides a dramatic view of Tranquillon Mountain, a prominent feature on the southwest horizon. Indians told John Peabody Harrington at the beginning of this century that this peak was *muy delicado*—"very sacred." The shelter itself is quite small and very low; it can accommodate only one person on hands and knees. Inside it is decorated with carvings, or petroglyphs. The pattern is complicated and difficult to make out, but a few of its elements seem clear. The left part of the design looks like a stylized vulva—a typical Chumash symbol for fertility and the

Window Cave is a low hollow in the sandstone ridges on Vandenberg Air Force Base. Its small size suggests very few people—perhaps only a single person—would occupy it and observe a display of sunlight near the petroglyphs within. *(E. C. Krupp)*

A small natural window in Window Cave opposite the larger entrance opens to the southwest for a view of sacred Tranquillon Mountain. On the winter solstice the sun sets behind the peak and drives a shaft of light through the window and against a wall of the shelter. *(Laurence W. Spanne)*

portal to creation; the point where new life emerges. On the right, a triangle points down to an eight-spoked wheel, a design often interpreted as a sun symbol.

A few feet to the right of the shelter's entrance, in the deepest recess of its womb-like space, the wall opens into a natural window that frames the mountain. From the shelter, which Spanne calls Window Cave, the peak is in line with winter solstice sunset. After hovering over the summit, now graded 25 to 50 feet below its original elevation, the sun moves halfway down the west slope, where it slips behind a spur. By observing sunsets from this location, it is possible to determine the date of the solstice to within a few days of its actual occurrence. The beam that enters the rock shelter's window streaks across a wall and ends where the wall turns, at the spot marked by the "sun symbol." The rock womb is penetrated and "fertilized" by the setting sun.

The themes at Window Cave seem congruent with the pattern seen elsewhere. Although the Chumash do not refer to the "death" of the sun, they say that the sun, the year, and the cosmos are reborn with the sunrise at the winter solstice. At Window Cave, winter solstice sunset is the key moment. Here, then, the *'alchuklash* experienced the prelude to renewal of the world's order. The eight-spoke disk may

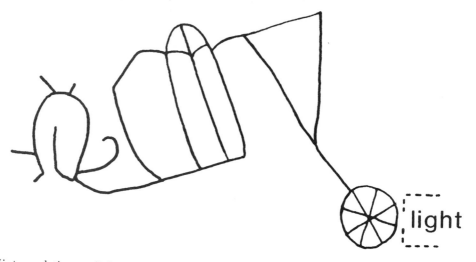

light

Winter solstice sunlight penetrates Window Cave and stops at a carved spoked ring that may symbolize the sun. The triangle nearby may represent the female sexual zone or the piercing beam of light. The image on the far left also resembles stylized female genitalia, and the entire design may equate the imagery of procreation with the renewal of the year. *(Griffith Observatory)*

represent the new sun fathered by the old, or it may just be an emblem of the shelter's solar connotations. Either way, it and the vulva petroglyph restate the familiar metaphors: fertility and the sun. Both imply renewal—the end of one cycle and the start of another.

In making his mystical voyage, the *'alchuklash* used his special knowledge and experience to assist regulation of the cosmos and ensure balance of the forces that drive it. He was familiar with the cycle of cosmic order. He recognized it in the winter solstice sun and saluted it in his rock art. By organizing tribal activity and integrating it with the world's cyclical changes, he became an active participant in the maintenance of public and cosmic order.

We can get an inkling of how the shaman perceived himself at La Rumorosa, a painted rock shelter in Baja California. Shamanic power, fortified by the sun, still seems in force there when winter solstice dawn disperses the night. The site is in Tipai Indian territory, just across the U.S.–Mexico border. Ken Hedges, archaeologist at the San Diego Museum of Man, stationed himself inside the shelter before sunrise on December 21, 1975, and observed a display of light on rock art when the first rays reached into the cave. At sunup, a sharp triangle of light stabs its way across a wall, hitting first a white circle with a stem and then a white vertical line. As it penetrates further, the "dagger" seems to menace a human-looking figure, about 13 inches high. The figure is painted red, sports two wavy horns on its head, and has little black eyes. Suddenly, the sun dagger jumps across the red figure. The pattern of light is split in

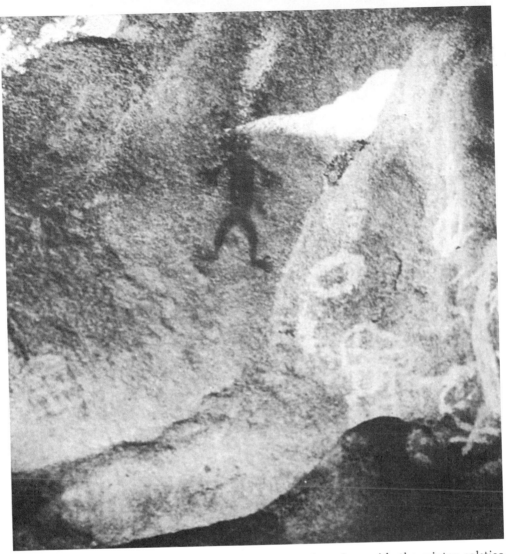

One of the most evocative rock art "light shows" takes place with the winter solstice sunrise in a Tipai Indian shelter in Baja California. The shrine is known as La Rumorosa, and on its walls a small red figure with horns "sees" the world's renewal in the sunlight that knifes across his beady black eyes. *(E. C. Krupp)*

two, and the painted figure is in the darkness in between. From the right the point of the knife seems to reform and stretch back toward the figure. When it reaches the figure, the dagger slices its face precisely across the eyes. The entire event seems to be a portrayal of a shaman's own participation in the winter solstice sunrise. Horns extending from the head are often thought to have something to do with supernatural power. As such, they mark the red figure as the sunwatcher: Its eyes actually catch the light of the winter solstice sunrise.

Journeys to the Sky

As explicit as the imagery at La Rumorosa or Window Cave may be, we really don't know many details about the shamanic rituals performed at California Indian sun shrines. This information was not shared with ethnographers, and it may not have even been preserved by the time ethnographers interviewed Native Californians. There are, however, other intriguing features of these sites that hint at the kinds of things that might have gone on. Deep mortars carved into the bedrock at some sites seem too large and too impractically placed for everyday use. They may have been used for offerings—ground acorns, for example. Paint for the rock art may have been readied in some of the shallow cuplike depressions. And perhaps the mortars were used to pulverize *Datura* seeds and other drugs. Leaves may have been soaked in some of the other cavities to prepare herbal infusions.

Even though these suggestions cannot be confirmed, we can look at the shamanic practices of other peoples and at least gain some understanding of the shaman's mystic techniques and celestial metaphors. Religion in ancient China, for example, despite considerable evolution, preserved many shamanic traditions, including the visionary ascent to the sky. In *Pacing the Void,* an elegant account of the celestial imagery and beliefs of the medieval Chinese of the T'ang Dynasty, orientalist Professor Edward H. Schafer describes the poetic accounts of Taoist mystics who roamed the stars. These initiates, following ancient shamanic traditions, practiced meditative techniques that involved concentrating upon the stars. In the mind's eye they traced out the stellar patterns or constellations, and from them derived the power to climb to the sky. Several routes were possible, but whether via the Big Dipper, the planets on the ecliptic, the Milky Way, or some other celestial trail, the goal was the same: The center of the cosmos, sacred and beyond space and time. This is the transcendental, religious experience of the mystic who enters, for a while, the realm where all is one and who melts there into the universe. This consciousness of unity is beyond the senses and beyond reason and typifies the mystic experience wherever it is achieved. In China, both the imagery of the soul's journey and the means of making it involved the sky. It is "pacing the void."

Shamanic ecstasy required shamanic knowledge. Intimacy with stars was a must. Hallucinogens may have been used as well; jimson weed, or *Datura,* was considered sacred by the Chinese, too, and its name was given to one of the circumpolar stars. Spirits from this star would transport its flowers to earth.

Prospective T'ang star treaders sometimes followed the "pace of Yu," a ritual dance step based upon the pattern and sequence of stars in one of the constellations, usually the Big Dipper. Yu was the semimythical founder of the legendary First Dynasty, the Hsia. Renowned for bringing civilization to China, Yu introduced agriculture, irrigation, mining, and metalworking. By T'ang times, he was regarded as divine and associated with shamanism as well. T'ang shamans danced the special step of Yu and found themselves Big Dipper bound. Through its circumpolar turns the Big Dipper was both calendar and clock. It was the sign of social order, an emblem of the king, and the chief reservoir of shamanic power.

The northern stars were the focus of many magical techniques. With eyes closed, an adept might try to visualize Alcor, the faint but visible nearby companion of Mizar, the second star from the end of the Big Dipper's handle. Others would recline upon a bed-size diagram of the Big Dipper and imagine its stars while reciting the names with appropriate prayer. Breath control and other trance-inducing exercises, in concert with the stars, transported the soul to heaven. One set of T'ang instructions is as explicit as a Michelin guide:

> When you wish to tread the Mainstays, first hold your breath, and turning left outside the seven stars, travel the three circuits above the Dipper's cloud- and white-souls. Only then may you ascend to Solar luminosity.

Nine sets of footprints mark the dance steps in the "pace of Yu," reminiscent of an Arthur Murray instruction book. T'ang Dynasty Taoist adepts, not modern initiates into the intricacies of ballroom dancing, tripped this light fantastic, however. The dance pattern was meant to carry the Taoist "star treader" around and through the stars of the Big Dipper and probably derives from a much older shamanic ritual of mystical ascent to the celestial realm of cosmic order. *(from Edward H. Schafer,* Pacing the Void, *after a twelfth-century source. Reprinted by permission of the University of California Press.)*

The "cloud-souls" and "white-souls" are the stars themselves. The "Mainstays" are imaginary celestial lines that bind the sky—meridians that extend through the stars from pole to pole, like longitude lines on earth. Only the proper path could permit one to amble up, star by star, to Dubhe, one of the brighter stars in the Dipper and the lead star in the bowl. Here Dubhe is called "Solar luminosity."

Some think the "pace of Yu" derives from impersonations of bears by archaic shamans. Throughout the northern world the bear is an important image in the shamanic tradition, and it is fairly easy to see why this might be so. As a powerful creature, the bear is a natural embodiment of the shaman's power, but the bear shares another aspect of the shaman's path. In winter, the bear hibernates. It emerges in spring. Hibernation is a kind of death, and it parallels the shaman's "death"—the obligatory trance. Through this transformation the shaman experiences a transcendental ecstasy associated with the sky. He is reborn, as is the bear when winter's death is done. Power and the cosmic cycle are locked in the habits of the bear. To reach them, the shaman shadows the bear, steps in its tracks, and is lifted to the sky. Although the Chinese never called the Dipper a bear, the link between the shaman's bear and the Big Dipper seems clear.

In the temperate latitudes, where the pole governs the sky, the shaman climbs to the northern stars. Those who live near the earth's equator, however, need not glorify the celestial pole. Yet their shamans, too, journey to the sky and to another zone of cosmic order. In the Colombian Amazon, for example, the Tukano Indians travel "beyond the Milky Way." Many equatorial peoples of South America use the Milky Way to organize the sky. Among the Tukano, the Milky Way is the sky's chief structure. It arcs up from the underworld and flows east to west over the earth. Because it intersects the horizon and reaches into the sky, it is the channel of communication between the powerful spirits of the upperworld and the people of the earth.

Tukano shamans also consume a hallucinogenic plant, a leafy vine known as *yajé*, to induce mystical visions. Anthropologist G. Reichel-Dolmatoff collected Tukano lore associated with the ritual use of *yajé*. Properly prepared and administered, it generates a feeling of ascent. Shamans and tribesmen, embarked upon the *yajé* vision quest, consider the Milky Way their first celestial destination. In their *yajé* "dreams" the Tukano experience mythic landscapes and events. They see the Milky Way and Father Sun. The myth of creation is reenacted. They enter the sacred realm of mythic time through *yajé*, and their dreamlike visions there restate the traditions of their culture. Their own metaphors for the drug-induced trance are death and the womb. Reborn in the sky or at the end of the trance, they say they have spiritually sampled the essence of what it means to be Tukano. The world's order is revealed in their creation myth, and they have gone to the sky to know it. They return transformed.

While the men drink *yajé* and experience these visions, the women encourage them and sing, "By drinking they will know all of the traditions of their fathers." Hallucinogenic ritual helps the participants feel in a cosmic context, orients them, and confirms their place in the world. In general, this transcendental experience integrates the personality with the world around it. The Tukano have a place in the cosmos, and they know where it is.

At the outset, *yaje* hallucinations involve changing, luminous patterns of repetitive geometric forms in vibrant colors. These sensations are called phosphenes, and they are triggered internally on the retina. They can occur spontaneously or be induced in a variety of ways. Psychotropic plants like *yaje* provide a chemical stimulus to release them.

Phosphenes are thought to be neurophysiological in origin—products of the circuitry of sight. They are universal, experienced by all people, and in some way relate to the structure of visual perception. An entire lexicon of simple shapes and patterns shows up in the phosphenes, and the Tukano transfer these designs to walls, dishes, baskets, gourds, musical instruments, and bark cloth paintings. Well aware of the source of their decorative motifs, the Tukano say, "We see these things when we drink *yaje*." Individual elements and designs are equated with various concepts, and their meanings are shared by the community. For example, a lozenge, or diamond, represents the female sexual zone and can symbolize many things connected with women. Fertility, sexuality, and the cycle of life and death are the dominant themes of the symbolic vocabulary. In a similar way, the rock art of California Indians seems to be closely related to the visions of shamans and their sense of cosmic order.

Tukano shamans seek power and knowledge in *yaje* visions and interpret the state of their communities in terms of the tribesmen's reactions during a carefully orchestrated ritual group consumption of the drug. It is the shaman's job to act as the community's "ecological broker," as Reichel-Dolmatoff phrases it, to maintain cosmic balance and natural order. He is interpreter and guide for the other tribesmen, who are less experienced with the drug and the celestial journey.

Here, then, in the lore and ritual of an isolated traditional people, the sky's essential meaning is confirmed: It is the door of perception to cosmic order. This involves religion because our sense of what is sacred is really our sense of what is real. Mircea Eliade, in *The Sacred and the Profane,* argues that consciousness itself requires us to sense an underlying fabric of reality and meaning. That is the cosmic order, and it is threatened by chaos—senseless events and capricious change. Achieving an experience of the sacred becomes, therefore, a cultural and personal necessity. Hallucinogenic substances, administered ritually under the guidance of an experienced shaman, expedite the process and keep it safely framed within the communal system of values. Phosphenes, with their order, pattern, and inventory of simple forms, reflect the ordered texture of consciousness.

Wheels of the Sun

Astronomical observation and calendar keeping are usually associated with settled, agricultural peoples. It is often argued, in fact, that the calendar really owes its invention to farming. This, of course, is not so. The Chumash were settled but hardly agricultural, and their shaman-priests kept track of celestial cycles, counted out the days in a lunar calendar, and established the times for important public events like the harvest festival and the winter solstice ceremony. Many people gathered for these holidays. Their impact upon the social life and the economy of the community was great. They provided structure, stability, and equilibrium.

This, then, is the real purpose of the astronomical calendar. It organizes our lives. Without doubt, a reliable calendar is of great importance to an agrarian society, because farming, like religion, commerce, and most other enterprises, must be organized—especially when it is undertaken on a large scale. Small farms and backyard gardens, however, can succeed without such institutionalized timekeeping. Folk traditions, nature lore, and informal skywatching provide ample cues to the appropriate times for planting and harvest. It is when farming—or any other activity—grows more complex that the calendar assumes greater importance. This is a reflection of society's increased complexity—not a legacy of discovering how to get crops to grow.

Now we usually equate complexity with civilization, but the intricacies of shamanism disclose the complexity of the spiritual life and social life of the early hunters. Ancient people on the Great Plains of the United States and Canada, for example, did not plant crops—or even settle down like the Chumash—yet they, too, built shrines that were aligned with the sky and probably used by their shamans to seek the sacred and experience the cosmic order.

These monuments are known as medicine wheels. The term "medicine" here refers to the supernatural, to sacred lore, and to shamanic power. The wheels themselves are usually circles of small rocks, centered on a large cairn, or pile, of stones of the same sort. They look like wagon wheels, but not all have spokes, and at least one, in Minton, Saskatchewan, is shaped more like a turtle—complete with head, tail, and four legs—than a wheel.

About 50 medicine wheels and related structures are known. Nearly all are found on the east flank of the Rockies or on the open plains below; most are in the north, on the grassy prairies of Canada. Some wheels seem to be only a few centuries old; others are very ancient. Excavations in Alberta, Canada, of the Majorville Cairn, a badly vandalized wheel with numerous spokes, a rim, and a 50-ton pile of rock in the center, imply an age of 4,500 years—making it about the same age as the pyramids of Giza!

The first of the wheels to be analyzed in astronomical detail was the Bighorn Medicine Wheel, high up on an exposed shoulder of Medicine Mountain in Wyoming. The wheel's central cairn is about 12 feet in diameter and 2 feet high. The ring itself is no higher than the scattered stones that define it, but its largest diameter is 87 feet—only slightly smaller than that of the large sarsen circle of Stonehenge. Radiating from the central cairn to the rim are 28 spokes, also made with small rocks. One

The Bighorn Medicine Wheel, near Sheridan, Wyoming, is a collection of small cairns and spokes at an elevation of 9,642 feet on an exposed windswept shoulder of Medicine Mountain. *(Courtesy Robert O'Connell, College of the Redwoods)*

spoke extends past the rim for 13 feet and ends in a small cairn. The ends of five other spokes also are marked with cairns where they join the wheel's rim.

Additional length suggests special importance for the extended spoke, and it was Dr. John A. Eddy, a solar astronomer at the High Altitude Observatory of the National Center for Atmospheric Research in Boulder, Colorado, who discovered what it is: The view from the cairn at the end of the spoke, across the center of the central cairn, and on to a low ridge to the northeast coincides with the direction of summer solstice sunrise.

By itself, the summer solstice sunrise alignment might be considered little more than the product of chance, but Eddy's analysis of the other cairns revealed an internally consistent set of alignments involving all but one of the remaining cairns. These include an alignment on the summer solstice sunset and three lines oriented on the rising points of three bright stars: Aldebaran, Rigel, and Sirius. The first of these, Aldebaran, in Taurus the Bull, rose just before the sun in the predawn sky, within a day or two of the summer solstice from A.D. 1600 to 1800. (Because precession has shifted the stars with respect to the solstices, Aldebaran now rises too close to the sun on June 21 to serve as a herald of summer solstice.) During the same period, two to four centuries ago, Rigel, in Orion, rose before the sun 28 days after the solstice, and

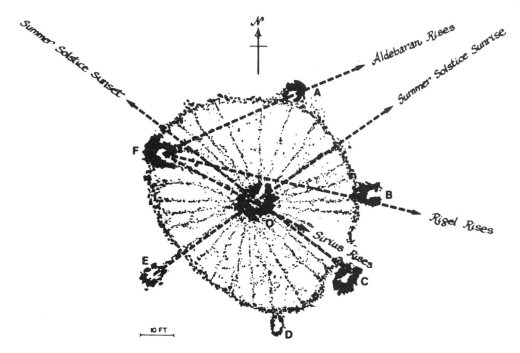

Perhaps the most important thing a plan of the Bighorn Medicine Wheel can tell us about the shape of the Wheel's rim is that like Newgrange it has a directed axis of symmetry. This axis coincides with the summer solstice sunrise alignment from the outlying cairn E to the central cairn O and emphasizes the astronomical significance of the line. Other alignments proposed by Dr. John Eddy are oriented on summer solstice sunset and three heliacal risings of bright summer stars. *(Dr. John A. Eddy)*

Sirius, in Canis Major, 28 days after that. An independent estimate of the wheel's age, based upon tree-ring counts in pieces of wood recovered from the central cairn, was consistent with the stellar alignments: It implied a construction date of at least as early as A.D. 1760. Whether the wood fragments really belong to the period of the wheel's construction is, of course, unknown.

Eddy was disturbed that he could find no astronomical explanation for the sixth cairn, but Jack H. Robinson, at the University of South Florida, showed that the backsight for the other stellar alignments also worked as a backsight with the leftover cairn to indicate the rising position of Fomalhaut, a bright star in Piscis Austrinus— the Southern Fish. Fomalhaut rose before the sun about 35 days ahead of the summer solstice in 1050–1450. This is slightly earlier than the period suggested by Eddy's analysis. Stellar dating has many uncertainties, and the alignments need not have been very accurate. The Bighorn Medicine Wheel could therefore be several centuries older than we think.

It is conceivable that the Bighorn Medicine Wheel was designed to give precise determinations of the date of the summer solstice. By using auxiliary posts and horizon features, the event could have been pinned down to the day. No evidence of the use of such techniques has been found, however. It is just as likely—perhaps more so—that only a rough accuracy was required. Alignments for the risings of stars do not have to be absolutely true; a sharp-eyed observer, facing more or less in the right direction, will spot the star when it becomes visible. Thus, the spoke and cairn for indicating Fomalhaut might be a few centuries off by our reckoning, but still adequate for the original purpose.

Arrangements for marking stellar alignments with cairns and rows of stones are more closely related to symbolic design than to astronomical instrumentation. The probability that the Bighorn Medicine Wheel's astronomical alignments are intentional, and not coincidental, was increased when Eddy verified that another ruin, 425 miles north in Saskatchewan, had the same basic plan as the Wyoming wheel.

Saskatchewan's Moose Mountain Medicine Wheel is but one of many stone and boulder designs on the plains of western Canada studied by two archaeologists on a National Museums of Canada grant, Alice B. and Thomas F. Kehoe. The Kehoes believe that many of the Canadian medicine wheels—those without astronomical alignments—were monuments to chiefs, and built not long ago. Thomas Kehoe had

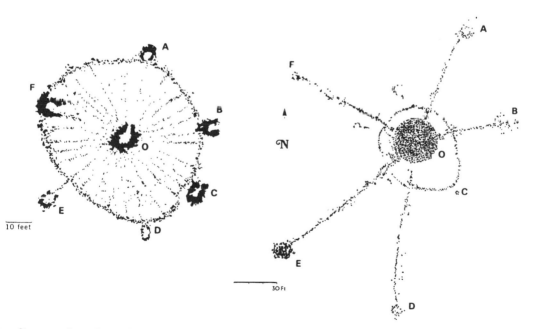

Support for John Eddy's ideas about the Bighorn Medicine Wheel seems to be provided by the similarity between its pattern of cairns and that of the Moose Mountain Medicine Wheel in Saskatchewan. Astronomically, the alignments are the same, and the greater age of the Moose Mountain structure suggests some of these wheels may be part of a millennia-old shamanic tradition. *(Dr. John A. Eddy, Moose Mountain plan by Tom Kehoe)*

noticed the similarity between the Bighorn and Moose Mountain wheels, however, when reviewing Eddy's original paper on the alignments on Medicine Mountain. Collaboration between the archaeologists and the astronomer in 1975 demonstrated that the Moose Mountain Medicine Wheel's cairns fit the same astronomical pattern. Both monuments could have been built and used by ancient predecessors of the historic Plains Indians.

Although built to the same plan, the two wheels differ physically. The central cairn of the Moose Mountain structure is much larger than its Wyoming counterpart. It is 30 feet in diameter, about 5 feet high, and contains perhaps 80 tons of rock. The egg-shaped ring around it is smaller than the Bighorn's flattened circle of stones, however. Only five spokes radiate from the hub, and all extend far beyond the rim of the wheel. The longest, the summer solstice spoke, ends in a cairn 123 feet southwest of the center, and the spoke itself is about four times as long as the radius of the rim. Next to the cairn at the end of this spoke, Eddy found a small "sunburst" arrangement of stones. Perhaps this signaled the solar alignment. Similar sunbursts are associated with summer solstice sunrise lines at two other sites, the Fort Smith Medicine Wheel in Montana and the Minton Turtle in south central Saskatchewan.

When the observer at Moose Mountain sees the sun rise above the central cairn at the solstice, it is actually about one solar diameter off its correct position. This is due only to the height of the central cairn, for the alignment between cairn *centers* is true. Many of the cairn's stones may have been added, long after the wheel was built, by various prehistoric pilgrims and visitors to the site.

Eddy's estimated date of construction for the Moose Mountain structure, based upon alignments of three other spokes with the rising points of the same three stars, makes the wheel almost 2,000 years old—considerably more ancient than the Bighorn Wheel. Alice and Tom Kehoe concluded the wheel was built around 440 B.C., although uncertainties in the radiocarbon technique leave the date open to question. It could be several centuries younger or older. In any case, there has been plenty of time for visitors to the site to build up the height of the central cairn, whose original purpose they did not understand.

Who built the medicine wheels? We really don't know. Similarities have been noticed in the plan of the Bighorn Medicine Wheel and the sundance lodge of the Cheyenne and other Plains Indians. To some, this meant the wheels belonged to the ancestors of the Plains tribes. In July, 1975, the Kehoes attended a sundance on the Sweetgrass Cree Reserve, in Saskatchewan, and after the ceremonies closed, they had the good fortune to examine the structures built for the occasion. They discovered the main axis formed by the components of this temporary camp aligned with the sunrise during the three-day ritual. In several respects, the layout of the Cree Indian sundance lodge resembled the Roy Rivers Medicine Wheel in Saskatchewan, also surveyed by the Kehoes. They pointed out the features the wheel shared with the temporary Cree structure: a marked central space, an entrance on the south, a surrounding ring, and stations analogous to the places of dancers marked in stone. Despite these resemblances, the other astronomically aligned medicine wheels don't really seem similar.

Interviews with tribal elders produced tantalizing, but inadequate, hints of the medicine wheels' origins. The elders believed the medicine wheels belonged to people who lived there before the Cree, to people who "had their ceremonies." From that time "these circles [of rock] are left." They thought of the ancients and their stone rings as belonging to another age. According to one elder:

> My old people said these were before our time. There was another creation, then a shuffling of this earth, you might say, then we were created. Sure, some Indians use them [boulder configurations] today, pray at these sacred grounds, but they were here before us.

Use of the Bighorn Medicine Wheel by Crow Indians was reported by ethnographer R. H. Lowie, in 1922. One of Lowie's informants, a Crow medicine man named Flat Dog, said the wheel was the "sun's lodge" and a place of fasting and vision-questing for the Crow. What Flat Dog described as roofless sleeping shelters for vision seekers may have been the rim cairns.

Familiar with Lowie's report, Eddy also suggested that the wheel may have been used as a place for vision-questing. He pointed out that occupation of the site by more than one or two persons at a time was improbable. The medicine wheel is remote. At an elevation of 9,642 feet, it is above the timberline. Water and wood are not readily at hand, and, because of snow, the wheel is inaccessible most of the year. (This is another reason for thinking the stellar alignments, which imply a stay on the site of up to two months after the summer solstice, are not coincidental. They are reasonable for the time of year during which the wheel could be reached.)

We don't know if the later Plains Indians—or their shamans—who made use of the medicine wheels had detailed knowledge of the wheels' astronomical potential, nor do we know if the original builders of the wheels used them as sites for vision quests as well as solstice vigils. But in some respects the medicine wheels do resemble the solstitial rock art sites in California. Both have solar orientations, both seem to have been used by only a few people—perhaps just a single person—at a time, and both appear in many variations.

Our knowledge of the builders may be tenuous, but the same themes persist: shamanism and the solstice sun. Whatever the astronomical medicine wheels may really be, they seem best designed to tune the shaman to the cycle of the sky. This medicine man and mystic conferred with the cosmos and returned to the tribe bearing the message and power of the world's order in trust for his people. This is important. These skywatching shamans and medicine men tell us our relationship with the sky is

From the summer solstice backsight (cairn E) of the Moose Mountain Medicine Wheel, the central cairn appears to sit upon the horizon. *(E. C. Krupp)*

an old, old religious response. It is not a byproduct of culture but something that makes culture the way it is. It started long before we became farmers, is at least as old as the cave paintings of paleolithic shamans, and probably reaches back into the origin of consciousness itself.

Priests of the Sun

Ceremonial societies rather than shamans dominate the religious life of the Pueblo Indians of the American Southwest, but the progress of the sun is closely monitored, for ceremonies must be held at special times and the dates of the calendar depend upon the sun's position in the sky.

Like shamans, the Pueblo priests regarded their sunwatching as a religious act. In traditional societies calendar keeping is more than a useful habit. It is a sacred ritual. Priests and shamans are technicians of the sacred who apply their special knowledge to maintain the stability and well-being of the communities they serve.

Although most Pueblo Indians were themselves familiar with the sun's movements and with the relation of the sun's position to the calendar and the schedule of ceremonies, authority to observe and establish these things was conferred upon an official sun watcher or sun priest. It was his responsibility to count out the days in terms of the shifting position of sunrise. This was a civil appointment in a sense, but the sun

Responsibility for the Zuñi calendar fell to the *pe'kwin*, or Sun Priest. In 1896, the Sun Priest was dismissed for making errors, and his successor appears in this vintage photograph. *(By permission of the Smithsonian Institution Press from* Annual Report of the Bureau of American Ethnology, 1901–02, *Plate xviii [opp. p. 108]. Smithsonian Institution, Washington, D.C.)*

watcher occupied an integral place in the community's religious structure. For example, the autobiography of Don Talayesva, the head Hopi Sun Watcher at Oraibi village, includes descriptions of rituals and prayers the Sun Chief must perform. His position also involved some esoteric and inviolate lore, for he declined to reveal certain secrets and activities to the anthropologist who collaborated on his life story. His autobiography contains no evidence of any sort of shamanic transcendence, but his duty was sacred nonetheless. His communion with the sun ordered Pueblo ceremonial life.

The methods of the sun watchers have been described by various ethnologists. According to Alexander M. Stephen, who lived among the Hopi in Arizona from 1891 to 1894, sunrise at Walpi village was monitored from a point south of all the houses, while sunset was viewed from the roof of an important clan house near the center of the village. Another early observer, Frank Hamilton Cushing, visited the Zuñi pueblo with a Smithsonian expedition, at the age of twenty-two. A remarkable man, Cushing stayed on in New Mexico as an active participant in the tribe instead of passively observing the Zuñi. He learned the language, lived with the governor of Zuñi, joined the tribal council, was initiated into the Priesthood of the Bow, and even participated in a run-in with an Apache raiding party. His accounts of Zuñi life are still lively, and one passage in *My Adventures in Zuñi* describes the activities of the Zuñi sun watcher:

> Each morning, too, just at dawn, the Sun Priest, followed by the Master Priest of the Bow, went along the eastern trail to the ruined city of Ma-tsa-ki, by the river-side, where, awaited at a distance by his companion, he slowly approached a square open tower and seated himself just inside upon a rude, ancient stone chair, and before a pillar sculptured with the face of the sun, the sacred hand, the morning star, and the new moon. There he awaited with prayer and sacred song the rising of the sun. Not many such pilgrimages are made ere the "Suns look at each other," and the shadows of the solar monolith, the monument of Thunder Mountain, and the pillar of the gardens of Zuñi, "lie along the same trail." Then the priest blesses, thanks, and exhorts his father, while the warrior guardian responds as he cuts the last notch in his pine-wood calendar, and both hasten back to call from the house-tops the glad tidings of the return of spring. Nor may the Sun Priest err in his watch of Time's flight; for many are the houses in Zuñi with scores on their walls or ancient plates imbedded therein, while opposite, a convenient window or small port-hole lets in the light of the rising sun, which shines but two mornings in the three hundred and sixty-five on the same place. Wonderfully reliable and ingenious are these rude systems of orientation, by which religion, the labors, and even the pastimes of the Zuñis are regulated.

Cushing's report is explicit. A pilgrimage and vigil are kept by the Sun Priest. Specially constructed for its special purpose, the site is symbolically decorated and involves direct observation of the sunrise as well as a visually dramatic display of shadow. A calendar tally is kept, and the people are informed of the outcome of the Sun Priest's watch.

The other Zuñi know well the date and the progress of the sun, but by investing authority in the Sun Priest, the Zuñi ritualize the sun watch. This puts the whole community on the same beat. The Sun Priest announces the date of the winter solstice, and everybody celebrates it together. To function at all, a community must have a shared perception of organized time. The Sun Priest's observations punctuate the passage of time and provide a framework that organizes Zuñi life. His calendar meshes Zuñi ceremony with the pattern of the year, and this puts the people in touch with the cyclical changes the seasons impose on farming, dress, food preparation, travel, and all of the other aspects of everyday life. The sun watch assumes a sacred dimension, for it harmonizes, unifies, and orients the Zuñi.

The function of the Sun Priest seems to date to prehistoric time—to the ancient cliff dwellers, pueblo builders, and basket makers, who lived in the Four Corners region, where Utah, Colorado, New Mexico, and Arizona meet, between A.D. 400 and 1300. These people are thought to be the ancestors of the modern Pueblos. Although these prehistoric Southwestern Indians are best known today by the name Anasazi, the word is Navajo, meaning "the Ancient Ones," and the Hopi prefer to

Bearing the sun's face, the upright slab in this Zuñi sun shrine at Ma'tsakïa resembles the station of the Sun Priest described by Frank Cushing in *My Adventures in Zuñi*. *(By permission of the Smithsonian Institution Press from* Annual Report of the Bureau of American Ethnology, 1901–02, *Plate xviii [opp. p. 108]. Smithsonian Institution, Washington, D.C.)*

call them by a Hopi name, *Hi-sat-si-nam.* They are, after all, their ancestors and not the Navajo's.

One of the great centers of the prehistoric pueblo builders was Chaco Canyon in northwest New Mexico. The canyon once harbored many settlements and pueblos, each incorporating numerous multistory apartments and sunken, round ceremonial chambers known as kivas. Traces of irrigation systems and great road networks are still to be found, indicating a complex, well-organized society. Chaco Canyon seems to have been occupied seasonally by people from distant pueblos. This suggests it was a ceremonial center for pilgrimages from a wide geographic area.

One of the smaller pueblos in Chaco Canyon is Wijiji, at the canyon's eastern end. Above it on the mesa a cliff wall marked with a white disk with four rays faces the southeast. A stone stairway leads up to the decorated face of the cliff and to a ledge. From this point, the horizon is dominated by a natural rock chimney, roughly 1,600 feet to the southwest. At the winter solstice, this stone column aligns exactly with the rising sun when observed from a point on the ledge about 50 feet north of the rock painting.

The rayed disk could represent the sun, and Dr. Ray Williamson, an astronomer with OTA for the U.S. Congress and an investigator of the ancient Southwest sites, believes it identifies the spot as a sun-watch station. He surveyed the site and determined that the width of the rock chimney, as seen from the ledge, is just slightly less than the apparent size of the sun. His photographs of the winter solstice sunrise show the sun's disk gleaming around the sides and over the top of the chimney's dark silhouette. They show how the site could have been used in this way, even though the exact viewing point is not marked by the rayed disk. Williamson offers the un-provable suggestion that the sun disk was intended only to mark the ledge, at its entrance, as a sacred place. Knowledge of the proper observation post remained—unadvertised—in the mind of the specialist who used the shrine, the Sun Priest.

Spirals of the Sun

Considerable publicity accompanied the discovery of another astronomical marker carved upon a cliff face 405 feet above the floor of Chaco Canyon. It is unique and impressive. Anna Sofaer, an artist from Washington, D.C., happened to be up on Fajada Butte, a prominent landmark at the south entrance to the canyon about a week after the 1977 summer solstice. While researching Indian rock art, which may be found all over the butte, she climbed up to have a look at a pair of spiral petro-glyphs known to be sheltered behind three slabs of sandstone, propped upright against the decorated wall.

With the edge of the southeast face of the butte making but a narrow ledge in front of the slabs, just 39 feet below the summit, the spirals remain remote, difficult to reach, and unsafe for casual visitors. There is no water on the butte. Rattlesnakes, steep walls, precarious climbs, and edges slippery with loose rock rightly discourage most climbers. A thousand years ago, when the Anasazi Indians inhabited Chaco Canyon, the Fajada Butte spirals were well isolated from the large population below.

On June 29, near noon, while Sofaer was examining the spirals, the shade beneath

Fajada Butte is an obvious feature in New Mexico's Chaco Canyon, a well-populated site of the ancient Anasazi Indians between 700 and 1,600 years ago. Three slabs that permit daggers of sunlight to interact with spiral petroglyphs are propped against a vertical cliff face near the summit on the southeast (right) side of the butte.   *(Robin Rector Krupp)*

The larger spiral beneath the slabs on Fajada Butte is bisected by a descending dagger of sunlight for 18 minutes in the hour before local noon on the summer solstice. This is reminiscent of what goes on in some of the California Indian shrines and, for that matter, at Newgrange. *(Photograph by Karl Kernberger. © The Solstice Project/Anna Sofaer)*

the slabs was pierced by what Sofaer called a "dagger" of sunlight. It very nearly bisected the larger spiral, taking about 12 minutes to pass through it. Aware that the summer solstice had just passed, Sofaer thought the event might have been an intentional component of the rock art, designed to mark the solstice. Subsequently, she returned to the site at monthly intervals, including the equinoxes and solstices, as well as some intervening dates. Allied with architect Volker Zinser and National Science Foundation physicist Dr. Rolf Sinclair, she became convinced the Fajada petroglyphs comprised an accurate and precise calendar marker.

On the exact day of the summer solstice, sunlight first slips between the middle slab and the one on the right at about an hour before midday. A spot of light glows at the upper rim of the larger of the two spirals, grows into a thin dagger, and slices down through the carved turns, splitting them through the center. Eighteen minutes after first light, the wall is dark again, and the performance is over. Another spot of sunlight appears near the small spiral throughout the year and also develops into a dagger.

Six months later, at the winter solstice, two light daggers exactly frame the large spiral, on its left and right edges. This occurs in the course of a 49–minute midday period when sunlight enters both openings between the three uprights. The space between the left-hand slab and the one in the middle is responsible for this second blade of light. It, like the other, is visible throughout the year, but it makes only a brief appearance at the summer solstice. In ancient times it may have not appeared at all on that date. If so, its failure to show up on that one day of the year also could have signaled the summer solstice.

At the equinoxes, in March and September, the display of light is also a distinctive voucher of the date. There are nine turns on the right side of the large spiral, and the larger, right-hand dagger cuts between the fourth and fifth of the grooves. This divides the nine in "half," just as the equinoxes split the time between the solstices in half. While the equinoctial slicing of the right spiral's right side takes place, the smaller, left spiral is lanced through its center by the other slender wedge of light.

I remember receiving a telephone call from Anna Sofaer some weeks before the 1978 summer solstice, as did several others actively interested in archaeoastronomy. I was intrigued by what she had discovered, but couldn't really understand it in detail. It was quite difficult by phone to visualize exactly what she had seen and how it worked. Her emphasis on the unexpected vertical movement of the sun dagger made the whole thing sound overcomplicated, unlikely, and confusing. Although interested, I remained a bit puzzled—and dubious. Was the sun dagger designed into the site or a coincidence, unknown to the Anasazi rock artist?

I noted the weaknesses in her theory—which she already knew herself—and suggested what appeared to me would be the most appropriate problems to solve. These obvious concerns were expressed by others she consulted as well. Could she show, for example, that the edges of the slabs that formed the daggers had been tooled? And it seemed important to demonstrate—if possible—that the slabs had been placed intentionally in their present positions and not simply fallen there.

It has proved difficult to reach any conclusions about rock surface tooling, though Sofaer and her colleagues have since explained how the curved surfaces of the slabs translate the horizontal movement of the sun into a vertical play of light against the face of the butte. They have been able to identify the probable original locations of the slabs, however. Geological analysis indicates that these 2-ton megaliths once hung horizontally, as a single block on the cliff face just to the left of their present position, with the top surface of the original block at about the same height as the top of the uprights today. It is difficult to imagine how the slabs could have fallen into their present positions, although U.S. Geological Survey researchers have managed to do so.

Now that more is known about the California solstice shrines, proof of artificial construction actually seems less crucial. They look like less precise examples of a similar technique. Meanwhile, Dr. Ray Williamson has found another example of this sort of solstice marker in the Four Corners region, at Hovenweep. There, where

These petroglyphs—two spirals and a set of concentric rings—near Holly House, an Anasazi ruin in Hovenweep National Monument, are split by two needles of sunlight that start their horizontal cut of the rock face at opposite ends of the decorated surface and meet in the middle during the morning of summer solstice.   *(E. C. Krupp)*

southeastern Utah meets southwestern Colorado, the Anasazi occupied the numerous canyons that cut into the high, rolling mesa. They built multistory pueblos and distinctive towers. In the canyon below the Holly House group, an inner wall of a rock corridor is decorated with a few fine petroglyphs, including a set of concentric circles and two spirals.

On the mornings near the summer solstice, sunlight penetrates the crack between the overhang and the huge block of stone that forms the opposite wall of the corridor. In fact, the corridor is the space once occupied by this mass of stone before it split and fell from the parent rock. When the light enters, two sun daggers appear on the decorated panel. Both extend horizontally across this south wall. The one on the left cuts through the spirals; the one on the right bisects the rings. As the morning progresses, the points of the two daggers meet.

Such dramatic lighting effects at significant times of the year and their interaction with rock art suggest, again, the handiwork of the shaman. With its inconvenient, precarious location, the Fajada Butte sun shrine is best explained as the site of a sun priest's sky vigil. In his role as sun watcher, he observed, in detail, the orderly passage and cycle of time. This knowledge was one source of his power, and he wielded it to keep his community in time and balance with the world. Each display of light upon the carved designs was a revelation—a personal experience—of celestial power. He might find in it his own personal vision; either by counting days or climbing the sky, he met with the sun. His astronomical vigils transported him to that realm of power. He returned with some of it, a custodian of celestial order, and he channeled it to earth. His simple tallies of the days were charged by his vigils with the power of the sky.

7

The Days We Tally

Our awareness of time and our need to organize it are fundamental underpinnings of consciousness. The calendar is one of the tools we use to arrange our time and activity, of course, but it is such a common and constantly used tool, we take it for granted and forget why it is important.

The calendar is much more than an array of numbers in sequence accompanied, perhaps, by a set of diverting pictures. Our ancestors understood its importance. They knew it had power: religious power, economic power, political power. Its influence was felt in nearly every undertaking. The calendar sustained society and for that reason was expressed in sacred ritual, in the imagery of agriculture, and in the responsibilities of kingship.

Our ancestors first started noting the passage of time long ago. It began, of course, in ancient observations of celestial cycles. Long before we became farmers or built civilizations, our brains must have focused on the rhythmic changes in the sky and measured the behavior of the world in terms of them. Survival depended on it. But is there any evidence of prehistoric calendar keepers? When and how did they first make their counts?

The oldest known calendars—or, at least, records of the passage of time—seem to be systematic notations carved on bones by Ice Age people in the Upper Paleolithic period, 20,000 to 30,000 years ago. A research associate at the Peabody Museum of Archaeology and Ethnology, Alexander Marshack, examined Upper Paleolithic markings under a microscope and photographed them. This simple act, which had never been done before Marshack tried it in 1965, led him to a discovery. The scratches were not simple decoration or hunting tallies, as others had guessed. They were much more complicated.

Marking Time in the Old Stone Age

One 30,000-year-old piece of bone, retrieved in the early part of this century from the Blanchard rock shelter in the Dordogne district of France, was engraved on one side with 69 distinct pits, all arranged in a snaky line that curved back and forth and around itself with five turns. Marshack's microscope told him that the marks had been made in sets. Twenty-four different tools—or types of stroke—had been used, and Marshack guessed each group was carved on a separate occasion. To Marshack, it seemed unlikely that someone would have applied a purely decorative pattern in this way. Instead, the marks looked like notations. As notations, they must have stood for something. Marshack thought it was the changing phases of the moon.

Something similar to the Blanchard bone had already turned up at Ishango, an 8,500-year-old mesolithic site in equatorial Africa. Ishango is near the shore of Lake Edward, one of the sources of the Nile and far to the south of Egypt. In 1962, Jean de Heinzelin, one of the excavators of Ishango, reported an unusual find—a bone, probably a tool, engraved with 16 groups of notches. The number of strokes varied from group to group, but de Heinzelin thought the pattern was intentional and in some sense a *numerical* sequence. Marshack, however, contrived a lunar interpretation for the 168 marks on the Ishango bone. He showed that visual changes in the design—breaks in the pattern—fall at proper intervals to coincide with new and full moons over a 5½-month period.

Although the impression of a fit is subjective, the idea remains possible. Marshack tested this approach on two other engraved artifacts, a piece of bone from Czechoslovakia and a fragment of mammoth tusk found at Gontzi in the Ukraine. Both belong to the Upper Paleolithic period. The Czechoslovakian bone's marks seemed to be collected in three groups of 15, 16, and 15 strokes respectively. Although there are some weaknesses in the details of his reasoning, the fact that tallies like 15 and 16—about half a lunar cycle—repeat themselves suggests Marshack is on the right track. His analysis of subdivisions on the Gontzi ivory gave him another, four-month-plus record of the moon, but here again "month-to-month" variations in the details proved puzzling. There seemed to be no exactly duplicated pattern of lines with a periodicity of 29 or 30 days, as we might expect. Marshack had had to work from drawings of these relics, however. Real objects, viewed under his microscope, might provide better information. So, however limited his success may have been, it was enough to send him to France in search of more examples of Upper Paleolithic notation. That brought him to the Blanchard bone.

To avoid selecting evidence prejudiced by his initial conclusions, Marshack adopted a healthy "either-or" policy toward the old stone age material available for examination: "Either notation was common in the Upper Paleolithic, or there was no notation." A random sample of engraved bones therefore would have to include some

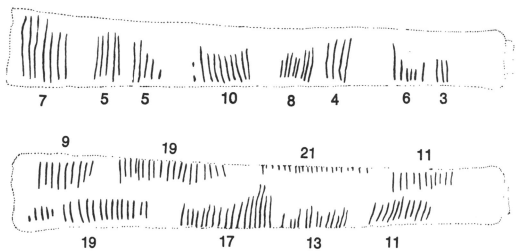

| 7 | 5 | 5 | 10 | 8 | 4 | 6 | 3 |

| 9 | 19 | 21 | 11 |

| 19 | 17 | 13 | 11 |

Altogether there are 168 marks on both sides of the Ishango bone. Alexander Marshack related these to the periods between new moon and full moon and between full moon and new. These alternate, of course, as the moon completes its cycle of phases, month after month. Changes in the angle or size of the strokes could correspond, he guessed, to "turning points"—or key phases—in the lunar cycle.   *(Alexander Marshack,* The Roots of Civilization)

Paleolithic marks on a piece of bone from Kulna, Czechoslovakia, may represent half-month intervals tied to the phase of the moon. The bone itself is shown in the upper drawing, while the lower picture depicts the marks schematically. According to Marshack, the 15 scratches on the top count the days from the day after the last visible crescent to the day before full moon. The 16 long lines then represent the days from full moon to the next first day of invisibility. Finally, the 15 marks on the bottom edge stand for the days from first crescent to full moon. We think of the moon's cycle as invariant—always 29½ days and always starting with the same phase. But here, the phase that starts one month is not necessarily what starts the next, and that makes Marshack's interpretation less explicit.   *(Alexander Marshack,* The Roots of Civilization)

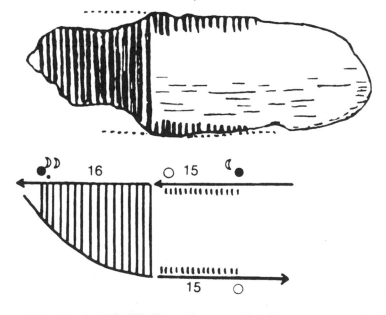

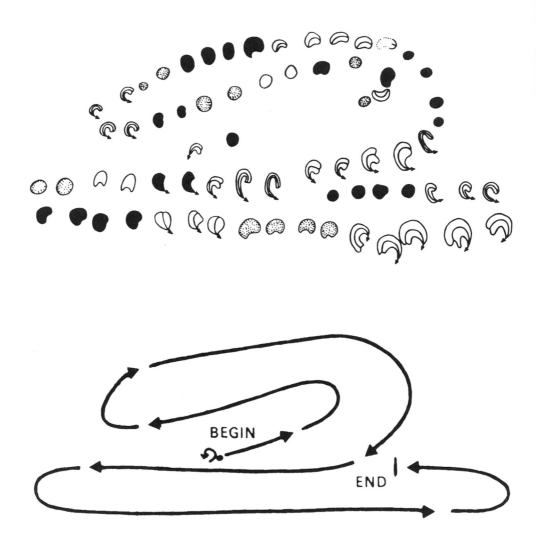

A remarkable collection of different notations was tooled into the Blanchard Bone perhaps 30,000 years ago, during the Aurignacian period. If Marshack is right, the sequence begins with the two marks near the center. These, he judged, marked the day of the last visible crescent and the first day of the invisible new moon. Continuing up to the right and then back down to the left again, he reached full moon at the first four-stroke group. This was also the point of the design's second turn, with two of the four hook-shaped marks above the turn and two below. As the line progresses back to the right, four black dots on the third curve coincide with the next new moon. At the fourth turn, on the lower left, the count adds up to another full moon, and the last stroke in a five-mark group, falls near the fifth, and last, turn. *(Alexander Marshack,* The Roots of Civilization*)*

symbolic sequences, or Marshack would have to abandon the idea. In the Upper Paleolithic material stored at the Musée des Antiquités Nationales in Paris, Alexander Marshack found at least three possible examples of notation, and all fit the pattern of the moon's phases, at least to his satisfaction.

The Blanchard bone's winding sequence was the first pattern Marshack noticed. After satisfying himself that the marks had been applied on several different occasions with several different tools, he took the same approach that convinced him the Ishango bone was a lunar record. By grouping the strokes in sets made with the same tool, he saw the following sequence of counts:

1ST TURN   2ND TURN

1 – 1 – 1 – 1 – 1 – 1 – 1 – 2 – 2 – 2 – 4 – 2 – 4 –

last   *new*   *full*
crescent

3RD TURN   4TH TURN   5TH TURN

5 – 4 – 8 – 2 – 2 – 2 – 4 – 3 – 4 – 5 – 3 – 4

new   *full*   *new*

For Marshack, the small, delicate, and detailed design was a 2¼-month tally of days arranged in symbolic congruence with the changing appearance of the moon.

Paleolithic Notations: Engraving the Ice Age Moon

The same museum cabinet that contained the Blanchard bone also housed several distinctive bone implements known as *batons de commandement*. These decorated, scepter-like instruments have been variously interpreted as ritual symbols of authority, as tools for bending and straightening spear shafts, and as the cheek fixtures of prehistoric bridles. Two of the batons examined by Marshack were scored with stroke patterns that fit his theory of moon-factored notation. Both were found at the French site known as Le Placard and are 7,000 to 10,000 years more recent than the Blanchard bone. One of them, with the head of a "smiling fox" carved on one end, may document four months of the passage of time. The other, marked by ten different tool points, covers 59 days, or two lunar cycles.

On many Upper Paleolithic implements, Marshack found totals and subtotals that seemed to have derived from the repeating cycles of the moon phases. If that is what the tallies represent, it is probably safe to assume those who inscribed the marks were eyewitnesses and recorded the days in terms of the moons they actually saw. In the short run, the interval of invisibility at new moon and the periods between various

phases can vary by a day or two. Cloudy weather or pressing matters may also interrupt or modify a count. These variations should average out over an extended period
of time, and the longer, continuous, and cumulatively correct tallies support Marshack's interpretations.

Two eagle bones from Le Placard each appear to register about a year's worth of
lunar months. Their notations seem to be more systematic and ambitious than the
others Marshack had seen before. Tiny vertical strokes represented one sequence of

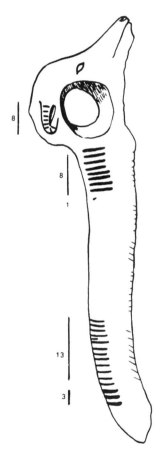

There are 132 marks on the "smiling fox" bâton from Le Placard. The artifact belongs to
the Magdalenian III period of the Upper Paleolithic and is 7,000 to 10,000 years more
recent than the Blanchard Bone. If these marks are sequenced according to the phases of
the moon, they represent about four months. Twenty-five different points were used to
tool them. In this view, only the marks on one side of the bâton are visible. The rest are
on the other side and on the "belly" between them.   *(Alexander Marshack,* The Roots of Civilization*)*

An eagle bone from the late middle Magdalenian period of Le Placard, about 15,000 years old, is also inscribed with what could be a moon modulated sequence of days. Here the technique is more complicated, however, for a small "foot" was added to each of the primary engraved lines. Marshack believes these "feet" had become, by this time, a standardized form of notation to continue the tally represented by the marks already applied. *(Alexander Marshack,* The Roots of Civilization*)*

days, and to each of these was later added a small "foot," or angle stroke, that continued the tally. These engravings are about 15,000 years old. They and others like them appear to be evidence that a conventionalized tally symbol and a less personal technique had evolved by this time.

There is probably no way to verify that the Upper Paleolithic notations are related to the moon's phases, but the idea is not far-fetched. Wands marked in similar ways by inhabitants of the Nicobar Islands in the Indian Ocean are known to be calendar sticks with lunar tallies. Their major intervals record lunar months, and the subintervals are related to the key phases in the monthly cycle. The records are not always astronomically exact, and the pattern from one month is not duplicated in the next. They are, however, cumulatively correct. They parallel what we see from the Upper Paleolithic. Perhaps, at least in some respects, Marshack is right.

Turning Time into Number

Marshack rightly emphasizes that we are not seeing scientific records of the moon's behavior in the old stone age carvings. They are symbolic and quantitative—although not necessarily exact—references to the passage of time.

This indicates that an idea we take for granted—the idea of a number—may have been used to mean more than its literal value or a quantity. In a sense, we can find an analogy in the title of the old song "The Twelve Days of Christmas." The number 12 in the title refers unambiguously to an interval of 12 consecutive and specific days. Here, 12 has the same superficial meaning as 12 tally marks scratched on a bone to represent 12 consecutive and specific days. Just as there is much more to the song, however, than its title would suggest, Marshack attributes more meaning to the marks. Everyone who has heard the song knows that it is a story about the gifts one lover gave to another. These 12 days of Christmas were special days, punctuated with a partridge, pear trees, French hens, gold rings, maids a'milking, and so on. It is a narrative of courtship, exchange, and seasonal celebration, but we would know none of that from the title.

The paleolithic notations are like the song title. They are related to a "story" that involves the passage of time, but the details were not recorded. Marshack believes the paleolithic notations have more to do with narrative or myth than with calendrical records. He is not contradicting himself but trying to clarify how the tallies were used. A change in stroke or a set of pits may have reminded the owner of the baton or bone slate of *what* happened when, rather than *when* something happened.

Also, the real meaning of the tally marks to the people who tooled them may not have been their individual positions, one by one, but the overall visual pattern of the placement. Perhaps only an appropriate sequence of numbers would permit the assembly of a pattern with the desired appearance. The face of a clock, for example, has 12 numbers that stand for the hours, but we can read the clock in terms of the positions of the hands. The various arrangements of hands acquire their meaning from the numbers that ring the face, but we can register the hour by the visual appearance the hands create without ever reading the numbers.

Marshack has also analyzed the content of the representational cave art from Europe's Ice Age and found that much of it can be interpreted in terms of seasonal indicators drawn from the environment of the Upper Paleolithic hunters. Spawning salmon, molting bison, a bison in rut, a pregnant mare, and a pair of mating snakes: All of these images confirm the intimate understanding these ice age people had of their world. Various calendar schemes are still used today by the shamans of hunter-gatherer societies. These shamans help maintain a dependable food supply by limiting the hunting to seasons the calendar and experience indicate will avoid depletion of the game. Rituals that celebrate these seasons and guide the activity in them may, in part, be symbolized by the notations of the Old Stone Age.

Celebrating Time in the New Stone Age

Although Marshack's most recent work traces notational systems right into the mesolithic, we have not yet found anything from the neolithic, or New Stone Age, we can be sure is a calendar tally. Some neolithic symbols may incorporate astronomical and calendric imagery, but genuine calendars seem to be missing. Their absence could mean we haven't recognized them yet or that they were recorded on materials too fragile to survive.

The moon's cycle is the backbone, the fundamental structure, of many traditional calendars, but we don't really know that the neolithic and bronze age peoples of northwestern Europe calibrated time in terms of the moon. It is likely they did, but Alexander Thom's surveys of stone alignments and rings of stones in Britain, carried out during the last three decades, argue in favor of a calendar based at least in part on the solar year.

The alignment of Newgrange on the southernmost sunrise implies that neolithic people were aware of the winter solstice and the sun's annual cycle, but a systematic set of other solar alignments would have to be found in the architecture of other megalithic monuments if we are to conclude their builders based their calendar on the yearly cycle of the sun. Alexander Thom's on-site measurements of hundreds of pre-historic monuments demonstrate the existence of solstitial alignments, equinoctial alignments, and alignments to the sun on dates that further subdivide the intervals between these significant yearly events.

These calendric alignments of the stone circles do not mean that the circles are in some way calendars or observatories at which calendar-related astronomical phenomena were monitored. Although the purpose of the stone rings and alignments is not known with certainty, most of them appear to be shrines or sanctuaries—sacred places dedicated to ritual and ceremony. Their calendrical components reinforce the idea that the calendar was regulated and spotlighted by those who directed the religious life in neolithic and bronze age communities.

Long Meg and Her Daughters is an early stone ring, flattened on one side. Thom believes the geometric rules of its construction were already worked out by the ring's designers 5,000 years ago, the most probable time of its use. The large stone outside the ring is "Long Meg" herself, and it is part of an alignment to the winter solstice sunset.

In addition to the numerous winter solstice alignments that are known, many monuments were oriented by the summer solstice. The main axis of Stonehenge, at least in its later stages, probably was intended to agree with the bearing of summer solstice sunrise. Woodhenge, a set of six concentric ovals of prehistoric postholes, or egg-shaped rings, shares the same orientation. It is only 2 miles northeast of Stonehenge, and the main axis of Woodhenge, common to all six rings, points to the northern limit of sunrise, occupied by the sun in June. Far to the west in southern Wales, coincidentally near the Preseli Mountains where the Stonehenge "bluestones" were quarried, a stone ring known as Gors Fawr is accompanied by two tall uprights about a quarter of a mile away. Summer solstice sunrise seems to be the target of the line formed by the two outliers.

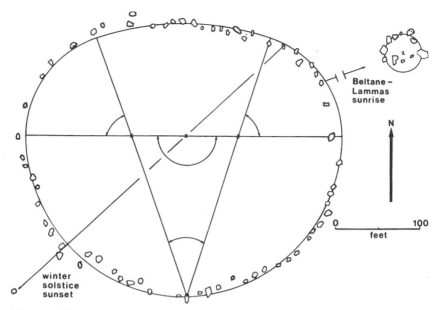

Long Meg and Her Daughters is a noncircular ring of 70 or so stones in Cumbria, England. The centers of curvature that permit fitting the stones to four circular arcs are given by the geometric construction devised by Alexander Thom. A line from a stone on the northeast through the center, and to the 12-foot-high outlier 75 feet beyond the southwest segment of the arc indicates winter solstice sunset and implies this may have been an important date in the calendar of prehistoric Britain.   (Griffith Observatory, after A. Thom)

A look along the winter solstice line to Long Meg, the standing stone beyond the ring, carries the eye up a sloping field to a relatively close horizon. This alignment does not offer many possibilities for exact determination of the solstice but seems instead to represent a symbolic alignment for the December event.   (E. C. Krupp)

Summer and winter solstice split the year in two, and the equinoxes quarter it. Equinoctial alignments can also be found at Long Meg and Her Daughters, where two conspicuously larger boulders, on opposite sides of the ring, establish the east-west line; in a line of 18 stones at Eleven Shearers, in southeast Scotland; and among the parallel lines of small stones on Learable Hill, near the border between the Highland counties of Sutherland and Caithness. One of the five lines extending from the main ring of Callanish, the "Scottish Stonehenge," is oriented equinoctially.

But many of these equinox indicators vary just slightly from true east-west orientation. Those that deviate usually indicate a sunrise a day or two later than the one that occurs on the true astronomical equinox in spring when the sun crosses north of the celestial equator and rises due east. The "megalithic equinox" sunrise is a little north of due east, and in the fall it marks a date a day or so earlier than the true autumnal equinox. This inaccuracy may seem to be nothing more than error, but it is what we would expect if the alignments were once related to a practical calendar rather than to the abstract notion of celestial geometry we use today. Although the earth circuits the sun, its orbit deviates slightly so that it is not truly circular. The orbit is an ellipse that carries the earth a little closer to the sun at one time of the year and farther from the sun half a year later. Gravitational attraction between the earth and the sun accelerates the earth when it is nearer to the sun, and over the last several thousand years this phenomenon has occurred in winter. Perihelion, our closest approach to the sun, now occurs on or about January 1. Although variation in our distance from the sun is relatively small and has little effect on the average temperature, it does affect the way we count the days in the year.

During one half of the year, from fall to spring, the earth moves a little faster around the sun. Less time elapses between the true autumnal equinox and the true, vernal equinox. The count of days between these two events is four or five less than half of the 365 that comprise the whole year. About 187 days elapse between the time of the spring equinox and the equinox in fall. Winter's half of the year takes about 178 days. The difference between these "seasons" was not as great 4,000 years ago, but the difference was large enough to notice. We might expect, however, a prehistoric calendar to divide the year into equal intervals of days. If this were done, the calendar's "equinox" would occur a little later in spring and a little earlier in fall in order to even out the two periods. Calendar alignments on the sunrise would be set a little north of due east, and this is what Thom seems to find.

Subdividing the Seasons in Standing Stones

Even a year quartered by solstices and equinoxes needs more subdivision into smaller and manageable intervals. Alignments on intermediate positions of the sun that divide the year into eighths, each 44 to 47 days long, led Thom to believe the early bronze age calendar keepers had adopted such a plan. In our own calendar the four additional breaks in the year occur on May 5, August 6, November 2, and February 2.

Alignments with the sun on these dates are found in many monuments. Long Meg and Her Daughters has one. Castle Rigg, an early ring, is also located in the Lake District, and its alignments include dates from the solar calendar. At Sheldon of Bourtie in northeast Scotland, sunrise lines for the May-August date and the November-February date involve outlying stones and the circles' common center.

Additional evidence favors the use of the intermediate calendar dates in prehistoric times. They are preserved in traditional holidays still observed in Britain: Candlemas (February), May Day (May), Lammas (August), and Martinmas (November). All of these derive from an earlier Celtic tradition and marked the beginnings of each season. For example, what is called the first day of summer in the United States, the summer solstice, is considered to be midsummer in Britain. There, early May marks the traditional start of summer, and the Celtic festival that set it off was known as Beltine. Once the seasonal cue to drive the cattle from their winter quarters to their summer grazing lands, Beltine evolved into our present May Day. It is suspiciously close to the May 5 date in Thom's megalithic calendar.

The sun occupies the same position on an old Celtic holiday, Lugnasad (the "Feast of Lug") on August 1. Lugnasad was an early harvest festival, a celebration of first fruits. In the Celtic calendar, Candlemas and Martinmas were originally Imbolg (Feb-

Alexander Thom has proposed a number of astronomical alignments and a specific geometric plan to account for the flattened circle arrangement of stones known as Castle Rigg. One line doubles for the summer solstice sunset and the November-February midcalendar (Martinmas-Candlemas or Samhain-Imbolg) sunrise. *(Griffith Observatory, after A. Thom)*

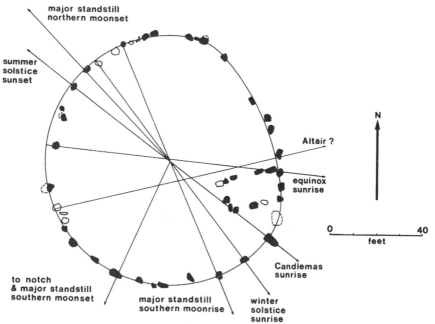

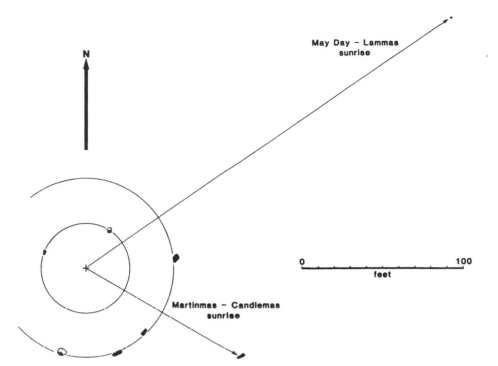

Sheldon of Bourtie, a pair of concentric stone rings in Aberdeenshire, Scotland, is still accompanied by two outliers. From the center, each indicates a sunrise for two of the quarter days of the traditional Scottish calendar. *(Griffith Observatory, after A. Thom)*

ruary 1) and Samhain (November 1). The name *Imbolg* seems to mean "sheep's milk" and refers to the start of the lambing season. Samhain is preserved in Halloween, and Imbolg, or Candlemas, persists in the United States as Groundhog Day (February 2).

Alexander Thom believes he has evidence for further subdivision of the solar year into 16 "months," but the case for them is not as clearly demonstrated and depends upon alignments alone.

Whether the megalith builders used a solar calendar or a lunar calendar or both is not what the astronomical alignments in these prehistoric monuments tell us, however, and is not what is important about them. They do say that certain cyclical celestial events filtered the flow of time for these people into an orderly series of dates. Because these events were enshrined in stone, we conclude they were significant. They reflect the megalith builders' awareness of stations in the passage of time. Through sacred ritual these moments might unify and calibrate the life of the community.

Measuring Time by the Moon

Even though we have no written calendar records from prehistoric Britain, we know its farmers and herders dealt with time by structuring its passage in terms of a calendar. Key events in the celestial cycles were symbolized by alignments they built into their megalithic monuments. The purpose of most of these monuments was ritual, and the passage of time, revealed by the sky, was part of those rituals. By identifying the alignments, we confirm our prehistoric ancestors' sense of the sacred aspect of cyclical time. The calendars they used to schedule their ceremonies are gone. But even though the procedures by which they actually measured time are lost, and the places where this went on uncertain, procedures and places and calendars must have existed. The megalithic monuments themselves show that calendrics were applied to religion.

An entirely different megalithic site, unrelated to those of prehistoric Europe, allows us to see one way calendar keeping can be done. The monument is in northwest Kenya, in Africa, and is known as Namoratunga II. Radiocarbon analysis dates a related burial and rock art site 130 miles south, Namoratunga I, to 300 B.C. The burial customs of the sub-Saharan people who used the cemetery resemble what is done today by the Konso, who live in the same general vicinity, in southern Ethiopia. The calendar scheme of peoples related to the Konso may, in turn, help us confirm the intent of the builders of Namoratunga II.

At the northern site, one setting of upright stones encircles a grave, like those at Namoratunga I, but the 19 other megaliths at Namoratunga II are arranged in an unusual pattern unconnected with any other burials. B. N. Lynch and L. H. Robbins, anthropologists at Michigan State University, measured the positions of the stones and established among them a plausible set of astronomical alignments.

These stones at Namoratunga II are not actually devices used to measure the positions of stars. They are too big and too close for that, and calendar keeping by the

This isolated collection of upright megaliths is in Africa, in northwestern Kenya. The stones were erected in this curious pattern in about 300 B.C., and the site is known as Namoratunga II. Its alignments are consistent with the traditional calendar of the people in the region. *(B. M. Lynch and L. H. Robbins,* Science *200: 766–768. Copyright 1978 American Association for the Advancement of Science.)*

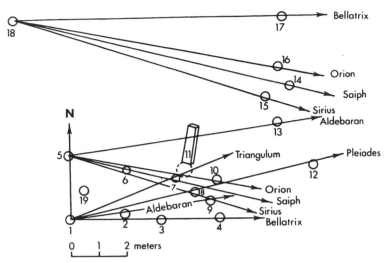

The stones of Namoratunga II are relatively close to each other, separated at the most by 35 feet, usually less. Alignments of the stones indicate the rising positions of bright stars, stars that have been adapted to calendar keeping by a variety of traditional peoples and ancient civilizations. For example, an observer stationed by stone #1 would sight eastward over pairs of stones or single uprights to the star Bellatrix in Orion, Aldebaran in Taurus, the Pleiades, and the small constellation of Triangulum. Stone #5 works similarly as a backsight for Aldebaran (again), the center, or "belt," of Orion, Saiph (also in Orion), and Sirius. These stars are the same as those in the luni-stellar calendar system of the Eastern Cushites, who still live near this region. *(Griffith Observatory, after B. M. Lynch and L. H. Robbins)*

stars doesn't require detailed mapping of their positions. The dates of their appearances and disappearances are significant, however, and the Namoratunga megaliths could have been used as reminders and guides to the stars of calendric importance.

Lynch and Robbins strengthened their case by showing that the same stars are still used by Eastern Cushite peoples in complex calibrations of a 354-day lunar calendar, 12 "months"—or lunations—long. The positions of key phases of the moon were intricately combined with zones of the sky that correspond to each star or asterism, in order of their use (Triangulum, the Pleiades, Aldebaran, Bellatrix, Orion's Belt, Saiph, and Sirius). Other months in a complex lunar cycle were coordinated with the position of Triangulum with respect to staggered phases of the waning moon. The Konso speak a Cushitic language, and it is reasonable to believe the Namoratunga sites were erected by the ancestors of the present-day Eastern Cushities, who still employ a calendar now likely to be at least 7,000 years old.

Namoratunga II is a modest site, relatively simple, with little there besides the alignments. While it could have been used ritually, its layout and the multiplicity of its alignments seem more appropriate for systematic observation.

Some of the same astronomical objects observed by the Namoratunga calendar keepers were used more than two millennia earlier by the Egyptian calendar priests.

Richard A. Parker, a renowned authority on Egyptian astronomy and calendrics, established that the earliest Egyptian calendar was based on the moon and calibrated by the stars.

Organizing Time and Keeping It Straight

Nearly everyone who develops a calendar depends on the moon to keep track of time, but any calendar marching in step with the moon's phases must eventually skip a beat—or it will slip out of time with the seasons. There isn't a whole number of lunar months in a solar year; the two cycles are not congruent. Moon calendars, therefore, cannot guarantee a year will close with the end of a month or open with a month's beginning. Anyone who uses a lunar calendar and wants it to conform to the annual seasonal cycle has to devise some scheme that will add extra months at acceptable times. This is one reason why calendars get complicated.

In Egypt the month began on the day the waning crescent moon disappeared from the predawn sky. The 29½-day lunar cycle made the length of some months 29 days, while others lasted 30. Over the long run they averaged out and kept the important days and feasts of each month in place with respect to the moon, but the Egyptians, like everyone else, had to deal with the fact that the lunar "year" of 12 cycles of the moon's phases is about 11 days short of the solar year, the year that keeps in step with the seasons. Without correction, any truly annual event—for example, the summer solstice—will occur later each year. The Egyptians solved this problem by adding an extra month as needed. Sirius (the "Dog Star") was the signal for the appropriate time to do it, and everything depended on the last month of the year. This was called *Wep-renpet.* The name means "Opener of the Year," and it refers to the heliacal rising of Sirius, its first predawn appearance. This event was supposed to occur during the last lunar month, but each year it would slip to a later date. If, for example, the celebration for the reappearance of Sirius fell close to the beginning of *Wep-renpet* in one year, by the star's fourth consecutive heliacal rising, the event would take place in Tekhy, the first month of the new year. To prevent this, a new "last month" had to be added. Of course, it would not do to add a new "last month" too often, or Sirius would start reappearing too early. So, when the heliacal rising of Sirius occurred during the last 11 days of *Wep-renpet,* an intercalary month—*Thoth,* named for the god connected with recordkeeping, writing, and the moon—was added. On the average, an additional month was added every third year.

Parker argued well in favor of the great antiquity of the first lunar calendar. He read the meaning of early symbols carved on an ivory tablet from the First Dynasty as "Sirius, the opener of the year, the inundation." This implies the basic elements of the calendric system were in place as early as 3100 B.C. The same calendrical imagery, the same celestial calibrations, endure all the way into the Ptolemaic period, 3,000 years later.

Sirius, the key calibrator of the Egyptian lunar calendar, is represented in a similar form by these two carvings, even though one was executed nearly 3,000 years before the other. On the left, a First Dynasty ivory plaque symbolizes Sirius by a cow, and Professor Richard A. Parker interprets the hieroglyphics as an unambiguous reference to Sirius, the beginning of the year, and the Nile flood. The astronomical ceiling of the Temple of Hathor at Dendera was decorated in Ptolemaic times with a cow, reclining in a boat, that also stood for Sirius (right). *(Griffith Observatory, after Richard A. Parker and Heinrich Brugsch)*

Sanctifying Time and Giving It Meaning

Calendars punctuate our lives with holidays, feasts, and ceremonies. These are guideposts. Through such events we are always closing out one period of time and starting another. This reinjects us with a sense of renewal, and we need these new beginnings. Our ancient ancestors needed them, too. For them, cyclic time and its renewal animated the world. It recharged the world with meaning and reconsecrated their lives.

Most of the days of the Egyptian lunar month had names that referred to feasts or priestly activities and emphasized the religious character of the calendar. It was the matrix in which Egypt's religion operated. It governed the sequence and timing of the festivals and events that defined Egyptian life and was administered by priests—a bureaucracy of scribes and astronomers who fashioned an ordered Egypt from the cycles in the sky.

Not long after the unification of Upper and Lower Egypt and the creation of the pharaonic dynasties, a second calendar was introduced by the Egyptian scribes. The earlier lunar calendar worked well enough as the metronome of religious festivals, but commerce and kingdom demanded a more uniform and constant calendar for civil affairs. The Egyptians quite sensibly based theirs on the sun. The Romans eventually adopted and adapted the Egyptian calendar, which therefore is the direct source of the solar calendar—with months no longer linked to the moon—that we use today.

This 365-day solar year came into use some time early in the third millennium B.C. It was subdivided in 12 "months," each by definition 30 days long and essentially independent of the phases of the moon. Each 30-day period was in turn split into three 10-day intervals. By convention these are called *decades,* and 36 of them, with five extra days at the end, metered out the civil year. Of course, the true length of the tropical year is closer to 365¼ days. By ignoring the ¼ day, the Egyptian calendar makers guaranteed the civil year would slip backward through the seasons and start a day earlier every four years. This means an annual event like the summer solstice would be dated a day later every four years. Similarly, the heliacal rising of Sirius, the herald of the new year, would occur on different dates until it had circuited completely through the calendar back to its original starting date. Egyptian administrators seemed untroubled by this circumstance and simply let the civil calendar run its course. As long as the system was clearly defined on its own terms it would serve business and government. The seasons and the moon made no difference. The religious calendar was a different matter. Because the religious calendar moved with rhythm of the moon, each of its months and each of its festivals recommemorated the cyclic order of life and time. It had to be kept in tune with the seasons of the sun, Sirius, and the Nile.

Eventually the Egyptians created a third calendar. It ran concurrently with the other two, but like the first was based on the moon. This later lunar year was linked to the civil calendar through an intercalation that brought the first day of the lunar year back into coincidence with New Year's Day in the civil year whenever it occurred too soon. Parker's review of a late text, *Papyrus Carlsberg 9,* reveals the existence of a procedure that allowed the astronomer to determine the civil calendar date of the first day of each lunar month. The system was based on a 25-year cycle, in which there are 9,125 days and 309 lunar months. In days, this number of lunations equals 9,124.95231. The near-congruence of the two cycles after the passage of this interval of time is what makes the scheme work. Ultimately, the motive behind it was the welding of some real celestial phenomenon to the arbitrary and artificial civil calendar. Through intercalation an attempt was made to give the civil months with their borrowed names some of the meaning measured out by the real moon overhead.

The lunar months were themselves named for the principal feast assigned to each. We have inscriptions from temples and tombs that provide us the names of the

A ceiling in the Ramesseum, the mortuary temple of Ramesses II at Thebes, portrays the months in the old lunar calendar of Egypt as gods. The jackal, in this detail, stands for the month of Phamenoth. *(Robin Rector Krupp)*

months in the early lunar calendar. At Thebes, a well-known "month table" depicts the gods associated with each lunar month in the lower register of a temple ceiling. The ceiling is in the hypostyle hall of the Ramesseum, the famous mortuary temple of the New Kingdom pharaoh Ramesses II. Although the 365-day civil calendar's months were not based on the actual behavior of the moon, their names are clearly borrowed from the earlier lunar calendar. Throughout the centuries, that old lunar calendar continued in use. The Egyptians could not abandon their oldest calendar. It kept the moon in tune with the seasons of the sun, Sirius, and the Nile, and that kept the Egyptians in touch with the world's sacred order.

Calendars, Corrections, and Kings

In Mesopotamia it was probably the Sumerians, the people who built the formative civilization of the region, who put the first formal calendar into use. The Sumerian calendar was lunar, but its months began when the first crescent was sighted in the west. A passage in the Babylonian creation myth echoes, in Marduk's instructions to the moon, a concern for the lunar cycle:

He bade the moon come forth;
  entrusted night (to him)
assigned to him adornment of the night
  to measure time;
and every month, unfailingly,
  he marked off by a crown.
"When the new moon is rising
  over the land
shine you with horns, six days to measure;
the seventh day, as half (your) crown (appear).
and (then) let periods of fifteen days be counterparts
  two halves each month.
As, afterward, the sun gains on you
  on heaven's foundations,
wane step by step,
  reverse your growth!"

The "crown" is the moon's fully lit disk, and the horns refer, of course, to the waxing crescent. On the seventh day a "half crown" describes the half-lit first quarter moon, and the rest of the text narrates the way in which the moon should continue to measure out the months.

Some of the Sumerian month names have survived in cuneiform texts and, like the Egyptian names, refer to the months' principal feasts: "the Month of the Feast of Shulgi" and "the Month of the Eating of Barley of Ningursu." Feasts were scheduled by the moon's phases, with regular celebrations at the first crescent, first quarter (seventh day), full moon (fifteenth day), and last day.

The Sumerians divided the year into summer, or *emesh,* and winter, or *enten.* We know the New Year holiday was consecrated by a symbolic "wedding" of the king with a high priestess. This ritual reenacted the marriage of Dumuzi, a god associated with the growth of grain and dates, and Inanna, a goddess identified with fecundity and sex, and was scheduled, most likely, in spring, when life seems to be rekindled in every blossom, seed, and fruit.

Of course, intercalation was the only way to keep the Mesopotamian lunar calendar in step with the seasons, and some inscriptions imply an extra month was added before the month of autumnal equinox. Other texts refer to a thirteenth month slipped in just prior to the vernal equinox. Whatever rule was followed in the early period, by 1000 B.C. or so Babylonian calendar priests were intercalating months according to an eight-year cycle. During this period three extra months were added. In Chaldean times, a "Metonic," or 19-year cycle with 7 extra months, was probably in use. This interval, which equates 19 tropical years with 235 lunar months, is named after the Greek astronomer Meton, who introduced its use in the Mediterranean world in the last decades of the fifth century B.C. Although it looks as though a numerical rule, and no observed celestial event, determined the years in which extra months were added, A. Sachs, a specialist in cuneiform and Mesopotamian astronomy, believes intercalations were designed to keep the annual heliacal rising of Sirius in a particular month. If this be so, it again stresses the important role of the sky's brightest star as a signal of the seasons and calibrator of the calendar for ancient societies. Its astronomical attributes—its brightness and the timing of its appearances—made it valuable wherever it could be seen.

No matter what method was used to keep the Mesopotamian lunar calendar coordinated with the seasons, only the king could declare when an extra month was to be inserted. In China, too, the calendar was a privilege of the king. The right to inaugurate and promulgate it was one of the prerogatives that actually defined what it meant to be king. These traditions were maintained until A.D. 1912, when the government of the Republic of China exercised the royal privilege to establish a new calendar. Throughout China's long history the calendar was an arm of government. It steadied society and prompted celestial prognostication. It was the job of the Imperial Astronomer to keep the calendar functioning properly and to inform the king of the state of heaven and earth. To this purpose, according to a Han dynasty text, the astronomer concerned himself:

> with the twelve years (the sidereal revolution of Jupiter), the twelve months, the twelve (double) hours, the ten days, and the positions of the twenty-eight stars. He distinguishes them and orders them so that he can make a general plan of the state of the heavens. He takes observations of the sun at the winter and summer solstices, and of the moon at the spring and autumn equinoxes, in order to determine the succession of the four seasons.

Inscriptions on oracle bones from the Bronze Age Shang Dynasty of ancient China confirm that a calendar based in part on the moon was in use at least by sometime between 1400 and 1200 B.C.

By adopting a lunar calendar the ancient Chinese also faced the problem of intercalation. They experimented with a few procedures and eventually settled on adding seven extra months over a period of 19 years. Later they refined this by omitting one day in four 19-year periods.

Of course, the Chinese also kept track of the sun. They split the sky into 24 equal segments and called the time it took the sun to move through any of them a *chhi* (or *qi jie*). These intervals helped determine when the extra months would be added.

Rather than use fractions, the Chinese defined some *chhi* to be 15 days long, and others were set as 16. On the average a *chhi* would actually span a little less than 15¼ days. Together, all 24 added up to a 365¼-day year. This solar cycle began, by convention, in early February, usually the 5th, about halfway between winter solstice and vernal equinox. The date nearly coincides with Candlemas. By starting the New

Year with the new moon closest to the first *chhi, li chun* ("Beginning of Spring") and adding a month, usually sometime in summer, in years it was obvious the fourteenth moon, and not the thirteenth (which should be the first moon of the next year), would fall closer to the start of *li chun*. This procedure kept the New Year close to the start of *li chun* year after year. The names of the other *chhi* refer to the weather and nature ("Cold Dew," and "Great Snow"), agriculture ("Grain Full," "Grain in the Ear"), and the progress of the sun ("Sun Equinox," "Winter Solstice").

Several other calendrical cycles were used in ancient China, including a 12-year cycle based on Jupiter's movement through the stars. Its progress provided each year with an emblematic animal corresponding to the station, or zone, of Jupiter in that year. This tradition persists, and today we still have the year of the rat, ox, tiger, hare, dragon, snake, horse, sheep, monkey, cock, dog, and pig.

Several of the rings of symbols on the back of this T'ang Dynasty mirror refer to cycles in the Chinese calendar. The innermost circle with the four cosmological animals signifies the four cardinal directions and the four seasons. The next ring of 12 animals represents the 12-year Jupiter cycle, and the outer ring is a set of 28 creatures: one for each *hsiu*, or stellar station. *(C. A. S. Williams,* Outlines of Chinese Symbolism & Art Motives*)*

Calendars, Cultivation, and Cosmic Order

Even though calendars originate before people became farmers, the imagery of food cultivation—like the marriage of Dumuzi and Inanna—naturally appears in the calendars and seasonal festivals of agriculturists. Their lives, after all, hinged on their ability to grow food, and their calendars reflect their life-styles.

The cyclical pattern of farming has great symbolic importance. What goes on in farming—sowing, cultivation, harvest, and the next new season of planting—follows the pattern of vegetation in general, the pattern of seasons, and the pattern of life. All these mimic the cycle of cosmic order—the birth, growth, death, and rebirth we see in the sky.

Two "calendar plants" are portrayed on a stone relief in a Han Dynasty (206 B.C.– A.D. 220) Chinese tomb in Shan dong province. One has 15 leaves on it, and they were said to accumulate on the tree, one a day, over an interval of 15 days. For the next 15 days, according to tradition, the tree lost a leaf each day. Symbolized, of course, is the waxing and waning moon. The other tree, shown with six leaves, was said to grow one leaf each month for six months. It lost them, one by one, in the following six months.

The leaves on the two plants depicted in relief on the wall of a Chinese tomb symbolize the days of the month (right) and the months of the year (left). The leaves appear, one by one, through half of their respective cycles, and fall away, one by one, during the second half. *(Griffith Observatory)*

We see real growth and decline in the vegetation as the seasons pass by. The seasons are linked to the year, and the "calendar plant" is a sensible symbol for the time and events that elapse. The same cycle seems evident in the ripening and dwindling of the moon each month, and so the metaphor of donning and shedding leaves is perfect for the passage of moon and month as well.

Even the Emperor of China participated in agricultural rituals that were tied to the calendar. On the proper day of the year he went to the Qi nian dian—the Hall of Prayer for Good Harvests—in Beijing's Temple of Heaven. The Emperor performed the ceremonies for which the temple hall was named, and the day he did so was chosen by the astronomer calendar keepers of his court. Nearby, in the Xian nong

In Beijing, the emperor performed the "spring" sacrifice on or near the New Year, around the date of February 5, which marked the first of the 24 *chhi* in the solar year. The ceremony took place at the Qi nian dian, or Hall of Prayer for Good Harvests, in the Temple of Heaven. Inside the round hall the architecture symbolized the year's cycles: Four large central posts stood for the four seasons, and 24 outer pillars, arranged in a double circle, represented the 12 months and the 12 "hours." *(Lois Cohen, Griffith Observatory)*

tan, or Temple of Agriculture, the Emperor, in another annual ritual, plowed the symbolic "first furrow" of the year in a sacred field and made sacrifices for rain. On April 25, the Chorti Maya of Guatemala begin an eight-day rainmaking festival that culminates on the day the sun passes through zenith. Two days later, on May 4, planting officially starts. Agriculture in ancient Peru also followed the annual pattern of the seasons and the sun and the rain, but the Andean farming cycle begins in August.

In Peru, the first plowing and planting seem to have been associated with a special—but somewhat surprising—event: the sun's passage through the nadir. This is the point directly below one's spot on the earth, and it is opposite the zenith directly overhead. For this reason, the day is sometimes called the antizenith date. The passage occurs at midnight.

Between the latitudes that mark the tropics, the sun can cross through the zenith. The dates on which this happens are well defined for any particular location, and the points of sunrise and sunset on those days are readily noticed, like those of the solstices. We know already that summer and winter solstice extremes are complementary. The winter solstice sunrise position is more or less opposite the summer solstice sunset. The four solstice sunrise and sunset points are related symmetrically. This same sense of symmetry is easily applied to the zenith passage sunrise and sunset, and so complementary calendar dates, defined by sunrise and sunset opposite those for zenith passage, would emerge naturally. These turn out to be the times of antizenith passage. Evidence of interest in the nadir sun seems to still exist among the descendants of the Inca. Calendrical and directional symmetry alone could create interest in the antizenith passage dates. Recognition of their relationship to the zenith passage dates could even prompt the notion that the invisible sun crosses the nadir at midnight on key dates, just as the daytime sun climbs to the zenith at noon on others.

At Cuzco, the Inca capital, antizenith passage occurs on August 18, a date probably connected with the August festivals and sacrifices that inaugurated the planting season. Astronomer Anthony F. Aveni and anthropologist R. T. Zuidema believe they have found evidence for observation of the position of the sun near and on August 18 and for association of this astronomical alignment with the first planting of corn.

The solstices were celebrated by the Inca as well, but because Peru is in the southern hemisphere, the solstices there are the reverse of what we would expect. *Inti Raymi,* the winter solstice festival, was held in June. The summer ceremony, *Capac Raymi,* occurred in December.

Several annual Inca festivals were actually timed by the moon. Its appearance as the new crescent marked the beginning of the *Citua* festival near the September equinox, and sacrifices for the ripening crops, part of the March equinox *Pacha-puchy* observances, also coincided with the first crescent.

Time and the Social Order: the Inca Calendar

The Inca calendar also measured time in terms of the moon. Details of the Inca calendar are still unclear, however, because most post-Conquest reports are incomplete and contradictory.

Although we do not know exactly how they did it, the Inca also must have integrated their count of the months with the solar year in a long-term cycle of congruence. Fragmentary references, now more than 400 years old, imply that custody of the calendar was divided according to strict rules of family lineage and social stratification. The so-called Anonymous Chronicler, the unknown author of a 1534 account of the Conquest and Peruvian life, tells us the Inca king split the people of Cuzco into 12 groups and assigned to each of them the name of a month and duties related to public announcement of the year's progress through the calendar.

Throughout the Inca capital were placed *huacas,* or shrines. These huacas were organized into *ceques,* which in general were straight lines across the landscape, marked at intervals by the huacas that belonged to each of them. Responsibility for the individual huacas and the ceremonies observed at them was related in a complex way to Inca social organization and kinship. Anthropologist R. T. Zuidema, an expert on Inca kinship, also thinks the huacas and their ceremonies were tied in with the Inca calendar.

Zuidema's careful study of the various post-Conquest reports led him to conclude there were 41 ceques and 328 huacas. He believes each huaca was associated with a day "... in a 328-day cycle, for the number 328 is astronomically significant." It takes

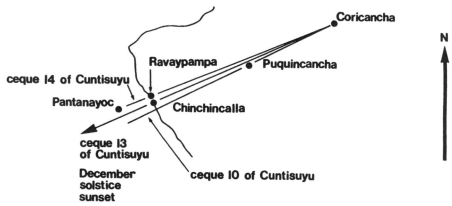

Huacas, or shrines, are marked by dots and identified by name here, and portions of three of the *ceques,* or lines of shrines, of Cuzco's southwest quarter (Cuntisuyu) are mapped. The Coricancha is the Temple of the Sun, the heart of Cuzco and the Inca empire. Ceque 13 coincides with the sunset on the December solstice, a calendrically important event. *(Griffith Observatory, after A. F. Aveni)*

the moon 27⅓ days to return to the same general position among the stars it occupied about a month before. The 27⅓-day interval is a *sidereal* month, and it, too, can be used to monitor the passage of time. The probable total of huacas—328—is equal to 12 times 27⅓.

It seems likely the Inca kept track of both the synodic month of 29½ days and the sidereal month and added intercalary days or months to keep both lunar intervals in pace with the sun and the annual cycle of agriculture and seasons.

In Peru, then, and in China, Mesopotamia, and Egypt, the civilizations that developed each fabricated their own unique calendar scheme. In each case the calendar evolved from observations of cyclical celestial events. Because these events organize and regulate the passage of time, they put order into life and society. Traditional calendars reflect, therefore, the sacred nature attributed to the sky, and we see this in the intimacy shared by the calendar and sacred ritual. The calendar schedules rituals. Rituals are named for calendric events. And the celestial phenomena that structure the calendar are celebrated in the rituals.

Although the sky is the obvious reservoir of cosmic order, its cycles—when examined in detail—are not congruent in a *simple* way. The belief that celestial phenomena nevertheless reveal the true and sacred character of the world prompted a variety of calendric compromises. Intercalations and multiple calendars were necessitated by competing needs in complex societies. A central authority had to balance and coordinate these needs, and so responsibility for the calendar fell to the pharaoh, the king, the emperor, or the Supreme Inca. His power was enhanced, accordingly, because he was in league with the sky.

Time and Divination

By observing what transpired overhead, shamans and astronomer-priests fashioned calendars and scheduled ceremonies. They had access to the domain of the gods and the source of cosmic order; this allowed them access to "knowledge" of the state of the cosmos. They could, then, communicate the celestial signs of the gods' intent to earth. Calendric divination made soothsayers, for example, out of the ancient Mesopotamian moonwatchers. In 1900, Assyriologist R. Campbell Thompson compiled hundreds of astronomical omens into a book with the engaging title *The Reports of the Magicians and Astrologers of Nineveh and Babylon*. Many of the reports involve the moon: "When the Moon at its appearance stands in a fixed position, the gods intend the counsel of the land for happiness."

This text refers to the first crescent ("appearance") occurring on the expected date ("stands in a fixed position"). "When the Moon out of its calculated time tarries and is not seen, there will be an invasion of a mighty city..." Unusual or unexpected behavior was regarded as a message. The views might be bad, but a proper word or spell recited by a knowledgeable priest could avert the threat. "When at the Moon's appearance in the intercalary month Adar its horns are pointed and dark, the prince will grow strong and the land will have abundance."

These texts tell us that the Babylonian prognosticators evaluated the match between what the calendar predicted and what the sky actually did. Departures from the expected order were viewed with concern.

One of the calendar cycles of ancient China was linked to divination. It was a day count, used as early as the late second millennium B.C., and based on the number 60. Each day was identified by the name of one of ten "heavenly stems" (*tian gan*) and the name of one of 12 "earthly branches" (*di zhi*). After the tenth "heavenly stem" was reached, the first was named again, but this time in combination with "earthly branch" number 11. All possible combinations were produced in 60 pairings, and then the cycle started up with the first "heavenly stem" and first "earthly branch" again. The number of unduplicated pairs is given by the lowest common multiple of 10 and 12. Both divide evenly into no number smaller than 60.

This 60-day cycle is the basis of an entire system of correspondences. It numerologically integrates many symbols into a network of associations and metaphors, and through these the ancient Chinese diviner interpreted the world. Although it may seem like an unusual way to organize the passage of time, the Mesoamericans did something very similar to this with their 260-day count, and with a similar divinatory intent.

Complexity and Congruence: the Mesoamerican Calendar

Of all the ancient calendar systems, that used by the Aztec and Maya and their neighbors in Mesoamerica is most often awarded the prize for intricacy and accuracy. Certainly this reputation is earned. The calendar of ancient Mexico and Central America was remarkable. One of the cycles they counted was, enigmatically, 260 days long. They also tracked a 365-day "year." Calculations of the congruence of such cyclical spans of time by the Maya calendar-priests equate with what we would call a solar year (365.2550 days). The correct length of the tropical year, according to precise, modern measurements, is 365.2422 days. Although the Maya did not measure the exact length of the year directly and did not conceive of it in terms of the fractions we use, their numerical manipulations of calendrical periods imply a year only 19 minutes in error. This is quite respectable, but not beyond the observational and computational techniques available to them. The Chinese, in fact, established the length of the tropical year as 365.2428 days by the fifth century after Christ, more or less the same time as the middle of Classic Maya civilization. Obviously, big telescopes are not required to obtain this kind of information. They didn't exist in ancient China or Mexico. But shadow-casting devices or careful horizon observations of the sun could do the job when coupled with accumulated records of the dates of celestial events. It is not, however, the accuracy of the Mesoamerican calendar that is most interesting, but rather its structure; the sacred dimension of time is more emphatically demonstrated there than in any other calendar scheme.

To understand the structure of the Mesoamerican calendar system, we have to identify its components. As we might expect, a yearly count of days was kept. For this purpose, the total was set at 365, and it was reached through a sequence of eighteen 20-day intervals (360 days) with 5 special days added at the end. The number 20 was the basis of the counting scheme among the Mesoamericans, just as 10 is the underlying principle of our decimal system. Probably the number 20 derives from the total number of human fingers and toes. In any case, this 365-day cycle was called the *haab* by the Maya and the *xiuhpohualli* by the Aztec, and in it each day was named by a number, from 0 to 19, and the name of the 20-day interval, or *veintena*. In our own calendar we adopt a parallel approach by assigning to each date a numeral and the name of the month. The Maya called the first day of a 20-day interval the "seating" of the interval. This is, in a sense, like numbering the first day 0. Then, as we might call out a date, like April 12, the Maya would identify a *haab* date as, say, 2 Zip—the third day in the third 20-day period of the year.

Recently a door was discovered in use as a tally board for the *haab* cycle in a Maya village in Chiapas. Strokes on the board were organized in groups of 20, and every twentieth stroke was bolder. Anthropologist Gary Gossen found that the tally included 18 groups of 20 marks plus 5 in a group of their own to total 365. Alexander Marshack's infrared photography revealed a century's worth of earlier *haab* tallies, the calendrical record of the current shaman's forefathers.

Just as we give a number to the year counted from an historically and culturally significant event (the presumed year of the birth of Christ), the Maya years were numbered from a date endowed with religious and cosmogonical significance. According to the most accepted correlation between our dates and theirs, the Maya starting point was in the year we call 3113 b.c. It is believed to be the year the Maya considered as marking the creation of the present world order. Naturally, the march of time would begin there. These notations of the year are called Long Count inscriptions, and in them the years that have passed are bundled into multiples of 20, just as our own dates are organized in multiples of 10. The year 1982 really means the sum of 1 thousand-year interval, 9 hundred-year intervals, 8 ten-year intervals, and 2 one-year intervals. Of course we call these successive multiples of ten a millennium, a century, a decade, and a year, respectively. The Maya units worked the same way:

$$1 \; baktun = 20 \; katuns$$
$$1 \; katun = 20 \; tuns$$
$$1 \; tun = 360 \; kins, \text{ or days}$$

They carried the sequence further, through the 20-day interval to the day itself.

$$1 \; tun = 18 \; uinals$$
$$1 \; uinal = 20 \; kins, \text{ or days}$$

A *tun* was an approximation to the true solar year and so 18 twenty-day intervals, rather than 20, were built into the sequence. The Maya chose to use 360 instead of 365, and their reason most likely was the numerological usefulness of 360. It can be divided and manipulated in many ways.

The Maya also kept detailed records of the moon, but these do not seem to constitute a formal lunar calendar. We find hieroglyphic inscriptions on commemorative stelae that report, for the dates registered on them, the number of the day in the current lunation (or the "age" of the moon) and the position of the lunar month in a sequence of five or six lunar cycles that was noted because it relates to the recurrence of eclipses. Glyphs that tell whether the lunation sequence for this date has five or six months and whether the month in question has 29 or 30 days also accompany the calendrical inscription.

Numerological significance is probably the key to the most enigmatic day count used in ancient Mexico: the 260-day *tzolkin,* or Sacred Count, of the Maya. In central Mexico the Aztec called it *tonalpohualli,* and it worked, in a way, like the 60-day cycle of the ancient Chinese.

The *tzolkin* was a ritual calendar in which days were numbered from 1 to 13 and named by a sequence of 20 emblems, or day-name glyphs. Most of the Aztec symbols represented animals, plants, or other things in their environment. In order, they were:

Cipactli	alligator	Ozomatli	monkey
Ehécatl	wind	Malinalli	grass
Calli	house	Acatl	reed
Cuetzpallin	lizard	Ocelotl	jaguar
Cóatl	serpent	Cuauhtli	eagle
Miquiztli	death	Cozcacuauhtli	vulture
Mázatl	deer	Ollin	movement
Tochtli	rabbit	Técpatl	flint knife
Atl	water	Quiáhuitl	rain
Itzcuintli	dog	Xóchitl	flower

By proceeding through the 13 numbers and the 20 names, a specific sequence of designations for the days is obtained. The first few days are, therefore, 1, Cipactli; 2, Ehécatl; 3, Calli; and so on. By the time 13 is reached, the day is 13 Acatl. The numbers then start up again, but the names continue in sequence. After 13 Acatl comes 1 Ocelotl. The 20 names have all been used once by the time we reach Xóchitl, and then they are recycled. Next comes 8 Cipactli. By the time all possible combinations are used, 260 days (the lowest common multiple of 13 and 20) have passed, and then the cycle starts over again.

There is no obvious astronomical cycle that takes 260 days, and we know of no one but the Mesoamericans who used this count. The motivation behind it is puzzling. Explanations, of course, have been tried. The interval could be related to the 260 days between May and August zenith passages of the sun at Copán, an important

Stela 3, in the Main Plaza of the great Maya ceremonial center of Tikal in Guatemala's Peten, includes a hieroglyphic inscription that established the date of this monument in the Maya Long Count calendar. In addition, the fourth glyph from the top, on the left side, is 4 (four dots) Ahau, the date in the Maya 260-day count (or *tzolkin*). The date in the 365-day count (or *haab*) is 13 Kayab, shown as the seventh glyph from the top, also on the left. Usually these calendrical inscriptions included a second series of glyphs that supplied detailed information about the moon at this time. Here, this lunar series begins with the fifth glyph from the top, on the left again. Part of that symbol is three bars (3 x 5) and two dots (2 x 1), the Maya way of writing the number 17. This stands for the seventeenth day of the lunar month. The glyph on the right side may be saying where among the background stars the moon was located on this night. Below it is a glyph that looks like a crescent enclosing a dot within its horns. The bar and four dots that accompany it stand for the number 9, and this is the Maya way of saying this was a 29-day month. Sometimes a lunar inscription like this would also include a glyph to indicate which moon cycle in each 6-cycle count this particular month represented. *(Robin Rector Krupp)*

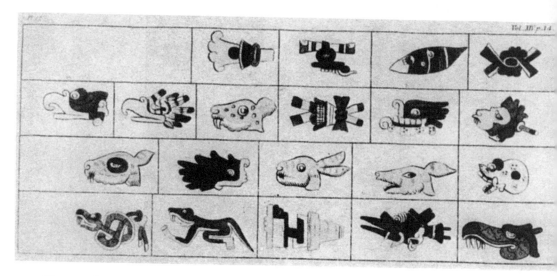

Twenty day-names and 13 numbers combine in 260 ways to complete the Aztec *tonalpohualli*, or 260-day sacred count. In 1814, Alexander Humboldt provided these versions of the day-name glyphs, and they can be identified by matching them with the entries in the list in the text. The first one on the left in the top row is Xóchitl, or "flower." *(From Researches Concerning the Institutions & Monuments of the Ancient Inhabitants of America)*

Maya center, or at Izapa, a site that shares the same latitude but was occupied by an earlier people with a Maya-like style of art and writing. Other proposals include 1½ times the period of 173.3 days between eclipse seasons (or 3 x 173.3 = 2 x 260); ⅓ of 780 days, the time between configurations of Mars (or 780 = 3 x 260); and the approximate number of days between the first and last appearances of Venus as either a morning star or an evening star (263 days). Even the interval between the conception and birth of a human baby has been suggested, an idea that may not be so far-fetched.

We know the 260-day cycle was used in divination, and the Maya approached this kind of thing with enthusiasm for numerology and the congruence of many cycles and events. Personal names were given to the Aztec newborns by a specialist, the *tonalpouque*. This individual, a priest and an astrologer of sorts, consulted the ritual almanac for omens—good or bad—associated with the day of the child's birth. In favorable circumstances the child took the number-name of the ritual calendar day of his or her birth. There may have been some sense of rightness, then, in the day of conception corresponding, at least schematically, to the day of birth.

These calendar priests still exist in the Maya region. They fulfill functions very similar to those of their ancient predecessors and keep the 260-day count for divination and other shamanic activity. Anthropologist Judith Remington, in interviews with Highland Maya in Guatemala, learned that although most people knew about the names of the Sacred Count days, only the shaman was aware of their sequence and the name of any particular day.

Probably no one relationship of the 260-day interval accounts for its important role. It ties several astronomical cycles together, and through more complicated relationships it acts as an elegant unifying factor in the structure of the Mesoamerican world and mind. The counting system, the 365-day count, motions of the planets and the moon, and the timing of human birth—all acquire more varied nuances of divinatory and cosmological meaning through the agency of certain numbers, like 18, 13, 20, and, of course, 260. Floyd Lounsbury, an expert in Maya hieroglyphics, has shown that an even more elaborate system of symbolic numerology and cyclical resonances lay at the core of Maya astronomy and calendrics. Many of its components originated in the sky, but they were manipulated by astronomer-priests to apply the sacred structure of the cosmos to affairs on earth. These priests juggled cycles of time and calculated when several would coincide. Theirs was a search for congruence—another form of order.

8

The Rituals We Perform

When most of us celebrate on December 31, we have little awareness that our New Year's revels are any kind of ceremonial observance of cyclical time. The old year is about to close out; the new year is about to begin. The dining, dancing, and drinking and the obligatory countdown to midnight are a kind of annual ritual that makes us active participants in the rhythm of time's passage and world order.

New Year celebrations are, of course, holidays, and such holidays are important destinations in the cyclic itinerary of time. New Year's Eve is the boundary between the year past and the year to come; congruence and cyclic order in the passage of time are most sharply focused at such junctions, for they mark the completion of one cycle and the start of another. These holidays reverberate in the human psyche because, as moments of time's renewal, they create hope. They are the times that contain the promise of new beginnings and generate the vitality that comes with renewal.

For traditional peoples, the myth of the cycle of cosmic order is dramatized in the arrival of the new year and the departure of the old. The old year dies, and the new year is born. This moment is the thin membrane between an end and a beginning, between a death and a resurrection. It is like the time of creation, when cosmic order first emerged from primordial chaos. In that sense, New Year's Eve, as the death of the last cycle of time, represents the intrusion of chaos. New Year's Day marks the rebirth of time and the restoration of order. Even in our secularized world we sense the symbolic importance of the New Year and continue to celebrate it. Every time it arrives, we unconsciously reenter the mythical time of the world's creation and play out the old conflict between original chaos and world order. Our ancestors saw this in the calendar and in the sky. And so they interrupted the flow of everyday life and celebrated the renewal of time.

Renewal is one of the key themes of ceremonialism, and we encounter connections between the sky and the idea of renewal most directly in seasonal ceremonies. The seasons, in turn, are reflected in the annual cycle of vegetation and—for farming communities—in the yearly pattern of agriculture. For this reason, elements of the growth cycle of plants and the various stages in cultivation become symbols in ceremonies that commemorate the completion of time and renewal of the world.

Reviving the Year

Hopi Indian farmers ritually revived the year with ceremonial symbols borrowed from the world around them: the sun and the seasonal growth of plants. Their year began in November and involved a "New Fire" ritual, designed to resurrect the spirits of the dead and, ultimately—by winter solstice—the sun itself.

Winter observances begin with Wúwuchim, a 16-day ceremony determined by the position of the setting sun and announced in accordance with the proper phase of the moon. After the ceremonial underground chambers, or *kivas,* have been purified and prepared, the new fire is kindled, before the sun comes up. The ritual involves prayers to the god of the nadir, or antizenith, who is associated with death and the underworld. Coal, in the Hopi view of things, belongs to his realm and is used to maintain the fire. This part of the New Fire Ceremony acknowledges the energy the sun transmits to the earth. At sunrise more ritual activity reenacts the cycle of germination, growth, and harvest, and then the new fire is carried from the kiva of the Two Horn priests to the kivas of the three other religious societies.

Because the Two Horn Society is considered to have the fullest knowledge of Hopi origins and the Creation, it is responsible for the New Fire and has the final authority to verify that all the ceremonies have been conducted properly. This is very important, for the rituals can fulfill their function—the maintenance of the cycle of world order and in particular, its annual renewal—only if they have been correctly performed. Following the exact prescription is what sanctifies these acts, and their sacred status is what makes them effective.

Songs in the last kiva ritual of Wúwuchim are timed by the appearances and passages of some of winter's bright stars—the Pleiades, Orion's Belt, Castor and Pollux, and Procyon—in the opening at the top of the kiva's roof. The Hopi singers start with the ritual smoking of tobacco, in a pipe lit from the fireplace directly below the roof window. The smoke ascends to the zenith, where the right stars are shining, and carries the songs skyward. The songs continue nearly to dawn. Finally, Wúwuchim ends with a public dance and the ceremony's last emergence from the kivas. This emergence is a reenactment of the Hopi creation.

As a theme, emergence harmonizes with the rest of the meaning of Wúwuchim. It marks the critical time when the sun, the world, and time itself are rekindled. Through dramatization of the creation myth, the identity of the Hopi themselves is reaffirmed, and their place on earth reestablished. Because shoots of new plants, hibernating animals, and rising celestial objects all emerge from the earth, Earth is a mother and a womb. Creation, in the metaphor of birth, is an emergence, and the Hopi rituals treat it that way.

However, creation is part of a cycle that ends in death. At the end of its growth, a plant goes to seed, and returns to earth. When winter's time comes again, the animals that hibernate reenter their burrows and sleep a seasonal death. The sun, moon, and stars slip into the earth's bosom as they set. All of these natural phenomena also make a tomb of the womb that is earth, and so the crucial junctions in the year are associated as much with death and the supernatural as with life and this world.

Near the end of Wúwuchim, all the roads to the village, save one, are closed by "blocking" each with a line drawn in cornmeal. This keeps contaminating influences—particularly those with evil power—from entering the village at such a sensitive time. The one road left open is a small trail. It permits the dead and the supernatural spirits, or *kachinas,* to return. Indeed, Wúwuchim marks the first reappearance of the kachina impersonators since the summer Niman ceremony in July-August.

By focusing on the New Fire, the emergence-creation theme, and the cycle of life and death, the Hopi prepare themselves for the next significant event in the year: the winter solstice. Winter solstice, among the Pueblo Indians, is an affirmation that the cyclical order of time and the world order will continue intact. The sun must turn back from its progress south, and the ceremonies of Soyál, held in the month of Kyamuya (December), "sacred but dangerous moon," are designed to guarantee the sun's return north. The season of Soyál is determined by the end of Wúwuchim and the phases of the moon, but its primary event, the winter solstice, is anticipated by making horizon observations of the sunset from the village of Walpi.

By standing on the roof of the ancestral house of the Bear Clan, the First Mesa Sun Chief commanded a clear view of the ridges on the horizon and the progress of the sun toward a notch (known as Lühavwü Chochomo) formed by the contours of the San Francisco Mountains far to the southwest. The sun's arrival there meant the winter solstice would occur 11½ days later. A clever analysis of the dates the Hopi chose for their festivals, undertaken by historian of science Stephen C. McCluskey, demonstrated that the Hopi observations were accurate to 4 minutes of arc, about one-seventh the apparent size of the sun. Hopi sunwatching techniques are quite good, but it is important to keep in mind that the estimated error of 4 arcminutes is an average error. The true date of the solstice is a challenge to determine, and the Hopi allowed themselves a few days' leeway in the actual scheduling of the main Soyál ritual. Accordingly, the rules that governed the scheduling of the ceremony were a little flexible.

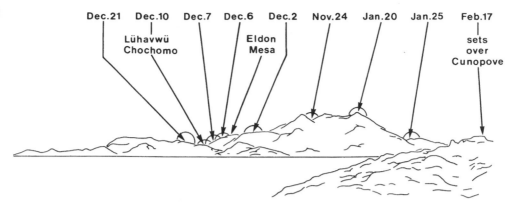

Dec.21 Dec.10 Dec.7 Dec.6 Dec.2 Nov.24 Jan.20 Jan.25 Feb.17

Lühavwü   Eldon   sets
Chochomo   Mesa   over
                                                            Cunopove

The sunwatcher at the Hopi village of Walpi helped set the start of the winter solstice Soyál festival by observing when the sun set in a horizon notch known as Lühavwü Chochomo. The next morning, those responsible for organizing the ceremony declare the Soyál season will begin in four days. The main dance usually occurs on the winter solstice, about 11½ days after the Lühavwü Chochomo alignment. *(Griffith Observatory, after Alexander M. Stephen)*

At the beginning of the Soyál period, the Soyál Kachina impersonator arrives in the village, moving unsteadily like a young child who is just starting to walk. A day later the Mastop Kachina moves about the houses. On each side of his face are painted three stars—the Belt of Orion. The Soyál Kachina is the symbolic announcer of the coming ceremonies, and Mastop is associated with human fertilization.

As the winter solstice approaches, the kivas are again prepared. By the twelfth night of the ritual, important ceremonies begin. Stars are again sighted through the kiva roof. The Pleiades and Orion's Belt schedule the songs, and the altars are sanctified. At dawn the participants walk to the mesa's edge, where prayers are said to the sun, and a path is drawn toward it in cornmeal. For three more nights the priests continue with the rituals and then assemble in the main kiva on the fifteenth night of the season. One of them, dressed as Star Priest, and carrying cornstalks to symbolize germination and growth, remains at the kiva of the Flute Clan until Orion's Belt culminates. Star Priest then races to the main kiva, jumps down the ladder, and announces he is there to assist in the ceremonies.

Star Priest wears a large four-pointed star on his forehead, which probably stands for Morning Star, the herald of the sun, and his body is painted with white dots that suggest stars. White corn leaves are fastened to the hub of his star headdress. In the main kiva he moves to each semi-cardinal point, dancing, throwing consecrated cornmeal, and making declarations about the meaning and intent of the ceremony. Later in the ritual, Star Priest receives from the Soyál Chief a shield painted with the face of the Sun. Partitioned into upper quadrants—one yellow, the other red—and a lower half that is blue, the face of the Sun is encircled by eagle feathers and has long strands of bright red hair dangling across it. Dancing with the sun shield, which is actually the face of Taiowa, the Creator, Star Priest spins it clockwise and jumps again around the semi-cardinal points. The ritual mimics the sun's motion and is meant to help Sun continue on his normal path by turning back from his southern house. He may linger there a few days, but he must turn back north or the world and the Hopi will die. Ears tuned to modern science and to a secular world will hear in this account anxiety, superstition, and naïveté, but, of course, a failure of the sun to continue on its accustomed course would signal catastrophe. Underlying the Hopi concern for the physical survival of their world is a sense of what is really at stake—the entire cosmic order. Southwestern writer Tom Bahti astutely noted that superstition "is the *other* man's religion."

The Soyál rites continue through the night, paced by the stars in the kiva roof. If all has been done properly and the Two Horn priests approve, the celebrants leave the kiva and take the prayer sticks to the shrine of the underworld and the shrine of the sun. During the ritual a young girl, dressed as the hawk maiden, "incubated" the prayer sticks by sitting on them in the kiva. The prayer sticks invoke the theme of germination, or birth, and after the ceremony more prayer sticks, and prayer feathers, too, are brought to all the families waiting in their homes. It is still before dawn, and everyone walks with the prayer sticks to the eastern edge of the mesa and plants them in the ground. Paths of cornmeal are traced to the east. The day begins, the sun heads back north, and the year is on its way.

Surviving the Year

Food production ensures the survival of a community, and we expect this concern to be expressed in celestially timed ceremonies linked to the agricultural year. Often there are other aspects of these rituals that have nothing to do with farming. The details of the rituals depend upon the culture of the people doing the celebrating, and they may have something to do with kingship, sacrifice, or with the perception of time. However these seasonal ceremonies may be fabricated, they have the same foundation: a sense of the structure of the cosmos, and the images they do incorporate reflect the same anxiety for continuity and survival. For that reason, we can find the same theme of renewal in ceremonies that invoke the seasons and the sky and yet contain no explicit references to farming or food.

The idea of rekindling the sun at a time of solar vulnerability and at a time of transition in the calendar, then, is certainly not restricted to the Pueblo Indian farmers of the American Southwest. When the Irish Celts occupied their capital, Tara, from about 300 B.C. to about A.D. 500, they, too, celebrated the new year with a "new fire" ceremony, also in November. Neither the date of the Celtic New Year nor the sun's situation on that holiday duplicated the mid-November conditions of the Hopi Wúwuchim, but the idea of the renewal of the sun and the renewal of time through the completion of a cosmic cycle was the same.

The Celtic New Year took place in early November, on Samhain. This falls about midway between the autumnal equinox and the winter solstice and was traditionally the start of winter in the British Isles.

Samhain was a crucial and dangerous time in the minds of the ancient Irish. It was one of the two hinges that held the year and its seasons together. As the portent of winter, it announced the sun's impending death, which would occur at the December solstice. Real fear of fairies, witches, and spirits of the dead kept people close to home and fire on the eve of Samhain. During the night, dangerous forces were abroad; supernatural beings and the dead were free to roam. This sounds a lot like the old tradition of All Hallow's Eve, and indeed that holiday evolved from Samhain and is still celebrated today as Halloween.

Because Samhain marked the decline of the sun and the close of a cycle, it symbolized the cyclical decline of cosmic order and the intrusion of chaos. All the old forces of primordial chaos caught the world at its weakest moment and upset the normal, ordered environment. Supernatural disorder, in the form of ghosts and demons, broke through. Somehow this threat had to be met. The Celts magically ensured the sun's renewal, and therefore the world's renewal, with their ceremony of the "new fire."

Tara was the headquarters of the Irish king and the site of the major general assemblies and important feasts. Today only the earth platforms of its wood and wattle buildings and earthwork enclosures survive to be seen on the Hill of Tara, about 23 miles northwest of Dublin. Even less remains on the Hill of Ward, 12 miles beyond, but that was the site of the kindling of the new fire, a ritual center known as Tlachtga, named for a famous sorceress in Irish legend. With the approach of Samhain, all the household fires and most of the ritual hearths of Ireland were extinguished to mimic the dying sun and the chaotic darkness to follow. Priests, Druids, that is, rekindled the fire on the night before Samhain sunrise and sanctified it by burning sacrificial victims in the new bonfire.

In the Celtic Samhain ceremonies we can see much more clearly the link between the cosmic order and the social order, as embodied in the capital, the king, and the law. All three were threatened when the year died. The Feast of Tara was also held at this time, and just as the Tlachtga fire renewed cosmic order, the Feast of Tara renewed the social order.

This great general assembly and major festival was a time for enactment of new laws and for the settlement of accounts, debts, and litigations. In this way the outstanding business was wrapped up and bundled away with the old year. Genealogies, records, and histories were brought up to date. Even the old laws were reenacted, and so reinvigorated, for another year's service.

As the seat of government, Tara on the Samhain was a fortress against the threat to world order. The myth of Aillen mac Midhna, a hostile fairy spirit from the Other World of cosmic chaos, makes this clear. Every Samhain, Aillen mac Midhna burst from a dark cave accompanied by many other destructive spirits of the Beyond. They intended to devastate the land, but Aillen's special target was Tara itself. Every Samhain he burned it down. Finally, Fionn mac Cumhaill (or Finn), a prominent hero in the Irish tales, drove Aillen mac Midhna from the tables of Tara and beheaded him.

Aillen menaced the world order by launching his attack on the day that marked the sun's decline into winter and struck the heart of the social and political order by burning down Tara. Aillen's cave is not only the dark route to chaos, it is the source of the darkness of winter, night, and death.

This myth was echoed in another Samhain fire ceremony performed at Tara. A pyramid of stacked timbers, called *torc tenned,* or "fire boar," was burnt to the ground. Its embers were placed, however, in a special Samhain hearth and tended for a year. At the next New Year, another "fire boar" was lit from the low flames of the last year's fire. Tara, itself, was symbolically burned and built anew each year in this ritual.

Even the protocol at the tables of the Feast of Tara was designed to defend society against the challenge of chaos. The social order was incorporated into a cosmically oriented and symbolic arrangement of the important guests. The High King dined at the center of a square formed by the tables of the four provincial rulers, each on one side. These sides were oriented toward the cardinal directions, and each belonged to the chief of the corresponding geographic quarter of Ireland. The tables also symbolized the four seasons and the orderly circuit of time. The High King acted as the center, the point of stability. By hosting the feast, the High King reaffirmed his authority and enhanced the social order. The very act of holding court at a commemorative feast—with a cosmologically organized seating of the guests—on the night when mischief, confusion, chaos, and death were on the move, was a stand against the darkness. By igniting their fires and holding their feasts, the Irish Celts rekindled their world and reenergized the sun. And this, in turn, held the kingdom together.

Harmonizing the Year

Celebrations of celestial renewal permitted ancient peoples to participate in the rhythm of cosmic order. They harmonized themselves and by their sacrifices helped maintain the world's structure and stability. A proper act at the proper time oriented them in the landscape of space and time, and that process itself, in their minds, contributed to the equilibrium of the whole cosmos and helped it on its way.

This way of thinking—and acting—is one of the things we encounter among peoples whose belief systems preserve a sense of the sacred. Understanding it can help us who, with our democratic, secularized view of the world, have trouble comprehending the true meaning of the divine kings of our ancient ancestors. One function of kingship is the maintenance of terrestrial order: the social, political, and economic structure of the society. Usually, the king acquired his authority through the mandate

Ceremonial sacrifices to Heaven occupied the emperor of China at the winter solstice. Each year at that time he was obligated to climb to the top level of the Round Mound in Beijing's Temple of Heaven. There he faced north and the pole of the sky and paid homage to the symbolic center of the cosmos. *(Griffith Observatory, Lois Cohen)*

of heaven, the source of cosmic order. As the representative and conduit of order, the king was part of a feedback loop. His rituals to heaven embodied the responsibility people sensed they had. Celestial order would flow from heaven to earth, but people, for their part, had to reenergize the sky. This made a priest of the divine king.

In imperial China, ceremonial renewal of the world order was the direct responsibility of the emperor. His annual sacrifice to Heaven on the winter solstice made him the link between cosmic harmony and the sun's yearly course. Heaven, in the official state religion, was an impersonal—but supreme—power. It preserved the structure and stability of society by mandating the emperor's rule, and it created order through cyclical celestial change. The Chinese did not think of Heaven as the actual sky, but as the sky's essence—the principle of cosmic order.

By performing the appropriate—and required—ceremonies, the emperor participated in the orderly flux of the universe and so helped the cosmos to continue. Such visions of a participatory universe are sustained by the concept that human beings have a specific, sacred role to play, a cosmic job to do.

Three days before the winter solstice the emperor and some of his retinue moved into the Hall of Fasting, or *Zhai gong*, where they abstained from the company of women, music, certain foods, and other activities. This pavilion is situated in the large ceremonial area known as the Temple of Heaven, just south of the Forbidden City, or central Beijing. On the winter solstice, two hours before sunrise, a procession led the emperor the short distance to the Round Mound (*Huan qiu tan*), also part of the Temple of Heaven complex. At times as many as a thousand people assembled in this parade: musicians, singers, dancers, soldiers, royalty, and high officials. Fans, parasols, and banners cut the air. Flags for the 28 lunar stations, the five planets, the four rivers, the five peaks, and other cosmological emblems put the whole universe into the pageant.

The emperor, closely associated with the north celestial pole, was used to being worshipped while he faced south. In this way he imitated the celestial pole, which shined in the north. To see it one must face north, and the ancient Chinese turned north to face their emperor. At the winter solstice, however, the emperor had to humble himself before the Supreme Emperor, Heaven, and he approached the Round Mound from the south. The Round Mound is a stack of three terraces, its summit nearer to—and open to—the sky. The circular shape of the Round Mound symbolized Heaven, and the square enclosure of its lower court represented the Earth. The emperor mounted the south stairway, climbed closer to Heaven, and faced north, as his own subjects faced him. He was the intermediary between Heaven and Earth and exalted above all others. Only he was permitted to climb to the highest terrace. But once there, he prostrated himself before the power that was presumed to know all, the power that moved the world—the sky.

At the top of the Round Mound the emperor lit a fire. Its rising smoke was his invitation to Heaven to join the ceremony. He read an account of all the significant doings of the past year and offered gifts of incense, silk, and jade, while music was played below. Libations followed and then a portion of roasted human flesh (taken from one of the sacrificial victims that had been killed and cooked in the Sacred Kitchens during the night) was presented. In closing, the emperor bowed low nine more times and descended to the lower courtyard, where at a porcelain furnace he attended the burning of a last victim.

Just north of the Forbidden City, near the An ding men Gate, the Ming emperors also built the Altar of the Earth, or Di tan. It, like the Round Mound, has a stairway at each cardinal direction and three terraces. The emperor's ceremonies there were counterparts to the rites at Heaven's altar, the Round Mound, but the terraces were square for the Earth, instead of round for the sky. Sacrifices took place on the summer solstice at the Altar of the Earth, and the victims were not burned but buried. For Heaven, the ashy smoke of the sacrifices ascended to the sky. For Earth, their remains were placed under the ground. These two ceremonies—in timing, style, and intent—complemented each other. One was intended to stimulate the celestial, male, active principle—*yang,* in winter when it was weakest. *Yin,* on the other hand, was the terrestrial, female, passive principle and needed to be strengthened in summer, when *yang* prevailed. Together, in alternating strength, they forged the cyclical and ordered pattern of the world. The solstices marked the crucial transitions from one phase to the other, and it was the emperor's business to assist the change.

Honoring the Year

In three different examples so far, we have seen three unrelated cultures treat the theme of the weakened winter sun and revival of the year with similar ceremonial metaphors. All three of these people inhabited the earth's northern hemisphere and shared the same seasonal pattern of the sun. The perception of cyclic time is universal, however, experienced by every culture with a life-style still tied to the sky. We can be certain, then, that these rituals of renewal will take place wherever people still have a sense of the joints of time, those transitions from one cycle to the next. We can verify this by looking for those same ceremonies in the southern hemisphere, where the seasons are reversed but the response is the same.

Nothing demonstrates the universal meaning in the theme of the seasons and cyclical time better than the calendrical ceremonies of ancient Peru. Chroniclers who wrote about life, customs, and the history of the times prior to the conquest of Peru verify that important ceremonies were held at both the summer and winter solstices. *Inti Raymi,* the "Feast of the Sun," paralleled in theme and purpose the winter ceremonies celebrated by other peoples, but in Peru, south of the equator, the winter solstice occurred in the month of June, not December. For the Inca, it was the month of Cusqui Quilla, "the Moon of Hard Earth." With the first breaking of the soil still a month away, and the time of sowing corn a month after that, the earth was still hard and dormant at the time of Inti Raymi. And the low sun, its path far to the north in this southern latitude, was ready to return south with life for the world.

As in China, the emperor officiated at the great winter solstice ceremony. The *Sapa Inca* occupied the peak of a vast and highly organized pyramid that was Inca society. When the time for the feast drew near, the *curacas,* who were the heads of the basic kinship groups, arrived in Cuzco from throughout the empire. Dressed in their finest clothes, they brought offerings for the sun. Some came in costumes or in masks. Others carried symbolic weapons. Drummers and trumpeters were brought to the capital, and the entire assembly advertised the greatness of the empire.

As we have seen already in these ceremonies of renewal, the rituals must be preceded by some kind of purification. This purification is not simply something one does, the way a physician washes his hands, in order to perform one's duties. It represents the fresh, purified state that renewal brings. The year dies, and the world is born anew, free of the encumbrances of the past. At Inti Raymi purification took the form of ritual fasting that began three days before the solstice. The Sapa Inca and his retinue were permitted only water, some uncooked corn, and handfuls of chucam, a native legume. The company of women was avoided, and throughout the city the fires were put out.

Before dawn on the day of winter solstice the Sapa Inca and the curacas in the royal lineage went to the Haucaypata, a ceremonial plaza in central Cuzco. There, they took off their shoes as a sign of deference to the sun, faced the northeast, and waited for the sunrise. As soon as the sun appeared, everyone crouched, much as we would get down on our knees, and blew respectful kisses to the glowing gold disk. The Sapa Inca then lifted two golden cups of *chicha,* the sacred beer of fermented corn, and offered the one in his left hand to the sun. It was then poured into a basin and disappeared into channels that conducted it away as though the sun had consumed it. After sipping the blessed chicha in the other cup, the Sapa Inca shared it with the others present with him and then walked to the Coricancha, or Temple of the Sun.

Again without shoes, the Sapa Inca went into the Coricancha, where sacrifices

At the June solstice the Inca celebrated the solar ceremony of Inti Raymi. In this drawing from Poma de Ayala's sixteenth-century chronicle of Inca life and Spanish rule, the Sapa Inca drinks *chicha* out of one goblet while the other is transported by a flying spirit and offered to the sun. *(Det-Kongelige Bibliotek, Copenhagen)*

were underway. Among the victims were llamas. Their entrails were removed, examined for omens, and burned in offering to the sun. This fire was not lit, however, until the sun came into view. To be ready for this, the Sapa Inca positioned himself in one of the rooms that opened on the inner court of the Coricancha. He was accompanied by no one but the seated mummies of his predecessors. We learn from Pedro de Cieza de León, who wrote *The Chronicle of Peru,* of two benches in the Coricancha, for the Sapa Inca's use only, upon which the light from the rising sun fell.

The Inca's "sun room," like much of the Coricancha, was shingled with gold plates. Garcilaso de la Vega still remembered the emeralds and the turquoise mounted in the gold when, as an old man, he wrote his *Royal Commentaries of the Inca.* Although much of the Coricancha is now gone, some of its walls and rooms are preserved in the Church of Santo Domingo, built on top of the old Temple of the Sun. The remains show that it was all oriented northeast-southwest, in alignment with the June solstice sunrise. Rolf Müller established this many years ago, and Tony Morrison, a writer and filmmaker for the BBC, affirmed this with recent on-site measurements.

A new fire was lit by focusing the sun's rays on wisps of dry cotton with a concave mirror. The mirror was worn as a bracelet by one of the priests, and it provided a fresh fire for the sacrifice, given, in Garcilaso's words, "by the hand of the sun." If it were cloudy, two sticks were rubbed together to ignite the fire. Felipe Huamán Poma de Ayala, an Inca of noble descent who wrote *The First New Chronicle and Good Government* in the late sixteenth or early seventeenth century, mentioned that sacrifices included the burial of offerings of gold, silver, different colored shells, and children. Children, we know, were sacrificed but only in certain significant ceremonies. Several post-Conquest writers mention that children were sent to Cuzco for this purpose. Those not killed in the capital returned home and were sacrificed there. Their route from Cuzco was governed, it seems, by sacred ritual, for they followed the *ceques*—the lines of shrines—and not the normal roads back to their villages.

The *ceques* and *huacas* (shrines) were inextricably entwined with the calendar, the *ayllu,* or kinship groups, the Inca political organization, and the Inca religious life. Their use as ritual paths—crossing even great distances—may help explain the enigmatic Nasca lines and figures drawn on desert plateaus of Peru's south coast. These huge ground drawings, or geoglyphs, were rendered by clearing the darker, loose surface rock to reveal the light desert soil beneath. A long straight line was produced then simply by removing the rock by hand along a line surveyed by eye. The drawings are more than just lines, however, and comprise a complex display of animals, plants, triangles, trapezoids, "avenues," and narrow lines. Mathematician Maria Reiche has been measuring and studying them for decades, and back in 1939 the Peruvian archaeologist Mejia Xesspe proposed they might be something like the *ceques,* despite the seven hundred years or so that separate the period of Nasca culture from the time of the Inca.

Such lines could have been used in a large number of different ceremonies and in different ways. Although we are unlikely to sort the system out entirely, we are not forced to take seriously the notion that ancient astronauts assisted in the layout of the designs and made use of them as runways or navigational aids. Lines are still in use. Until recently, at least, some people in remote Andean villages, in Bolivia as well as Peru, drew lines to hilltop shrines where religious ceremonies were held on certain days. In *Pathways to the Gods,* Tony Morrison reported his discovery of a hilltop sacrifice containing burned llama bones. From this hearth a line extended straight to the flat below. Morrison reported further encounters with lines that terminated in small piles of rocks. Beneath the stones he found fresh coca leaves. The coca, a traditional Inca offering, left no doubt these were *huacas.* Elsewhere, he collected accounts of offerings made at a station, or *huaca,* on the route of a line to sacred ground. The line and shrines were part of a ceremonial complex linked to the calendar.

We don't know that the Nasca people conducted winter solstice ceremonies, but the chances are good that they did. For now, however, the accounts of Inti Raymi are our best clue to what the winter solstice meant in ancient Peru. Although most of the details have dispersed like the smoke of the burnt offerings, Inti Raymi was officially reestablished as an Indian fiesta in 1930, and is reenacted every June 24 at Sacsayhuaman, the Inca fortress above Cuzco. Despite being a tourist attraction, Inti Raymi still seems to project some of the power the winter solstice commands over us.

Straight lines drawn by the Nasca on the pampa may turn out to be some kind of version of the system of *ceques* and *huacas.* *(E. C. Krupp)*

Starting the Year

Time, measured out in celestial tallies by skywatching shamans and calendar priests, eventually rounds the last turn in the cycle of cosmic order and begins the cycle anew. These technicians of the sacred punctuated that joint in time with ceremonies that consecrated the moment and mirrored the pattern of the sky. Such moments can occur, however, at various times of the year. They are not necessarily married to the winter solstice. And they can also occur after various intervals of time, not necessarily restricted to the passage of the year. Just when these moments are celebrated by any particular group of people depends upon where they live, their way of life, and their particular perception of cosmic order.

Babylonian priests performed a kind of ritual drama at the New Year ceremony in ancient Mesopotamia. It, too, initiated the cycle of ceremonial renewal and involved a recitation of the *Enuma elish,* the Babylonian creation myth. The priests also reenacted some of the key events in the story of Marduk's victory over chaos and Marduk's assembly of an ordered cosmos. Unlike the other rituals of renewal we have considered, however, the Babylonian New Year did not take place in winter. It was called the *akitu,* and it was held at the equinox, either in spring or in fall. Records of intercalated months suggest that in Old Babylonian times the autumnal equinox started the year. Later, the New Year was celebrated in spring. Which date does not really matter. What counts is the choice of a turning point in time that was significant to the Babylonians. More than one reason must have suggested an equinox, and only hints of those original reasons remained in the ceremonies that continued to commemorate them.

In the first few days of the ceremony, Marduk was symbolically confined in what texts call "the Mountain." For three days Marduk remains in this underworld, a realm of chaos and the dead. The term "mountain" also refers to the tall, multileveled temple-towers (or ziggurats) the Mesopotamians built of clay bricks on the flat flood plain of the Tigris and Euphrates rivers. It is possible that this part of the ceremony was connected in some way with the ziggurat. On the fourth day of the *akitu,* the *Enuma elish* was repeated, and this activity, accompanied perhaps by others, brought Marduk back to life and allowed him to "emerge" from the Mountain, or the underworld. We have already seen how such metaphors equate with sunrise and with the start of the New Year.

Marduk was not Shamash, the sun, but he assumed many attributes of the sun as part of the elevation of his status in Neo-Babylonian times. Marduk's emergence from the Mountain at the equinox and New Year, in any case, represents the creation of world order. We already know that is Marduk's role in the creation epic. By staging this myth in ritual terms at a turning point of the seasons and the year, the Babylonians recognized the cyclic nature of the world. The end of each year is a reentry into the time before creation of the world. The previous world must break down before it is refabricated, and that is why Marduk is imprisoned and slain in the Mountain.

Some of the mythological scenes portrayed on cylinder seals may relate to these ideas. When the Mesopotamians wanted to put an official stamp on a clay document or protect the integrity of the contents of a container, they impressed a design in the soft clay by rolling a small stone cylinder in it. The cylinder was intricately carved, and one of these seals, from the Akkadian period (2360–2180 B.C.) and now in the British Museum, portrays the sun god, Shamash, brandishing a saw and emitting undulating rays of light as he emerges in a gap between two mountain peaks. The god at the right, with streams of water and fish flowing about his shoulders, is Ea. The goddess on the left, perhaps heralding the appearance of the sun, is the goddess Ishtar, who was sometimes identified as the planet Venus or as the morning star. Ea's waters here may represent the spring floods. We can't be certain, for no text accompanies the picture. But if the springtime is meant, the scene may symbolize the vernal equinox sunrise, and possibly the New Year.

More prayers and rituals continued the New Year ceremony, which lasted for 11 days. A ritual called "fixing of the destinies" and clearly involved with omen readings for the coming year took place. Also, the Babylonians perpetuated the "Sacred Marriage" ceremony of the Sumerians. This time the king represented Tammuz and a high priestess was Ishtar. But the message was the same: fertility. The passage of cyclical time meant in Babylon what it meant elsewhere: renewal—in the gods, in the king, in the fertility of the land, in the calendar, and in the sky.

In this Akkadian cylinder-seal impression, Shamash, the sun, makes his dawn appearance in the notch of a mountainous horizon profile. This may be a representation of the vernal equinox New Year sunrise, the event embodied in the *akitu* ceremony. *(Griffith Observatory)*

"Bundling" and "Bearing" the Years

We who count things decimally package the years into decades, centuries, and millennia. We speak of the "Fifties," "Sixties," and "Seventies" as though fashion takes a sharp turn in the first year of each interval of ten and continues in suit until the decade is over and another trend and ambience begin. It is also natural for us to compartmentalize history into hundred-year spans and think of the twentieth century as a different sort of place from each century that preceded it. And there is something special, too, about the passage of a thousand years. Today, our fixed reference in historical time is the year accepted as the time of the birth of Christ, and from the early days of the establishment of the Church the ideas of the Millennium, the Second Coming, and the Apocalypse also have been fixtures of Christian tradition.

Numerous and varied predictions of our imminent date with doom and the dawn of a new age have come and gone without incident. Now, as we approach the year 2000 and see in its string of zeros a kind of cosmic odometer that measures the maximum mileage the vehicle can endure, we are tuning up our instincts for apocalypse. Many books catering to our curiosity about doomsday have already appeared, and doubtless more catalogs of pseudoscientific catastrophe are on the way. Some contemporary religious movements emphasize the same impending cataclysm. We can expect to see more and more of this anticipation of disaster as the year 2000 rounds the bend.

Catastrophes do occur, of course, but rarely on schedule. All of this concern really originates in our sense of cyclic time and in the myth of cosmic order. For many peoples of the world, A.D. 2000 holds no particular magic. But they have their own appointed rounds with the destiny of the universe just as we have ours. The Aztec, for example, expressed a real concern for the world's stability—and their own continued existence—at the end of a 52-year calendar cycle called the "Bundling of the Years." They played their proper role at that time by performing the New Fire Ceremony on top of Cerro de la Estrella, the "Hill of the Star."

To get a 52-year cycle, the Aztec had to combine two of their calendar cycles: the 260-day count and the 365-day count. A "year" in this sense was the 365-day year. As both the 260-day count and the 365-day year continued through their appointed rounds, each day acquired a number and name combination from each calendar. For example, a date in the 260-day cycle might be 3 Técpatl (three "flint knife"), and the same day might be designated 1 Panquetzaliztli (one "Raising of Banners") in the 365-day, 18-"month" calendar. The complete name of that day would be 3 Técpatl 1 Panquetzaliztli, and that pair would not be seen again until another 18,980 days had passed.

This Calendar Round, or *xiuhmolpilli*—the "Bundling of the Years"—had great cosmological significance for all of the Mesoamerican peoples. They divided it into four groups with 13 years each and paid special attention when the 52-year cycle approached completion. In some instances, a new Aztec pyramid seems to have been built on top of what was already there after a 52-year interval.

Few people lived long enough to see more than one New Fire ceremony, and so anguish for the world's end was real. Although the Aztec knew other 52-year cycles had come and gone, few individuals had the perspective to suppose the chances were good for survival. Time was neatly sewn up when the two calendar cycles returned to their start. It seemed reasonable the world would end on a cycle's last beat.

When the end of the 52-year cycle was at hand, the priests of Tenochtitlán prepared for the New Fire Ceremony. The last four days of the old "bundle of years" were a time of great anxiety, for it was thought that in the congruence of the calendars and in the climax of the count resided the possibility of the world's end. Everyone knew the present age, or "sun," would end sometime and the finish was programmed for the completion of one of the 52-year cycles. Each time the cycle rounded its last turn, every Aztec mind wondered if this age's time was up. If so, the sky would cease to turn. The sun would lack the strength to rise. Darkness and its demons would descend to earth and devour everyone alive. Night would last forever, and the world order would dissolve. Here then, on the stage of the Aztec New Fire Ceremony, is a scene from the drama of cosmic order we have seen before. At the "Bundling of the Years" chaos threatened to intrude.

Worry for the future of the world showed up in curious ways. Pregnant women were fitted with masks and placed in *cuezcomates*—small, clay corn bins—the Aztec equivalent for a silo. Somehow this confinement was to have been effective should the women have turned into dangerous wild beasts. Small children were also dressed in masks and kept awake on the night of the New Fire vigil, lest they turn into mice.

Earlier in the day, the Aztec priests left the sacred central precincts of Tenochtitlán in order to arrive at the summit of the Hill of the Star, a hill they called Uixachtecatl, near midnight. Up on top of the Hill of the Star one is stranded in solitude by the sound of the wind blowing across your ears. Dust whips across the small plaza in front of ruined temple steps, which lead to an empty foundation, all that is left of the temple in which Aztec priests congregated for the New Fire Ceremony. High above the valley floor and closer to heaven, they prepared for the end of an age.

Once they reached the temple on the Hill of the Star, the priests watched the progress of the Pleiades, as that star cluster passed high overhead. Throughout the Valley of Mexico all fires had been extinguished, in the temples and in homes. Pottery had been smashed, houses cleaned, and votive statues and hearth stones pitched into the lake. All of these acts suggest preparation for destruction and perhaps renewal in the Aztec's anticipation of the world's end.

Fray Bernardino de Sahagún, the compiler of the *Florentine Codex,* told us in that book what happened next. The Pleiades, at their highest point, can go nearly through the zenith. When they continued, after midnight, to head toward the west, the priests concluded the world would last another 52 years. At that point the priest of

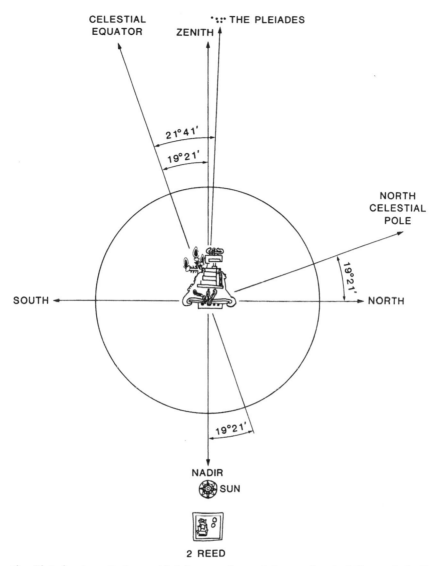

CELESTIAL EQUATOR

ZENITH •::• THE PLEIADES

21°41'

19°21'

NORTH CELESTIAL POLE

19°21'

SOUTH

NORTH

19°21'

NADIR

SUN

2 REED

When the Pleiades transited at midnight, nearly straight overhead of Cerro de la Estrella, the sun was at the nadir—straight below. The glyph in the center of the diagram comes from a post-Conquest Aztec codex and is part of a drawing that records key events of A.D. 1507 (2 Reed), the year of the last New Fire ceremony. The bell-shaped structure is Cerro de la Estrella itself, and it is shown with a temple at its summit. *(Griffith Observatory)*

Copulco took the firemaking sticks he had carried with him and kindled a fire on the breast of the victim. As the fire caught, a flint sliced through the victim's chest, and the priests seized the throbbing heart and dropped it in the fire. With a little tinder from the small fire on the victim's breast, a bonfire was lit on the temple altar. The populace, miles away, in the valley below, watched for that beacon. It was their sign the world would be preserved.

In that preservation was renewal. Torches of New Fire were carried down the hill to the Templo Mayor, to priests' houses, to clan chiefs, and to public braziers. All the hearths in the city were reignited, and runners carried flames to villages through the valley. New clothes, new mats, and new crockery commemorated the renewal of the sun, rekindled by the sacrifice of the victim, whose body was consumed in the first bonfire. The despair of the previous day was replaced by joy. The last "bundle of years," a symbolic bundle of 52 sticks, was buried.

The last 52-year cycle ended as the Pleiades climbed to the "top" of the sky, and the clue that the world would go on was seen in what they did next. The Pleiades continued on course, as a fire in the night, and signaled that the sun—the day's fire—would return renewed at dawn. The Pleiades reach their highest point, or culminate, once each day, and the time of day varies with the seasons. In Aztec times their culmination at midnight occurred in mid-November, about November 14 in A.D. 1500.

This date falls reasonably close to the time when the sun at midnight is at the nadir—or straight down, opposite the zenith above the Hill of the Star—on November 18–19. It is possible that the real astronomical motivation for the timing of the start of a new Calendar Round was the sun's "rebirth" from its entry into the deepest, darkest underworld. The New Fire ceremonies themselves suggest an association between the renewal of the sun and the renewal of time. The Pleiades, perhaps, were simply the visible sign of the sun's renewal at an hour when the sun started toward its rebirth but was still invisible. It is the day, or night really, of antizenith passage and the sun's vulnerability, then, that was symbolized by this event—and its date. The New Fire Ceremony took place at the time of year when the sun was most threatened and in the greatest need of rekindling.

The ancient Mexican calendar cycles were played out in a number of native ceremonies and dances. Even the 52-year cycle is embodied in the relic of a ritual still performed near Papantla, a market town in the state of Veracruz. Papantla is a major supplier of the world's vanilla, and in its *mercado*—a labyrinth of stalls, small shops, and tortilla griddles—one can buy a liqueur flavored with local vanilla beans. It is the label on the bottle that makes this product interesting to us however. Pictured is the Pyramid of the Niches at the nearby archaeological site known as El Tajín. This is where the ancient ritual "dance" of the *voladores* is staged.

El Tajín belongs to the Classic Period of pre-Columbian Mexico. Even though it is now in Totonac Indian territory, it probably was built and inhabited by the Huastec Indians, who occupied this region between 10 and 17 centuries ago. El Tajín incorporated many general Mesoamerican traditions into its otherwise distinctive regional style. Ruins have yielded evidence of the same calendar system used throughout Mesoamerica, and the six-level, temple-topped Pyramid of the Niches even seems to confirm that in its architecture.

Each level's façade is actually a row of small compartments, or niches. Excavation and restoration by the Mexican archaeologist José García Payón led him to conclude that the original number of niches had been 365, the number of days in the year. In discussions of the pyramid's calendrical system, this claim is usually repeated. Because no one in recent years had tried to confirm García Payón's result, it seemed appropriate, on one of Dr. Anthony F. Aveni's archaeoastronomical expeditions to Mexico, to do the obvious: Count the niches.

This is not as easy as it sounds, however, for the main stairway, although part of the original design, is built right over some of the niches on the pyramid's east side. With a team of "niche counters" drawn from the group of Colgate University students accompanying Tony Aveni's field expedition, I began a tally. We had to measure the width of the stairs and figure out how many niches were hidden, on each level, behind them, and we had to double check each other's counts—even on the sides unobstructed by stairs. In the muggy heat and high brush it is easy to slip a niche.

Everyone persevered, however, and the compartments were counted, even the unseen ones hidden by the stairway. By the time we had climbed and worked our way around the ruined sides and back of the temple on the top level, the total was 363. Unfortunately the front of the temple is gone, but it is likely that the front door was flanked by niche 364 on one side and 365 on the other. The length of the front wall foundation suggests this, and 363 seems too close to 365 to be coincidental.

While we were on top of the Pyramid of the Niches, our count completed, the *voladores*—or "fliers"—arrived and readied the 100-foot pole in the plaza in front of us for an acrobatic feat that would carry the calendrical symbolism even further. Although the original ritual had religious significance, today the voladores put on their show at El Tajín for the benefit of the tourists.

The Totonac voladores dress in bright red and white costumes, richly embroidered and decorated with yellow fringe. Their hats are conical and covered with mirrors, flowers, and sewn designs, and topped by a multicolor fan. The voladores represent ceremonial birds, and four of them "fly" from a square frame on top of the pole all the way to the ground, dangerously far below. A fifth volador plays music during the entire aerial ballet.

After all five men have climbed to the top of the pole, the musician plays a melody on a five-tone pipe and beats a rhythm on a small, two-headed drum. He faces each cardinal direction in turn, repeating the tune and stamping his foot, as if to punctuate the segment, and returns sunwise—or clockwise—to the east, the direction that started this cycle of the song. At the appropriate time the other four voladores push off and descend in a spiral path around the pole. Dangling by ropes tied to their ankles, they circle and glide lower as the ropes unwind. Up on top, the fifth volador takes an equal risk as he continues to play while standing on a small revolving platform. The

Suspended from a pole 100 feet high, the four voladores slowly circle 13 times until they reach the ground. At the top, a fifth participant plays a drum and pipe to provide a symbolic musical accompaniment. *(E. C. Krupp)*

"dance" is paced to reach the ground just as all four voladores complete their thirteenth turn. Here both numbers—four fliers and 13 turns—have important calendrical meaning.

There are 20 day-names in the 260-day count, but only four of them—"reed," "flint knife," "house," and "rabbit"—can fall on the first day of the 365-day year. This follows because 20 divides into 365 eighteen times, with five remaining. The remainder of five days then fits back into 20 four times. This means that the last day of the year, and therefore the first day of the year, must jump five names from one year to the next. As a result, the same four names keep showing up. These names are known as the year-bearers because they name the year, and they are associated with ceremonial birds.

Each year-bearer is repeated 13 times in the 52-year cycle. Now, even though most of the original meaning of the voladores' ritual flight is lost, we can still recognize the aerialists as the four year-bearers who bundle the years in their 13 turns through the

The Pyramid of the Niches at El Tajín seems to carry calendric significance in its compartmented façades. The pole nearby is used by the voladores in a ritual also based upon calendrical cycles. *(E. C. Krupp)*

sky. Appropriately, they continue this relic of a once-living calendric tradition in front of another relic of the old ways, a pyramid that puts all the days of the year on display in its 365 niches.

Resurrecting the Soul

Although people in traditional cultures take part in the renewal of cosmic order through seasonal ceremonies and other calendrical rituals that are locked in step with intervals of cyclic time, their ultimate participation is the journey of their souls. When, at death, they imitated the setting of the sun, the waning of the moon, the end of the year, or the completion of a cosmic age, they also prepared for the next life that must follow like the dawn, the moon's first crescent, the new year, and the next era. Ceremonies that called upon celestial imagery were devised, then, to revive the dead and provide them with new life. For example, the symbols of ancient Egypt's funeral ceremony, "the Opening of the Mouth," though not seasonal, still were borrowed from the sky.

"The Opening of the Mouth" ceremony was intended to restore the *ka,* or personality, to the body for existence in the afterlife. A touch on the eyelids and a touch on the mouth with an adze-like ritual instrument reanimated the senses. This adze was called by the same name as the word for the constellation of the Big Dipper. To the

Egyptians these stars represented a celestial bull (not Taurus) or just its leg or thigh, and they were portrayed as part of the bull's body, stylized as a haunch or an actual leg. The name of the constellation was Meskhetiu, which identifies it as the leg of Set, the Adversary of Osiris. Tethered to the sky's "mooring post," the circumpolar leg of Set swings around the celestial pole.

The ceremonial adze even looks like a dipper, as it appears, for example, in the painting on the north wall of the celebrated tomb of Tutankhamun. Tutankhamun's successor, Ay, dressed in the leopard skin cloak of the *sem*-priest, is portrayed here using it to perform the Opening of the Mouth on Tutankhamun. Between Ay and

On the north wall of the burial chamber in the tomb of Tutankhamun, Ay, the successor of the boy king, carries out the Opening of the Mouth ceremony upon the mummy of Tutankhamun. The instrument held out by Ay, a similar adze on the table, and the leg of meat with it all allude to the stars of Meskhetiu—the constellation we know as the Big Dipper. *(Griffith Observatory, Lois Cohen)*

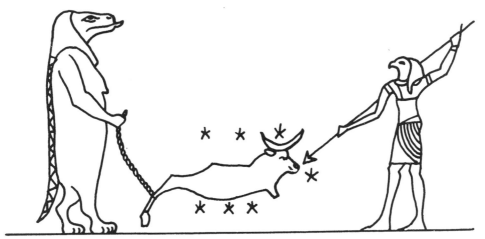

Meskhetiu, stylized as a leg of beef with the head of a bull at the joint, is tethered to the north pole of the sky (symbolized by the hippopotamus) and challenged by hawk-headed Horus, the son of Osiris. Surrounding the leg of Set are the seven stars of the Big Dipper that comprise it. *(Griffith Observatory, after Heinrich Brugsch)*

Tutankhamun, a bull's leg, another adze, and more funeral paraphernalia sit upon a small table.

In charge of the ceremony, the *sem*-priest impersonated Horus, the son and successor of Osiris. By spiritually reviving the deceased, the priest in effect reenacts the victory of Horus in his battle with Set and the reestablishment of the authority of Osiris. This authority was, of course, transferred to Horus just as Tutankhamun's authority was handed over to Ay. In the myth of the Triumph of Horus, Horus lost an eye. A judgment by the gods returned it to him, however, and the "Eye" became one of the trophies of the contest. Set's leg also became such a trophy and is sometimes called the "Eye" as a result. This symbol of the victory—and therefore of resurrection—also turns up as the actual leg of a bull that is slaughtered as part of the ceremony. This leg, like the adze, is used to give new life to the dead.

The leg of the slaughtered bull, the dipper-like adze, and the Big Dipper in the sky all represent the same thing: the symbolic renewal of life. But all three of them are associated with Set, the personification of chaos. Because chaos was defeated by Horus, these emblems of Set belong to the new order. They are powerful symbols to be used in what is a parallel renewal-resurrection, itself a victory over chaos. The Big Dipper was already associated with eternal life because its stars are circumpolar. They were the undying, imperishable stars. In death the king ascended to their circumpolar realm, and there he preserved the cosmic order.

9

The Space We Enclose

For the last eight decades, and probably in the nineteenth century, too, a procession of white-robed "Druids" has passed through a country turnstile in the last hour before summer solstice sunrise and entered the rings of Stonehenge. The participants assemble there, among the massive upright stones of a prehistoric monument about ten miles north of the cathedral town of Salisbury, England. They are prepared to perform a ceremony, to witness the sunrise, and to celebrate the summer solstice.

From the center of the main ring of stones, or Sarsen Circle, the year's most northerly sunrise appears within one of the circle's arches, on the northeast side. As the sun continues to rise, it also moves slightly south, and when its entire disk clears the horizon, more than half of the sun protrudes above the tip of a tall and massive outlier of Stonehenge, the Heel Stone, 256 feet from the center. This is the event the "Druids" have come to watch. By greeting the dawn with ritual, recitation, and music, these pilgrims believe they are perpetuating a tradition and a priesthood that reaches back to Britain's Iron Age and the Celtic peoples who populated it.

There really were Druids in prehistoric Britain, and although Greek and Roman accounts of them are relatively sparse, we do have some understanding of their role in prehistoric Celtic society. They were priests, poets, and seers. Julius Caesar mentioned their knowledge of celestial phenomena and expertise with calendrics. We can nearly imagine them conducting their sacrifices in the first light of the summer solstice sun—they seem to belong at Stonehenge. The Druids did not, however, build this monument. It was designed and constructed, in its earliest phase, around 2800 B.C., nearly 2,000 years before the Celts arrived in Britain. Even the last major phase of Stonehenge, which includes the Sarsen Circle and the five freestanding archways,

The view from the center of the Sarsen Circle along Stonehenge's main axis and through its "summer solstice" arch frames the Heel Stone in a position offset to the right of the main axis. *(E. C. Krupp)*

or *trilithons,* inside the ring, was completed about 1550 B.C.—still too early for the Druids.

The "Druids" of Stonehenge are a seventeenth-century romanticized vision of the past. In attempting to identify the builders of Stonehenge, an antiquary, John Aubrey, had cautiously suggested in *Monumenta Britannica: A Miscellanie of British Antiquities* (1666), that the stone rings of Britain were Druid temples. For Aubrey's time, this was not an outrageous notion. Aubrey did not have the benefit of our perspective on British prehistory; he took the best information available to him: details on the monuments he himself had discovered through thoroughly respectable field work and the oldest references to the inhabitants of pre-Roman Britain he could find.

Aubrey's manuscript remained unpublished until 1980, but a transcription of it was available, in the eighteenth century, to another, far less disciplined interpreter of Stonehenge, the antiquary William Stukeley. Stukeley embraced Aubrey's idea with enthusiasm and inflated it far beyond Aubrey's intent and—for that matter—the existing evidence. In *Stonehenge, a Temple restor'd to the British Druids,* published in 1740, Stukeley really started the Druid Revival. With far more imagination than fact, Stukeley called Stonehenge "the metropolitical Church of the chief Druid of Britain." He peopled Stonehenge with great congregations, assembled there at the times of the

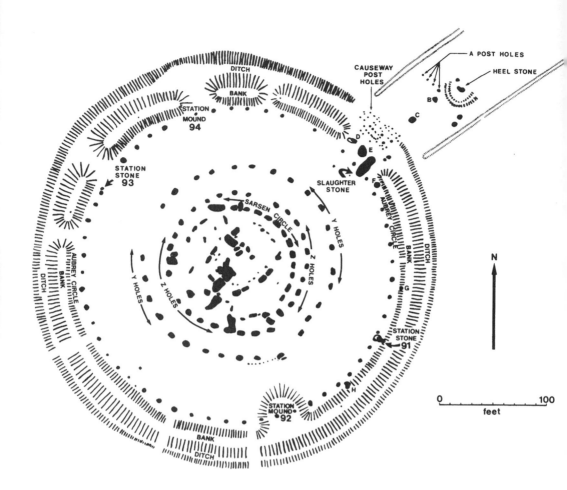

Stonehenge is several monuments, each built upon the relics of an earlier vision of enclosed and sacred space. In the first era, this space was enclosed by the circular earthen bank and ditch. In the last phase, the linteled circle of sarsen uprights and a horseshoe arrangement of trilithons helped define its major axis. Recently, the discovery of a pit near the Heel Stone, once occupied by a large stone itself, has complicated interpretation of Stonehenge astronomy. *(Griffith Observatory)*

year's important seasonal festivals. With considerable care he reconstructed rites and sacrifices no one had ever seen, let alone described. The popularity of Stukeley's book made Druids and Stonehenge almost synonymous, and since the eighteenth century several neo-Druidic orders, invented by those who believed they were only reviving Britain's ancient, pure, and true religion, have carried on at Stonehenge—especially at the summer solstice.

How Stonehenge was really used at the summer solstice by its prehistoric builders is still unclear. But this hardly seems to matter. In 1980, 2,000 people waited outside the barriers for the midsummer sunrise as the Druids stood their watch again. Whether the original builders of Stonehenge did this kind of thing is not what is important here. The lesson today is that the solstice and a ritual of cyclic time still speak to us. People seek and occupy the special ground—the sacred space—where cosmic order is revealed.

Sacred space is a realm where what we sense to be the basic organization and meaning of the universe is experienced and celebrated. Order and structure: when our brains perceive these, the world is sensible, manageable, and meaningful. It was the essence of the way our ancestors interacted with the world, and so it was a religious experience. They dealt with it in temples—places designed just for that purpose.

Even though we are removed by time and technology from the traditions of our prehistoric ancestors, we embrace with enthusiasm a romantic vision of Stonehenge astronomer-priests keeping a solstice vigil. We still respond to the monument as a place where we can encounter a visible sign of celestial order. We still want to affirm the pattern of cyclic time. Our response to Stonehenge folklore makes it clear that we still feel some kind of ceremony ought to take place on this special ground, in this special space.

Stonehenge and the Sky

Astronomical alignments funnel celestial order into prehistoric monuments and turn them into sacred space. Which alignments were intentional, what they meant in detail, and how they were actually used remain controversial questions, however. Several astronomical interpretations of Stonehenge have been proposed since Stukeley's day. Sir J. Norman Lockyer, the British astronomer who discovered helium in the sun, carried out the first systematic and scientific study of the monument's alignments and reported some preliminary results in 1901. Five years later, his ideas appeared in *Stonehenge and Other British Stone Monuments Astronomically Considered.* Lockyer tried to relate the alignments he measured, between certain stones and along the axis, to the summer solstice and to the four traditional holidays in the old Celtic calendar. He did not regard Stonehenge as an astronomical observatory or as some sort of a megalithic calendar. To him, the alignments were symbolic and suggested ritual use. He thought of Stonehenge as a temple and tried to prove its last features were built upon the remains of a much older temple dedicated primarily to worship at Beltine, the pagan holiday now saluted as May Day, and at other calendrical midpoints that separate the solstices and the equinoxes. Although Lockyer's approach was severely limited, and his alignments sometimes incorrect, he pioneered the study of astronomical orientation.

In 1963, many decades after Lockyer's efforts, a more thorough attempt to understand Stonehenge astronomy involved the moon in the monument's alignments as well as the sun. Dr. Gerald S. Hawkins, an astronomer and a naturalized American, analyzed the alignments between the features of individual phases of Stonehenge's development on an electronic computer and discovered that a virtually complete "set" of solar and lunar horizon extremes were marked. C. A. Newham, an amateur astronomer in England working independently, had surveyed some of the alignments himself and had reached a similar conclusion several months earlier. What was significant here was alignment on moonrises and moonsets—at the inner and outer monthly extremes—during the 18.6 year cycle. This implied not only a familiarity with astronomical cycles longer than a year but also the existence of specialists who kept track of them. Newham interpreted a set of posthole rows out by the Heel Stone as an actual record of the annual northernmost moonrise, and Hawkins developed a scheme for eclipse prediction involving the 56 pits, or Aubrey Holes, that form a circle 284½ feet in diameter, just inside the earthen bank that encircles Stonehenge. The significance of these and other discoveries was hotly debated by archaeologists, astronomers, and anyone else game enough to enter the fray. Stonehenge was identified, at turns, as a precise astronomical observatory, a computer of astronomical phenomena, a symbolic set of alignments built to commemorate an earlier observatory, a digest in stone of the astronomical knowledge of the day, and an astronomically aligned tribal center where bronze age warrior chiefs invoked heaven to buttress their dynasties.

Gerald S. Hawkins was able to demonstrate that many features of Stonehenge I and Stonehenge II aligned with astronomical events on the horizon. Here, his sun lines and moon lines for the first phase of Stonehenge indicate the solstices and the maximum and minimum monthly excursions, or "standstills," of the moon.   *(Griffith Observatory)*

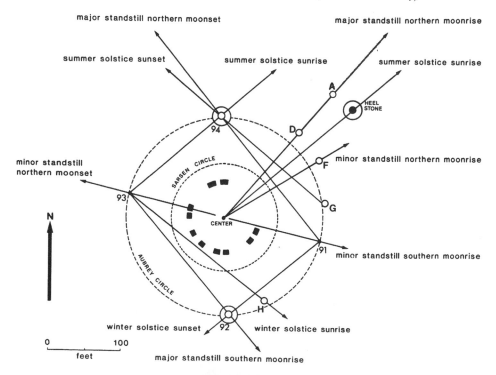

Today Gerald Hawkins and most other interested researchers emphasize the existence of an astronomical plan at Stonehenge. This plan may not reveal the purpose of Stonehenge, but it confirms the importance of the celestial phenomena to the people who built it.

The Rest of Stonehenge

There is more to Stonehenge than alignments, however, and the other things we find there shed some light on its purpose. Symmetry and enclosed space seem to have been part of the architecture of Stonehenge in all of its phases. Stonehenge was carefully set out to a well-considered plan. The Sarsen Circle, part of the third phase of construction, consists of uprights that support sandstone lintels that weigh 6 to 7 tons apiece. Originally 30 lintels completed a level circle suspended more than 13½ feet in the air, and although only a fraction remains in place today, it is obvious that all of them were worked into the curved sections of a circle. It would have been much simpler to cut the stones into rectangular blocks, but only a crude, unfitted ring of straight-sided lintels would have been the result.

The builders of Stonehenge had a more elegant sense of design. Professor Richard J. C. Atkinson, a British archaeologist and the world's foremost expert on Stonehenge, documented many other fine details of the layout and construction of Stonehenge as well. What looks at first glance like a rather rudimentary exercise in megalithic architecture becomes upon close examination an ambitious and sophisticated achievement of early bronze age technology.

When engineer and mathematician Alexander Thom surveyed the Sarsen Circle in 1973, the results led him to conclude that its dimensions and the spacing of its uprights were measured out in a prehistoric unit he has called the "megalithic yard." With a length of 2.72 feet, the megalithic yard is just about 8½ inches short of a standard English yard, but it compares in length with certain traditional units of European measure—the Spanish *vara,* for example. Most archaeologists regard Thom's megalithic units of measure skeptically, but whether we accept the megalithic yard or not, Thom's analysis of Stonehenge demonstrates that a conscious sense of geometry and measurement determined the design of Stonehenge. This shows up in the size and spacing of the Sarsen Circle uprights. Because the inner surfaces of these uprights were worked smooth by the builders, they define the circle that the stones fit rather clearly. Atkinson was able to show, in fact, that these dressed and polished faces missed the theoretical circle they all fit best by less than 3 inches on the average. This is remarkable when we consider that the inner diameter of the ring is nearly 97½ feet. From center to center, the uprights are spaced uniformly around the ring, and all but one of the remaining uprights share comparable thickness and width.

Although the purpose of the astronomy, geometry, and dimensions at Stonehenge remains obscure, there are other facts to guide us to the monument's meaning. Excavation of some of the 56 Aubrey Holes showed that they had been only 2 to 4 feet deep and held deposits of cremated human bone, bits of burnt wood, bone pins, and flint artifacts. The holes were filled up not long after they were dug, and some of them seem to have been dug up and refilled again. This suggests that some sort of ritual burial or consecration took place at Stonehenge. Other cremation burials were found by Colonel W. Hawley when he excavated the bank and the ditch outside of it, in the early 1920s. It is not the burials themselves that surprise us, for the area around Stonehenge is a vast bronze age cemetery. Barrows, or earthen mounds, from many prehistoric periods may be found there, and they contain human remains and the goods buried with their occupants. But the Aubrey Circle, Bank, and Ditch are all considerably older and belong to Stonehenge I, the earliest phase. And unlike the barrows, at least some of the burials at Stonehenge seem related to the actual features of the monuments—and therefore to its use, a use that may have continued into the period of Stonehenge II, when the first elaborate rings were erected. A piece of charcoal from one of the secondary Aubrey Hole deposits provided the radiocarbon date of 2240 B.C. These earlier, "bluestone" rings, made with smaller stones of spotted dolerite native to the Preseli Mountains in Wales, were later dismantled when the sarsen stone structure was put up as Stonehenge III.

Even the earlier arrangement, Stonehenge II, incorporated orientation and conscious design. The earthen Avenue that leads northeast from the Bank's entrance was added at this time. It and the axis of the incomplete bluestone rings were oriented southwest-northeast and may have been targeted on the summer solstice sunrise. Outside the Sarsen Circle, two mounds and two stones have long been recognized to form a rectangle with right-angle corners accurate to about 1 degree. These are the Stations of Stonehenge, and alignments among them Hawkins and Newham demonstrated as astronomical. Until 1978, however, the exact position of one of them, the north mound on the west side, was not known. After excavation pinpointed its location, the rectangle's alignments were remeasured. They are still roughly consistent with the proposed horizon phenomena, but there are too many uncertainties to conclude anything about their precision and use.

Colonel Hawley found an undatable but probably prehistoric burial inside the Sarsen Circle and on the solstice axis of the trilithon horseshoe. Like the Aubrey Hole and Ditch cremation deposits, it links death with whatever else may be present at Stonehenge. In 1978, another important burial—belonging to the Beaker People period of Stonehenge II—was discovered in the Ditch a little more than 30 feet west of the northeast Entrance Gap. The bones belonged to a young man, interred with three flint arrowheads and an archer's wristguard made of stone.

Stonehenge, then, is a confusing mixture of monumental architecture, astronomical alignment, geometric design, and burial through all of its phases of development. It is pointless to interpret Stonehenge without consideration of all aspects of the monument. Even its highly formalized architecture hints that something more than astronomical observation was going on here—the burials mark it as a very curious observatory. We don't know if these burials represent human sacrifices or natural deaths, but in either case they tell us that Stonehenge was sacred ground. Even if some of the burials belong to a later period and have no ritual function, their presence at Stonehenge suggests the site's special character was recognized and used for burials thousands of years later.

What the Rings Contain

Astronomical alignments are found in other prehistoric rings besides Stonehenge. The way they fit into the architecture of these other monuments helps to confirm the relationship between celestial alignment and sacred space already noted. Woodhenge, a monument built about 2300 B.C., precedes by a century or two the rearrangement of Stonehenge that led to the solstice avenue and the bluestone rings. We have already learned that the main axis of Woodhenge's forest of standing posts was also aligned with summer solstice sunrise. The posts may have been freestanding or, possibly, they supported a roof, but in any case Alexander Thom reports that the six rings of postholes that remain were measured out in megalithic yards. They are not true circles, but their circumferences increase systematically: 40, 60, 80, 100, 140, and 160 megalithic yards respectively. Only the absence of a ring with a 120-megalithic-yard perimeter breaks the pattern.

According to Thom, the shape of these rings originates in fairly simple principles of geometric construction. Circular arcs are struck from the corners of two adjacent right triangles—also set out in megalithic yards. One needn't believe in the megalithic yard to see the well-ordered, proportional increase in the size of six rings drawn to the same pattern, but it shows us the overall design in the monument and implies some system of standard measure was involved.

Just as at Stonehenge, evidence of ritual activity is found at Woodhenge. A three-year-old child was killed by a blow to the skull and buried near the center and on the solstitially aligned axis of the rings. Chalk ax heads, too soft to be of any practical use, were retrieved from Woodhenge postholes. Symbolic axes are known from other sites, including Stonehenge, and imply, as do the burials, some act of consecration or ceremony.

We have already encountered astronomical alignments at two of the oldest stone rings: Castle Rigg, and Long Meg and Her Daughters. In the first case stones on opposite sides of the ring established the lines of sight. In the second ring, an outlier—Long Meg—provides the celestial orientation. Neither setup works very well as an actual observatory: Castle Rigg's stones are large, irregular boulders and don't operate as astronomical instruments.

As an outlier, Long Meg gives us an accurate line to the winter solstice sunset. Long Meg, however, is too large and too near any place an observer might stand to

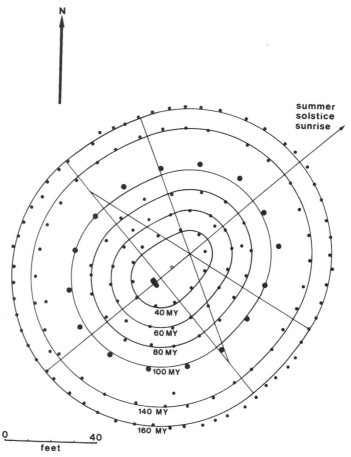

N

summer
solstice
sunrise

40 MY
60 MY
80 MY
100 MY

140 MY
160 MY

0 — 40
feet

Woodhenge has an axis of symmetry that is shared by six concentric egg-shaped rings of postholes, and this axis points to the summer solstice sunrise, a key date in many traditional calendars. The units noted for each ring give the circumference in the megalithic yard of Alexander Thom. *(Griffith Observatory, after A. Thom)*

be used as a precise foresight. Furthermore, the ground slopes upward in the direction of Long Meg. This makes the horizon seem close, the profile, in fact, of the low hill on which the ring was built. Despite the accuracy of the alignment, the arrangement is unsuited to astronomical observation.

Professor Thom began surveying megalithic rings in 1938 and noticed that some rings—Castle Rigg and Long Meg and her Daughters among them—were not truly circular. One side looks flattened compared to the rest of the ring. The exact degree of flattening can vary, but Thom found he could reproduce these shapes with a few general rules of geometric construction.

We may not understand why the builders decided to flatten one side of a ring, but their action does tell us the shape of a ring was taken seriously, for flattening a side

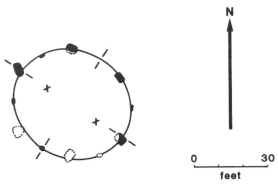

A careful survey of Machrie Moor's Circle #1 by A. E. Roy revealed it to be an ellipse, with major and minor axes and north-south and east-west lines established by stones in the ring itself. *(Griffith Observatory)*

gives the ring a single axis of symmetry. Another geometric figure, the ellipse, also has a principal axis of symmetry and obeys specific rules of construction. The first prehistoric ring recognized as an ellipse was noted by Scottish astronomer A. E. Roy, on the island of Arran. Its details of symmetry, orientation with the cardinal directions, and geometric shape would not be a part of a modest stone ring if they had not meant something special to its builders.

The stones in Circle #1 on the Isle of Arran's Machrie Moor alternate between large granite blocks and smaller pieces of sandstone. This view coincides with a north-south line defined by two of the ring's granite boulders. *(Robin Rector Krupp)*

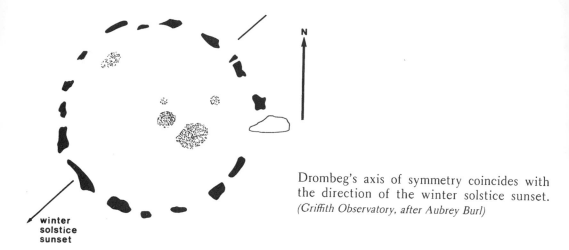

Drombeg's axis of symmetry coincides with the direction of the winter solstice sunset. *(Griffith Observatory, after Aubrey Burl)*

winter
solstice
sunset

Stone rings are found throughout the British Isles, and though they differ in detail from one region to another, the local architectural conventions each accomplish the same thing. They enclose space, set it apart, and make it special. Drombeg is a typical circle in the general regional style of Cork and Kerry, in southwest Ireland, and many of its features help define sacred space. The ring is attributed to the Early Bronze Age. Burned bone, bits of charcoal, and pottery fragments were recovered from one of the pits inside the ring. The builders leveled the enclosed area with care and put down a floor of gravel 3 to 4 inches thick. They contrived an entrance on the northeast by making portals out of the two tallest stones. The rest of the stones diminish in

The main axis of Drombeg, a bronze age stone ring in southwest Ireland, is created by the entrance portals on the northeast and a recumbent, altarlike stone on the southwest. Also, the heights of the stones are symmetrically graded. *(Robin Rector Krupp)*

height about halfway around on each side and then symmetrically grow taller until a massive stone, opposite the entrance, is reached. This stone lies on its side in contrast to the others, all upright.

The builders of Drombeg adjusted the sockets in which several of these uprights were set, in order to make their tops slope upward and draw the eye toward the horizontal, altarlike stone. A line from the entrance and across the center of the recumbent stone aligns with winter solstice sunset. Although this could be a chance orientation unintended by the designers, all of the features of the ring emphasize this axis. It is reasonable to think the astronomical orientation was also part of a plan, although nothing suggests that precise determination of the solstice was desired. The level floor, the graded stones, the altarlike recumbent, the architectural symmetry, the unambiguous axis, the cremation burial, the imprecise solstitial alignment, and the presence of quartz, often found associated with stone circles, create a symbolic alignment and give the place importance. It is an environment of symbols—suitable for sacred ritual because they invoke cosmic order and so sanctify the ground.

Rituals and Rings

Northeast Scotland has its own circles with recumbent stones. British archaeologist Aubrey Burl has inventoried the key traits of these structures, and all his evidence points to a ceremonial purpose and religious significance for the rings. They are astronomically aligned but not in a way we might have guessed. Their axes of symmetry are defined by the line from the center through the recumbent stone's midpoint. This massive horizontal stone is always flanked by a pair of tall uprights, usually the tallest stones in the ring. From the side opposite the recumbent, the standing stones gradually increase in height, and like the pillars that form the rings of southwestern Ireland, they focus attention on the big horizontal stone.

More careful review in 1980 led Burl to the discovery that the range of permitted alignments actually was more closely defined. Fifty rings were still in good enough condition to evaluate orientations reliably. Of these, no axis pointed more than 25 degrees east of south or more than 55 degrees west of south. In fact, the majority with westerly orientations did not exceed 24 degrees west of south, and the few beyond this limit were clustered at the 55-degree extreme. Burl realized that the points 26 degrees either side of south marked, in this region of Scotland, the greatest southern limits of moonrise and moonset, positions the moon reached every 18.6 years, at major standstill. At minor standstill, the minimum limits of moonrises and moonsets are seen, and the point 53 degrees west of south corresponds to the southern minor standstill moonset. The rings have a number of different orientations, but all fall within a fan bounded by the extreme positions of the moon. Because the rings fit so nicely within these well-determined, celestially meaningful borders, it appears certain these stone circles had something to do with the moon.

The Scottish rings were not lunar observatories, however. The alignments themselves offer no means of precise measurement and timing of the moon's position, and too many other features imply ritual use. Burl believes the passage of the moon through the space between the flankers and over the recumbent played some part in a prehistoric ceremony centered in the circle.

Other features of these rings reinforce this idea of ritual use. For instance, at Loanhead of Daviot the ground was carefully leveled and then surfaced with a layer of stone rubble. The recumbent, like those at similar circles, was meticulously engineered into position, its flat upper surface leveled true. After a body or bones were burned in a pyre at the center, the fire-baked surface was cleared and then penetrated with a small ritual deposit: a piece of burnt bone, a piece of broken pottery, and a piece of charcoal. Whatever these fragments meant, they were placed with specific intent. Other pieces of pottery, broken on purpose, were buried beneath small cairns near each of the circle's uprights. Soil, mixed with the burnt bone of adults and fragments of children's skulls, was laid down in the center, and a doughnut-shaped

Aubrey Burl's excavation of the Berrybrae recumbent stone circle revealed the fragmentary remains of a central ring cairn, inside of which three cremations were buried. The cairn was destroyed by a later group, around 1750 B.C., and the cobbles were rearranged by them into the circular wall in which the uprights of the circle seem to be embedded. The recumbent is the horizontal stone on the upper left, and the axis across it is oriented toward the minor standstill southern moonset. *(Aubrey Burl)*

cairn was built around this deposit, nearly to the inner faces of the circle's uprights and about knee high.

The stone just east of the recumbent's east flanker is decorated with five cup marks. These shallow, circular depressions are found on other megalithic monuments, sometimes with more elaborate designs. This cup-marked stone at Loanhead of Daviot completes a line from the center to winter solstice sunrise. Marked stones often are part of the celestial alignments at other megalithic sites.

Loanhead of Daviot belongs to the neolithic period, about 2500 B.C., but later, bronze age inhabitants of this region apparently recognized something consecrated about the circle and built their own cremation cemetery next to it. This late monument is more modest and consists of a low rubble enclosure with elliptical and circular elements in the layout. Thom observed that the axis of this later burial monument points to the summer solstice sunrise. The astronomical target of this alignment may differ from that of the earlier recumbent stone circle, but it still seems to restate the existence of a relationship between celestial alignment, funeral traditions, and sacred space.

The ingredients that make Loanhead of Daviot a site of ritualized astronomy are found in the other Scottish recumbent stone circles, too. Cremated bone and charcoal were recovered from the center of the ring cairn enclosed by the stone circle known as Sunhoney, about 11 miles south of Loanhead of Daviot. Sunhoney points to the minor standstill southern limit of the setting moon. Its recumbent, now fallen, has three cup marks on what was once its inner face. The circle's 11 uprights are shortest on the northeast and grow systematically taller to the flankers, which are tallest of all. All of the uprights are the same kind of stone, but the recumbent is distinctively different.

We have sampled only a few of the prehistoric stone circles of the British Isles. There are more than 900 of them, and they range considerably in age and design. We have no reason to think they are all part of any continuous tradition or systematic evolution, but they do share a number of properties that may help us understand the astronomical alignments sometimes incorporated into them.

The most obvious thing a stone circle does is enclose space. We can tell that this space had special meaning because permanent architecture and formal principles of design are what set it apart. Death, in some symbolic sense, is also a part of the meaning of these monuments, for burials of one sort or another are associated with stone circles. Stonehenge has them from all of its phases.

Most of the burials found at stone rings do not suggest that the chief purpose of these monuments was funereal. Instead, the bashed skulls, the cremated fragments of bone, and the purposeful but enigmatic deposits imply sanctification of the site, a place fortified with a richly symbolic substance: human bone.

The bones we find may have belonged to revered ancestors or to victims of sacrifice. Both make symbolic sense. Honored ancestors really represent a society's own identity. In them repose a people's sense of who they are. By placing these dead in the sacred precincts we affirm our place in the cosmic scheme.

Sacrifices, on the other hand, repeat the key sequence in the cycle of cosmic order: death. However they may be viewed, the dead are a vital component of sacred space because they have access to the transcendental realms of cosmic order. In one way or another, most cultures have linked the journey of the soul after death with the sky, the source of sacred order. The dead, then, operate as symbols of cosmic order and sanctify the places where its ceremonies are observed.

We still use the imagery of the skeleton to summon the concept of mortality. Bone can also imply, however, immortality or resurrection, for bone persists, though the body decays. In shamanism, bone is extremely important. The shaman, in his visions, may be reduced to nothing but his bones, which are then refleshed. His bones are, then, a symbol of death and rebirth. The same imagery of bone and death is also part of the initiation ritual by which a candidate becomes a shaman. The process is a spiritual "rebirth," and to achieve it—and so acquire a direct linkage to the reservoir of sacred power—the shaman-initiate must undergo a spiritual transformation, achieved only through a symbolic "death."

We can guess, with reasonable certainty, that there was a shamanic thread in prehistoric religion. Bone, burial, and sacrifice, in any case, sanctified the stone rings by invoking the fundamental rhythm of life, death, and, presumably, rebirth. This cosmic theme also appears in the astronomical alignments. Emphasis may be on the sun or the moon, on rising or setting, on summer or winter, or on some other celestial metaphor which had to do with the details and purposes of the ceremonies performed at each ring. Any astronomical orientation emphasizes, however, the sacredness of the space by bringing a visible sign of cosmic order into the area enclosed by the stones.

There are stone circles with celestial significance that is less ambiguous than the astronomical alignment of the stone rings of prehistoric Britain. We know that the Druid-like priests of the ancient Thracian people of Romania had knowledge of the heavens and calendrical skill. In Roman times this region, called Dacia, had its capital at Sarmizegetusa, near the town today known as Grădiştea Muncelului. Situated on the broad terrace above the walls of a citadel, Sarmizegetusa's earliest structures date to about 100 B.C., and it was also an important religious center. Of the several sanctuaries and temples, two are circular. The diameter of the larger is 96 feet 5 inches, equal to 100 Roman feet, implying interest in integral measure. Dacian temples were open to the sky. At Sarmizegetusa the large sanctuary's axis points to the winter solstice sunrise, and the solar character of the temple is restated in a circular altar with ten rays—a stone sun.

The cathedrals of medieval Europe were designed in the shape of a cross. Because we know the story that frames Christian belief, we understand the meaning of the cross. But imagine the trappings of Christianity having disappeared long before our

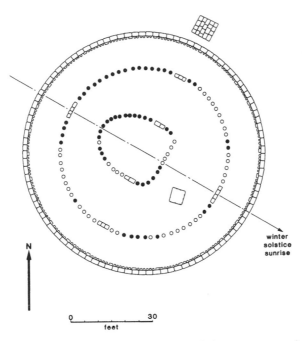

An astronomical alignment is clearly a component of the symmetry of this Thracian temple, in Romania. By bringing the sky's cosmic order down to earth, monumental architecture creates sacred space. *(Griffith Observatory)*

time. The loss would keep us from knowing why its temples were cruciform in design. We would never know, without the account of the miraculous birth, the crucifixion, and the resurrection of Christ, that we were dealing with a symbol which, like a sacrificial burial, spotlights the cycle of cosmic order. Despite that, we could recognize pattern, symmetry, and order in the layouts of the ruins, and we might conclude, with the help of other chance pieces of archaeological evidence, that these sites were sacred places.

The stone rings are not mathematical exercises. They are patterns contrived to restate our sense that the universe has pattern. This is achieved by imposing symbolic and visual order on the space enclosed, and the order may be expressed in measurement, geometry, and astronomical alignment. The steps of measured geometric construction and astronomical alignment are like natural laws. They impose a plan upon the placement of stones, just as the universe seems to require proper placement and timing of the events that give it its shape.

The evidence available to us today tells us that John Aubrey, and even William Stukeley, were in part right. Stonehenge and many of the other stone rings most likely *were* temples. Temples are the places reserved for ceremonies, and in an unsecularized age, ceremonial use can have broad meaning and involve activities that today might be considered political, social, or economic. What transforms a space into a

The gnarled stones of Rollright, a late neolithic or early bronze age site in Oxfordshire, set the space they enclose apart and thereby give it special meaning. We may not be able to revive the prehistoric ceremonies that took place here with accuracy, but we can recognize the stone rings as enclosures of sacred space—ground sanctified by geometric and, in some instances, celestial order. *(E. C. Krupp)*

temple is the significance we attribute to it. We endow it with cosmic order and so make it sacred.

It used to be possible to walk in the sacred space between the great megaliths of Stonehenge. This privilege Stukeley accorded only to the highest Druids, but during most of this century Stonehenge has been open to the public. Now, however, the central part of the monument is once again forbidden territory. The decision was made to protect an old monument from the wear and tear of nearly a million visitors a year. And as much as one dislikes the idea of being barred from the sacred inner precincts, it must be recognized that the old stones have regained their dignity. People no longer scramble over the stones or dart between them. Grass has grown back in the center of the rings. The ground seems sacred once more.

10
The Temples We Align

Temples link the living to heaven. Whether in the underground ceremonial chambers of the American Southwest, the sanctuaries of New Kingdom Egypt, or the halls and platforms of ancient Mexico, the temple's environment restated and emphasized cosmic order. The ceremonies performed there marked the pivots in the passage of cyclic time, those moments of reentry into myth and encounters with the sacred. Even if the temple's inner precincts are reserved for the high priests, the temple is a public building. It belongs to the community, and because its ceremonies organize and focus community life, the temple's cosmic order sustains the social order.

Temple ground may be sanctified by astronomical alignment and cosmological symbolism, as well as by the ceremonies that commemorate the ordered and rhythmic passage of time. Celestial alignment might tempt us to see an astronomical observatory in any monument oriented by the sky, but by recognizing such places as settings in which the sacred would be seen and celebrated, we are better able to understand the intentions of those who built them.

Mimicking the Myth

The Anasazi, prehistoric Indians of the American Southwest, performed their ceremonies in subterranean chambers known as kivas. For them—and for their Pueblo Indian descendants, as well—kiva walls marked the limits of sacred space. Although today the Hopi kivas are rectangular, most prehistoric kivas were circular. Covered by a roof, their usual entrance was a hole in the ceiling, and a wooden ladder provided access to the floor. Ordinary kivas were usually no more than 25 feet in diameter and deep enough to permit those inside to stand upright. Much larger kivas—known as

Pueblo kivas are entered through the roof by ladder. This restored kiva has a circular floor plan inside but is incorporated into the rectilinear design of the main building at Edge of the Cedars State Monument in Blanding, Utah. *(E. C. Krupp)*

great kivas—were more complex and are thought to have been used by an entire community. The smaller kivas belonged to individual clans. The kivas still in use among the Pueblo share a number of traits with the ancient chambers. Pueblo tradition provides, then, a window at least partially open to the past.

Rituals conducted in the kiva and the architectural symbolism of the kiva itself are both related to the Pueblo myth of creation. A key element of this myth is the idea of emergence. According to Hopi tradition, the first people lived deep in the Underworld, but through a series of migrations and emergences they passed from one underground realm to another, until they found themselves outside of the world below, on the surface of the earth, and under the sky. The earth became their new home, the fourth and last world.

Symbolically, the kiva stands for the place and process of emergence. Its roof entrance is the point of emergence, and the ladder is not just a way of getting in and out but a symbolic component of the chamber related to the myth. Inside the kiva, usually on the floor, are found several other hallmarks of this type of structure. A small opening, known as the *sipapu,* represents the place of emergence from the second world to the third, and usually the *sipapu* falls upon the main axis of the kiva. The opening of the ventilator shaft that brings air into the kiva, a firepit, and a screen to

Several interior features of a typical kiva: the fire pit, the deflector, and the ventilation shaft, are preserved in this prehistoric underground chamber in the Pueblo Ruins at Mesa Verde. *(Robin Rector Krupp)*

shield the fire from the current of air also define this line. Usually the axis is north-south. Archaeologist Jonathan Reyman analyzed the placement of niches in kiva walls and found they, too, emphasized the north, the direction in which they are most often found.

The kiva's connections with the Underworld associate it with the idea of creation and birth. Of course, we have already seen how the earth, through these same parallels, is part of the metaphor of the creation of life. Sprouting plants emerge from the ground. As an architectural space, the kiva parallels the idea of the womb, the place from which we are born or emerge into the world. Religious initiation of the young Hopi men is itself a kind of "birth" and so occurs at the Wúwuchim ceremony, when the emergence myth is ritually reenacted.

With the creation myth designed into its architecture, the kiva might be expected to incorporate alignments that reflect cosmic order. We can understand the north-south orientations, then, at least in this general sense. Other expressions of celestial order seem to be part of the layout of great kivas. Only 19 or so of them have been excavated, however, and they are not identical. One of the best known is Casa Rinconada, on the south side of Chaco Canyon, in northwestern New Mexico. It is a short distance—about ½ mile—from two of the major settlements in the Canyon, Chetro Ketl and Pueblo Bonito, famous for its exquisite masonry.

At floor level, Casa Rinconada has a diameter of 63½ feet. Its anterooms, peripheral chambers, bathtub-like floor vaults, and stairway entrances are great kiva features, not usually found in the smaller variety. Like the clan kivas, great kivas have wall niches, and strings of beads and turquoise pendants—possibly for ceremonial use—were found in the ten wall niches of a great kiva at Chetro Ketl. Around the walls of Casa Rinconada there are 28 regularly spaced compartments, and these are accompanied by six larger niches that are spaced irregularly, two on the east side and four on the west.

The function of these kiva niches is uncertain, for it seems possible they were sealed up. A coat of plaster could have made them invisible unless they were marked by wall paintings. This is an interesting problem because a dramatic astronomical event involving a window and a niche in the walls of Casa Rinconada has been observed and reported by astronomer Dr. Ray Williamson and his associates, Howard J. Fisher and Donnel O'Flynn. The walls of Casa Rinconada were built higher than the surrounding ground level, and so it is possible for sunlight to penetrate into the kiva. When it does so, it falls squarely upon one of the six "irregular" niches. The window on the northeast arc is quite narrow and from the niche just frames the sun, which can appear there about 15 minutes after sunrise during a four- to five-day period centered on the summer solstice. Summer solstice is ceremonially observed among the Pueblo, and it would not be surprising if their Anasazi ancestors did the same. There is no clear evidence for other alignments, but an irregular niche on the southeast is well positioned for summer solstice sunset. Unfortunately the upper part

Huge kivas were built for wider community use by the prehistoric Anasazi of the American Southwest. This great kiva is in Chaco Canyon, New Mexico, and is known as Casa Rinconada. On the morning of the summer solstice, the sun shines through the window in the wall at the far right, and its light falls upon the seventh niche to the left of the T-shaped entrance on the north. This niche is lower on the wall than the others adjacent to it. While it and five other niches are spaced irregularly around the kiva's circular wall, the other 28 niches are carefully and evenly placed around the perimeter. *(Robin Rector Krupp)*

of the west wall was in ruin when Casa Rinconada was excavated, and we don't know if a window would have permitted the setting sun to shine in the room.

Casa Rinconada's roof is missing. Its firepit is cold, and its chamber is empty. But the sun in the northeast window and its light upon the wall niche seem to put ceremonial life back into the kiva. Great kivas are thought to have been major assembly areas where important members of the larger community would gather for special meetings and ceremonies, and it is reasonable to imagine the dramatic effect of summer solstice sunrise played some part in the ritual at Casa Rinconada.

Casa Rinconada's general orientation was also carefully planned. Its main axis is well aligned with cardinal north-south (this is also true for the larger great kiva over at Pueblo Bonito), and each pair of Casa Rinconada's opposite "regular" niches defines a line that passes across the kiva. All but one of these go through the same point—the center—underlining how much care went into the kiva's design and construction. One pair of these niches establishes an exact east-west line. Through these various features, the plan of the kiva is tied to the order of space and the direction of time.

Four large posts, each about 2 feet in diameter, once held up the massive roof of Casa Rinconada. The sockets for these supports can still be seen there and in other great kivas as well. Usually the sockets form the corners of a square, but even when the figure is slightly rectangular, it still emphasizes the idea of cardinal directions. The square's diagonals suggest the intercardinals. At Aztec Ruins, also in New Mexico and about 65 miles north of Chaco Canyon, the great kiva has been restored, roof and all. Its four roof supports were rectangular and composed of alternating courses of masonry and cut logs. An opening in the roof was positioned over the firepit. We know by now that these various features of a kiva were both symbolic and functional. The Acoma, another Pueblo people, maintained that the four roof supports in the "first kiva" of the world were four trees that grew from the Underworld. Here, then, the idea of world structure and the roof's physical support merge into the components of a ceremonial room. The cardinal and intercardinal directions are embodied in the roof supports, while zenith and nadir are represented in the roof opening and *sipapu*, when present. Kivas are round, according to the Acoma, because the world's rim, or horizon, is round. As blueprints for sacred space, the great kivas are organized in terms of the Pueblo concept of the universe.

Lodging in the Universe

In a very similar way, the earth lodge of the Skidi Pawnee (who dwelt on the Great Plains of Nebraska) was also a mirror of the cosmos. Much of their religion involved the sky and the stars, and reports from Pawnee informants supply us with the meaning of at least some parts of the earth lodges. Their ceremonial sacrifice of a young woman captive to the "Morning Star," intended to renew the land's fertility, has stimulated considerable speculation over the astronomical identity of the Morning Star.

The Pawnee lived in earth lodges, but ceremonial dances, feasts, and other gatherings also took place there, and the designated lodge was specially prepared for whatever event might be scheduled. A typical earth lodge was circular and might have a diameter of about 40 feet. In the center of the room the ceiling was about 15 feet above the firepit in the floor. Directly above it the ceiling and roof, penetrated by a circular smoke hole, opened the chamber to the sky. There was a tunnellike entry at cardinal east, and opposite this entrance, on the west side of the room, the Pawnee had an altar. A bison skull and other ritual objects might be placed upon it, and above it, on the wall, they hung a bundle containing sacred objects.

Walls were built of sod and faced on the inside with poles. Two circles of upright posts, near the wall and in the center, provided support for the roof beams, rafters, and a covering of saplings, grass, thatch, and soil. Four of the posts in the central circle were sometimes specially decorated, each with a band of a particular color. The four special posts formed a square and from the center firepit marked the intercardinal directions. Each of these directions was associated—in the Pawnee vision of the world—with a particular tree, a particular animal, a particular phenomenon of weather, a particular season, a particular time of life, a particular star, and a particular color: northwest, yellow; southwest, white; northeast, black; and southeast, red. These are the same colors on the four specially marked central posts.

The Skidi Pawnee endowed four stars with special status as the four pillars of heaven. These were the four world quarter stars whose hands supported the sky while their feet were firmly planted on the ground. In terms of their Pawnee names we know them as Yellow Star, White Star, Big Black Meteoric Star, and Red Star. These stars have the same colors as the intercardinal directions and are the same as those directions. Here again the structural supports of the world—the pillars of heaven—are the directions which organize the world. It is even more interesting that these directions are explicitly linked to the stars. Von Del Chamberlain, an astronomical researcher at the National Air and Space Museum and an expert in Pawnee sky lore, has plausibly identified the four world quarter stars:

Yellow Star	Capella
White Star	Sirius
Big Black Meteoric Star	Vega
Red Star	Antares

All four of these are conspicuous stars in the night sky, and with the exception of Vega, each appears to the eye to be the color associated with it. The idea of a "black" star is, of course, contradictory, and the origin of this name remains undetermined. At certain times of the year all four of these bright stars can almost be seen "standing," simultaneously, in their respective intercardinal directions. At the Nebraska latitude, 41 degrees north, Sirius sets just a bit before Antares rises, but the four still serve as perfect signals for the Pawnee sacred directions.

Every element of the earth lodge was borrowed from the architecture of the cosmos. In fact, the Pawnee say that the floor of their lodge is the earth itself, and the ceiling is the sky. The "four pillars of heaven" literally hold up the sky-roof, and the top of the sky—the zenith—is the smoke hole. Corona Borealis, the constellation of the Northern Crown, was called the "Council of Chiefs" by the Skidi Pawnee and associated with the smoke hole and the zenith. It would not transit directly overhead, but according to their tradition, their original homeland was farther south. There Corona Borealis might have crossed the zenith. In the firepit the Pawnee saw a little spark of the sun. The east entrance of the earth lodge was supposed to allow the light of sunrise to fall upon the altar, but this cannot occur throughout the year. The alignment is symbolic, although it may have been connected with ceremonies on or near the equinoxes. In any case, that small portion of sunlight in the central firepit could illuminate the altar each morning after one of the two household heads went outside to examine the predawn stars and then came back in to kindle the fire.

Weaving on the Cosmic Loom

When a Kogi shaman builds a temple, he actually makes a model of his universe. The temple is, in fact, a loom on which the sun weaves the cyclic pattern of time and transforms its structure into organized space. The Kogi are a South American Indian tribe of northern Colombia. Although they have relatively little in the way of material things, anthropologist Gerardo Reichel-Dolmatoff has reported their rich tradition of cosmological symbols and their elegantly woven network of metaphor and myth.

For the Kogi the cosmos is a spindle, a concept that reflects the turning of the sky and the movements of objects in it. Male and female are essential in the cosmic scheme, for through sexuality the creation—new life—becomes possible. The spin-

The Kogi Indians of Colombia describe the cosmos as a spindle. Its nine levels, named above, are divided between heaven and the underworld, and the central disk—the one that corresponds to the earth—is oriented by the intercardinal directions, which embody, symbolically, the sunrise and sunset extremes at the solstices. *(Griffith Observatory, after G. Reichel-Dolmatoff)*

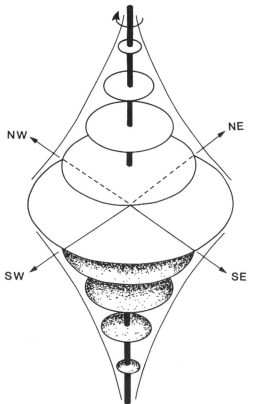

1. XATSÁLNULANG

2. NYUÍNULANG

3. MULKUÁKUKUI

4. MÁMANULANG

5. NÍNULANC

6. HABA SIVALULANG

7. HABA KANÉNULANG

8. HABA KÁNEXAN

9. HABA GUXÁNEXAN

dle's hardwood shaft is, therefore, male. Its disk is made of soft wood and represents the female. The shaft penetrates the whorl, and together they spin the thread which will be used to weave a fabric. On a cosmic scale the spindle of the universe reaches from zenith to nadir. Tapered at both extremes, it has nine levels—each imagined as a disk—comprising the cosmos, and the layer with the largest diameter, in the spindle's middle, corresponds to the earth. These nine layers derive, in part, from the nine months of a human pregnancy.

To construct a temple, the Kogi shaman must first select an appropriate spot, which will become the center of the building. Once the right place is found, strict rules of measurement and design determine the temple layout. With a rope, knotted to preserve key lengths, the shaman traces out a circle around a stake driven into the center point. The rope and its knots are used to determine the placement of other features of the temple. The height equals the diameter at the base, about 25 feet.

In this activity of the Kogi shaman we can see how standard units of measure might show up in ceremonial architecture. The act of design means imitating the

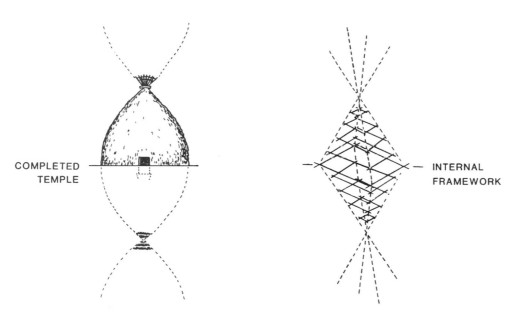

COMPLETED TEMPLE   INTERNAL FRAMEWORK

The Kogi imagine their temples to be patterned after the spindle-shaped universe they believe they inhabit. Although they build only the conical, above-ground section detailed on the left, they think of their temple as continuing in mirrored segments toward both the zenith and the nadir. These are counterparts of the celestial and underworld realms, and the floor they actually build corresponds, quite naturally, to the earth. The temple's internal framework also reflects their sense of cosmic construction, for each of the four-sided frames matches an upper realm of the universe. *(Griffith Observatory, after G. Reichel-Dolmatoff)* cosmos. Since the universe seems to be organized, measured, proportioned, oriented, and shaped, sacred architecture must echo the design and behavior of the universe.

Once a circle establishes the temple's outer wall, large posts are placed on it at the intercardinal directions. These four wall posts form a square, but the wall is built of smaller posts set along the circle. Pairs of larger posts mark the north and south points on the ring, and entrances are placed at cardinal east and cardinal west. As the walls are built upward, four successively smaller four-sided frames support the vault overhead. Each frame stands for one of the four levels of heaven. On the temple floor are four fireplaces also positioned as the corners of a square at the intercardinal points.

When the June solstice arrives, the sun travels north of the zenith, and at about 9 A.M. shines through a hole in the very top of the temple. Normally this opening is covered with a piece of pottery. But on the solstice, sunlight is allowed to enter the room, and the first light strikes the southwest fireplace. As the day continues, the beam traces out a line to the east and stops in the southeast fireplace. At that time, about 3 o'clock in the afternoon, the sunlight can no longer enter the roof hole, and the spot of light fades.

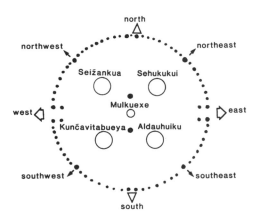

KOGI TEMPLE
PLAN

The floor plan of a Kogi temple is organized around four fireplaces (Seizankua, Sehukukui, Kuncavitabueya, and Aldauhuiku), which fall upon intercardinal lines drawn from the center (Mulkuexe) and are equated with the solstitial extremes. Light is allowed to enter through an opening at the top of the temple, and at the June solstice, the first light to enter strikes the fireplace on the southwest. During the day, the beam crosses over to the southeast fireplace, where it stops and disappears. Each day for six months, this horizontal line of sunlight moves a little to the north until, at the December solstice, the beam crosses from the northwest fireplace to the one in the northeast. This pattern is metaphorically described by the Kogi as weaving by the sun. *(Griffith Observatory, after G. Reichel-Dolmatoff)*

If we were to leave the roof hole uncovered and watch each day, we would see this line of sunlight gradually edge north until, on the December solstice, the first beam fell on the northwest fireplace and ended its display on the northeast fireplace. Now the sun is south of the zenith, as far as it will travel, and each day following, the threads of daily sunlight will weave the fabric on the floor back to the south.

Weaving is exactly the metaphor used by the Kogi to describe this hierophany that takes place in their temples. The four fireplaces are the four corners of a loom to which the sun adds a thread. Through the year, as the sun moves north, south, and north again, it is said to spiral about the world spindle. It weaves the thread of life into an orderly fabric of existence, and the cyclical changes of the sun's daily path are transformed into a cloth of light on the temple floor.

Just as the daytime sun weaves in white thread, from west to east, its nighttime alter ego, traveling in the underworld from west to east, weaves a black thread though the year's fabric between each pair of white threads. The female earth is the loom and provides the north-south warp of the cloth. The sun is male, and like the shuttle it penetrates the vertical threads of the warp. The shuttle is the "active" component of the process, and the loom is "passive." Accordingly, only Kogi men weave. The women do not participate. The loom itself is—through the imagery of the temple—a picture of the universe. Kogi looms consist of four poles lashed together into a square. Two thinner diagonals cross at the center. The four corners are the intercardinal directions. The Kogi regard the upper horizontal bar of the frame as the path traced by the June solstice sun. The lower bar belongs to December.

Weaving creates an orderly pattern. Whether the "cloth" that is woven is a garment, an individual life, or the universe itself makes no difference. All partake of the cosmic order, and so weaving among the Kogi is not just a practical enterprise: It is a sacred ritual act. For the Kogi, cosmic order originates in the pattern of the movement of the sun. This pattern is perceived in the mind, remembered on the loom, and played out in the temple, where the visible signs of celestial order make the space sacred.

A Temple on the Tropic

Throughout Mexico we find ancient buildings with astronomical alignments built into them. Any attempt to interpret all of these structures as observatories is doomed to failure. The alignments are present, but in most cases the architecture does not provide what would be needed for systematic and precise measurement. We know the ancient peoples of Mesoamerica were competent astronomers and established the periodic behavior of numerous celestial objects with considerable accuracy. But the structures we find oriented upon the sun, or even on Venus, don't seem to be the places where skywatchers engaged in scientific observation of the heavens. Instead, these astronomically aligned plazas, platforms, and doorways are components of ancient ceremonial centers. They are there, at least in part, to express in the language of architecture the sacred meaning of the sky. When a celestial object appeared at the proper time in alignment with a temple's design, the event itself was a revelation of cosmic order. The act of observing it was an immersion into the sacred.

Sometime in the middle of the seventh century after Christ, while the great urban and ceremonial complex of Teotihuacán dominated the political and economic life of central Mexico, someone influenced by the styles and traditions of the complex established a modest center of ceremonial buildings about 400 miles to the northwest. Here, just west of the present town of Chalchihuites, we find the site today known as Alta Vista.

Excavations by Dr. Manuel Gamio in 1908 and the more recent work by Dr. J. C. Kelley, an American archaeologist, certainly demonstrate that Alta Vista was relatively small—especially compared to Teotihuacán. Alta Vista's builders worked in adobe and rock and covered their structures with a coating of protective plaster. One of the most peculiar buildings at Alta Vista is known as the Hall of Columns, although Kelley prefers to think of it as the "Temple of the Sun." Its four walls form a square that encloses an array of 28 columns. These take up so much of the available space inside the temple that it is hard to imagine how the room would have been used. Additional structural support was later achieved by increasing the diameters of at least

A peculiar congregation of tapered pillars crowds the Hall of Columns at Alta Vista. The view is south, from the north corner.   *(Robin Rector Krupp)*

some of the columns—so much so, in fact, that their appearance is squat and grotesque to our eyes.

Kelley noticed that the orientation of the Hall of Columns appeared to be part of a cosmographical plan. The diagonals are oriented north-south and east-west. This puts a corner of the building in each cardinal direction and recalls again the idea of four world quarters—so prevalent in New World systems of space and time. Kelley's survey of pottery found at Alta Vista provided numerous examples of designs and symbols that also invoked this idea of four directions and four realms, and additional architectural evidence supported this emphasis on special directions.

A gallery or narrow hall, adjacent to the southeast wall of the Hall of Columns, bends east and becomes an even narrower passage that snakes and angles for about 80 feet until it reaches a straight channel, cut into the ground and oriented due east.

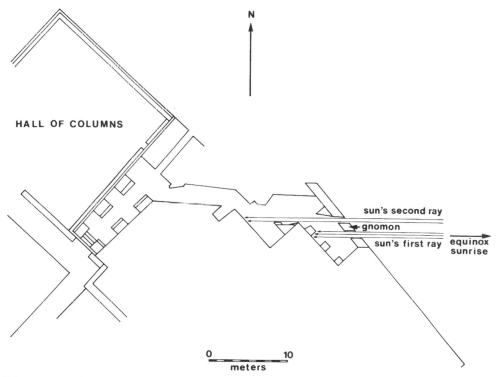

The corners of the Hall of Columns at Alta Vista, situated near the Tropic of Cancer in northern Mexico, point toward the cardinal directions. A strange zigzag passage begins on the southeast side of the Hall of Columns and eventually straightens out to point east and to the equinox sunrise. Archaeologist J. C. Kelley believes the builders of Alta Vista may have monitored the places where the first light fell as the beams passed on either side of a free-standing, diamond-shaped columnar gnomon. From the entrance to this "labyrinth" a pathway continues to the east, oriented upon Picacho Montoso, the horizon foresight for the equinox sunrise. *(Griffith Observatory, after J. C. Kelley)*

Kelley observed the equinox sunrise and found that the sun's first flash of light appears in a notch formed, in part, by the north slope of Picacho Montoso, a conspicuous peak. In silhouette at equinox sunrise this distinctive natural landmark commands even more attention. The pathway is painted in sunlight and becomes, in Kelley's words, "a true Camino del Sol," a road of the sun. The play of various beams of light upon the walls and edges of the odd angular passage may also be significant, but we do not know how to interpret them.

Oddly angled walls create the east end of Alta Vista's labyrinth. The freestanding pillar in the center of the picture is in front of the "Avenue of the Sun," visible here as an excavated trench. On the horizon the distinctive peak of Picacho Montoso marks the point of equinox sunrise. *(E. C. Krupp)*

Only by actually walking the zigzag route from the Hall of Columns to the gate of the "Sun's Road" can one really sense how strange the passage is. Perhaps it was used in some way to make actual measurements of the sun, but the form and purpose of such observations elude us. The architectural peculiarities of the passage may be better explained in terms of ritual and symbolic meaning. One can imagine, for example, a ritual "journey" by an initiate or priest, with stops at various corners and alcoves of the passage, culminating in a view from the east entrance of the equinox sunrise and a walk down the "Camino del Sol" to greet the sun; but we have no evidence to support these ideas.

Although no alignments were observed at Alta Vista on the summer solstice, shadows cast by the high, unroofed walls of the passage disappeared completely at local noon. Alta Vista is now only 2½ miles north of the Tropic of Cancer, and at this latitude the summer solstice sun passes through the zenith, straight overhead. A slow but steady shift in the angle of the earth's axis moves the exact latitude of the Tropic, but even in A.D. 650, when Alta Vista was founded, the Tropic was only about 14 miles north of the site. This is still a small error if we imagine the Alta Vistans were trying to establish the exact location of the place where the sun reached the zenith on the summer solstice. It is not at all farfetched to think this may have been on their minds, for the phenomena that determine this latitude are readily observed, and farther north the sun would never be seen straight overhead.

The only other information we have to assist us is Kelley's discovery of a burial in a crypt beneath the pyramid. Three bodies were placed there, all facing east. From their dress and grave goods, Kelley concluded they were important individuals. Authority and high status probably account for their presence in a special crypt beneath the pyramid. The face of one of the three was covered with a mosaic. It bore a rayed sun emblem, and the pyramid itself had once been ornamented with a similar design. A mosaic with the image of a turkey was also found in the tomb, and Kelley recalled the Huichol Indian tradition that the turkey is a sacred sun bird. There are enough associations to link the architecture at Alta Vista with the sun. As a ceremonial center, consecrated by the high-status burials found there, it most likely served as the stage for major community events, and it seems likely the equinox sunrise was one of them.

A Chamber for the Zenith Sun

Oddly shaped buildings seem to be judged possible observatories because no more obvious explanation is able to cope with their unusual designs. Structure J, a sort of pointer-shaped or "arrowhead" building at Monte Albán, was called the Observatory by the site's excavator, the renowned Mexican archaeologist Alfonso Caso. He thought that a doorway of a tunnel that cut through the "point" of the building opened southwest to allow observation of certain stars. Even though this interpretation is incorrect, Structure J's peculiar shape and its departure from the axial alignment of the rest of the site demand explanation.

Monte Albán was a hilltop center of ritual activity and public life built by the Zapotec of ancient Oaxaca; under its domination, the Zapotec established a sphere of influence of their own and maintained contact with Teotihuacán, about 230 miles northwest. Most of the buildings at Monte Albán are four-sided structures oriented 4 to 8 degrees east of north, but Structure J is noticeably skewed from the rest. Certainly there is something special about it, but the tunnel and southwest doorway do not seem to have anything to do with the sky.

Astronomer Anthony F. Aveni, one of the most active field researchers studying archaeoastronomy in Mesoamerica and Peru, reexamined Structure J's astronomical potential in 1972. Since then, in collaboration with Dr. Horst Hartung, an architect and expert on Mesoamerican architecture, these measurements and interpretation of them have been refined. Now we know that we are prejudiced by the shape of Structure J. As long as we think of it as an "arrowhead," we are inclined to believe it must point somewhere. Because Aveni's survey of the building's orientation shows no precise alignment, the arrow seems puzzling. Aveni did determine that the point was loosely oriented toward the setting positions, in 250 B.C., of the five relatively bright stars of the Southern Cross and two even brighter stars, the conspicuous pair known

Structure J is an unusual arrowhead-shaped building skewed with respect to the others at Monte Albán in Oaxaca. With Structure P, across the plaza in the background, Structure J seems to be part of an architectural complex devoted, in part, to the zenith sun. *(Robin Rector Krupp)*

as Alpha and Beta Centauri. Most of these stars fall within three degrees of the azimuth singled out by the arrow. We know the stars of the Southern Cross were important in ancient Mexico, and Structure J may have been intended to point them out. Even if it did, however, the building hardly qualifies as an observatory. Nothing in the alignment indicates the building itself was used to time carefully the setting of these stars. The alignment, if intentional, seems symbolic.

If instead we think of the shape of Structure J as like the homeplate on a baseball diamond, the point is no longer the front of the design, but the back. In fact, the stairway is on the northeast face of Structure J, the front of "homeplate." This actually is consistent with the general design of Mesoamerican pyramids and platforms. The temples these structures supported usually opened in the direction of the stairway; so do the ruins of a room on top of Structure J. Aveni measured the direction given by a line perpendicular to the base of Structure J's stairway. This line shoots through what was once an opening, or doorway, in a building on top of Structure P, a platform on the other side of the plaza. If continued northeast, the line indicates a position on the horizon where the bright star Capella would have been seen to rise in 275 B.C., about

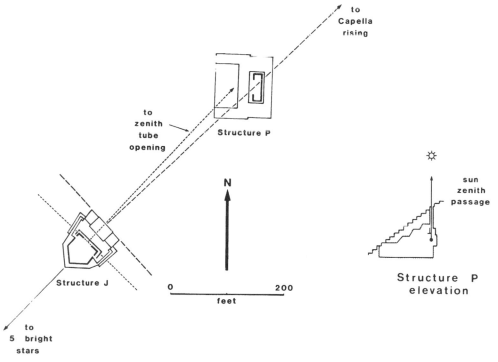

At Monte Albán, the perpendicular to the front stairway of Structure J points to Capella, the herald of the zenith sun. The perpendicular from the entrance of the temple on top of Structure J continues across the plaza and intersects the upper opening of the "zenith tube" in the stairway of Structure P. *(Griffith Observatory, after Horst Hartung)*

the time Structure J was built. There is special significance in the Capella alignment because in this period the star made its first reappearance in the predawn sky on the first of the year's two zenith passages of the sun at Monte Albán. Capella's heliacal rising could have heralded the first of these events for them. The alignment, however, may have been symbolic. Certainly astronomer-priests could have watched for Capella on top of Structure J. As experienced observers they would have known where to look for the star's momentary gleam, and the alignment may simply record in stone the principle behind whatever real observations may have been taking place, perhaps elsewhere.

In the 1960s, Horst Hartung had noticed that the foundations of the room on top of Structure J were skewed slightly north of the perpendicular from the stairs, and he found that a line from the center of the upper doorway and perpendicular to it intersected the stairway of Structure P where the open end of a vertical tube broke through a stair. The tube originates in a small chamber inside the pyramidal platform.

The most interesting thing about this chamber, which is buried well within Structure P, is the tube that points straight to the zenith. Aveni and Hartung see a connection between Capella's behavior, the stairway orientation of Structure J, and this shaft that would permit light from the zenith sun to enter the little room. The tube is too wide for determining the exact day of zenith passage. Although we cannot categorically state that the chamber in Structure P was an observatory for the zenith sun, a small niche in the back wall, at about chest height, could have held either additional equipment for making observations or ritual objects intended to be lit by the sun when it passed overhead. And there are other features of Structure J that fire the imagination and support these ideas.

A carved figure on one of the jambs of the doorway on top of Structure J looks like a skywatcher. His head is turned up, and he faces Structure P. Without question he, too, is one of the slain prisoners depicted elsewhere at Monte Albán, but Hartung has made the interesting suggestion that this stone slab was appropriated at a later time for use in Structure J because the figure resembled a celestially oriented priest.

Whether or not Zapotec astronomers actually gazed skyward from a temple on Structure J is not certain. But all of the architectural elements of this building, made a temple by ritualized observation, infer that sightings on Capella and the sun were part of ceremonial observance of a significant moment in the passage of cyclic time.

Stages for the Sky

Group E, a pyramid and set of temples at Uaxactún, a ceremonial center in the Guatemalan jungle, earned a reputation as an Early Classic Maya observatory as early as 1924. The claim has been repeated without critical review in many general surveys of Maya civilization. Only a new on-site survey could establish the accuracy and precision—and perhaps the purpose—of the alignments, and Anthony Aveni, who has

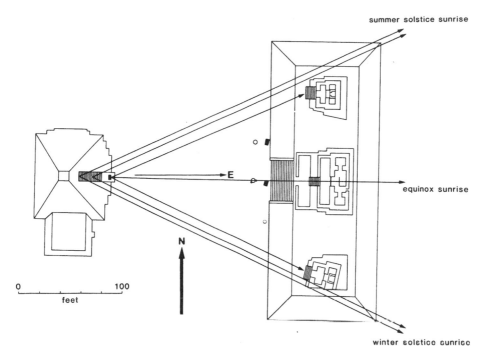

summer solstice sunrise

equinox sunrise

winter solstice sunrise

N

0   100
        feet

E

Three temples upon a platform just east of the pyramid of Group E at Uaxactún are placed in a way that permits them to mark—at least symbolically—the solstice and equinox sunrises. *(Griffith Observatory, after Horst Hartung)*

pioneered modern field measurement of astronomical alignments in pre-Columbian antiquities, visited Uaxactún to obtain the needed data.

From the top of Pyramid E VII sub or, better still, from a lower spot on its east stairway, solstice sunrises appear at the outer corners of the temples on the north and south. The line of sight over the center temple agrees very nicely with equinox sunrise. These are the sightlines, then, that would turn the group into an observatory. Although the alignments are reasonably accurate, within about a diameter of the sun as seen by an observer, they don't give us enough information to conclude that this architectural group was a functioning observatory.

At a genuine observatory the skywatcher obtains information about a celestial event: its timing, perhaps, or its placement. Armed with the new data, the astronomer-priest might correct a calendar, plan the design of a temple, predict another astronomical event, or prognosticate the future. By contrast, the skywatcher carries out ritual observation when the astronomical answer is already known. The celestial event fulfills a ceremonial role in those circumstances and is observed in the expectation that it will be seen. Certainly there were times when the distinction between these two types of astronomical activity was blurred, but Group E makes more sense as a stage for the seasonal drama of the sun. Ritual observation of the sunrise there could have been the focus of a public event.

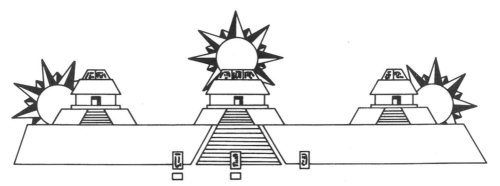

The middle temple of Group E is aligned with the equinox sunrise. The temple on the left coincides with the summer solstice rising sun, and at winter solstice, the right-hand temple indicates the position of sunrise. *(Griffith Observatory)*

A sightline from the Palace of the Governor at Uxmal extends nearly 3 miles to the ruins of another Maya center in this district of Yucatán. The site is known as Cehtzuc and its main pyramid, which protrudes above the otherwise featureless eastern horizon, created a marker for the southernmost rising point of Venus, over its 8-year cycle, at the time the Palace of the Governor was erected, about A.D. 800. Recently it

From inside the center room of the Palace of the Governor it is possible to see a horizon feature that coincides with the southernmost rising position of Venus for the eighth century after Christ. This small bump on the otherwise characterless horizon is actually a ruin, a 25-foot-high pyramid at Cehtzuc about 3 miles to the southeast. A fallen cylindrical stela and a platform that supports a statue of a double-headed jaguar both occupy spots on this line. The most recent research suggests that the best alignment with Venus is the view northwest, along this same line, from Cehtzuc, over the central door of the Palace of the Governor, to the northernmost setting of Venus. *(E. C. Krupp)*

The monumental upper façade of the Palace of the Governor at Uxmal contains hundreds of Maya Venus symbols. This view, looking back along the Venus axis, begins with the doubled-headed jaguar platform, continues across the fallen stone cylinder, and ends at the center door of the Palace of the Governor. *(Robin Rector Krupp)*

has been shown that the alignment is more accurate in the opposite direction, from Cehtzuc, over the middle of the Palace of the Governor, to the northernmost setting of Venus.

The upper facade is one of the marvels of Maya stonework. More than 20,000 individually cut and fitted pieces create a very deep mosaic of glyphs, patterns, serpents, rain god masks, and other symbols, including more than 350 emblems of Venus. An ancient Maya astronomer standing at Cehtzuc at the right hour, in the right season, and in the right year, would have seen Venus dramatically poised over the Palace of the Governor. This alignment, or the view from the Palace southeast to Cehtzuc, may explain why the orientation of Uxmal's Palace of the Governor deviates from the axis most other structures there share.

Although Cehtzuc and the Palace of the Governor at Uxmal could have been used to verify and date the position of Venus at its northern extreme, the extravagant vaults and rooms and facades of the building tell us that something more than a lone astronomer's observation of Venus went on there. The Palace is a monumental building on an even more monumental platform. The open area in front of it provides ample space for ritual, for sacred drama, or at least for an assembly of a large number of witnesses. In any of these circumstances the rising or setting of Venus at an important station in its celestial journey could have created an impressive and theatric moment in ceremonies at Uxmal or Cehtzuc. Also, the alignment's symbolic power might, at any time, ratify the power of the king.

Astronomical observations may have been made at Alta Vista, Monte Albán, Uaxactún, Uxmal, and other places like them, but the formal public setting of these "observatories" demands a different explanation. Everything is too well-staged. If genu-

Several astronomical alignments may have been built into the placement of major buildings at Uxmal in Yucatán. Although most of the site shares a single axis, one structure, the Palace of the Governor, is noticeably skewed from the rest. In fact, it faces close to the southernmost rising point for Venus during the period when most of Uxmal was erected. Recently it has been shown that the opposite direction of the line—northeast from neighboring Cehtzuc, over the Palace's center door, and on to the most northern setting of Venus—is better targeted.   (*Griffith Observatory, after Horst Hartung*)

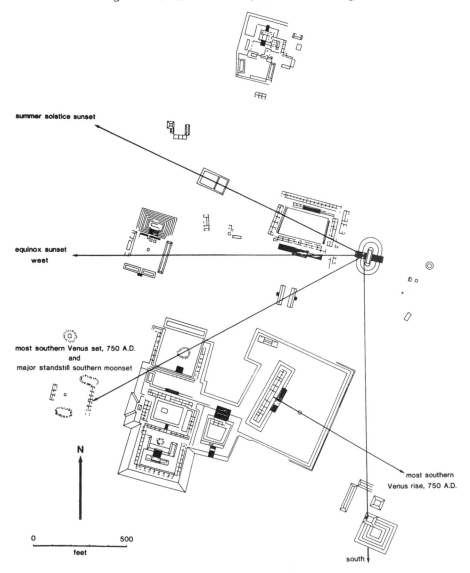

summer solstice sunset

equinox sunset
west

most southern Venus set, 750 A.D.
and
major standstill southern moonset

N

most southern
Venus rise, 750 A.D.

0   500

feet

south

ine astronomical observations took place, it seems they were treated as public events. The sighting of a star or a sunrise, the return of a planet to the place where its journey began, a beam of light from a zenith-crossing sun—all of these, when ritualized, reinjected cosmic order into the sacred spaces designed to receive it.

Lighting the Temples

Without written accounts of what went on at Stonehenge or at Mesoamerican ceremonial centers, we remain uncertain about the actual function of their astronomical alignments. We can only be guided by the religious aspects of the architecture. In Egypt we encounter similar alignments, but we have a better understanding of the celestial elements of the temples because reliefs and inscriptions describe the primary rites performed there.

Sir Norman Lockyer measured the orientations of Egyptian temples and in his book *The Dawn of Astronomy*, published in 1894, proposed that some of them had been aligned with the sun. New Kingdom temples belong to the period between 1567 B.C. and 1085 B.C., the time of the Eighteenth, Nineteenth, and Twentieth Dynasties.

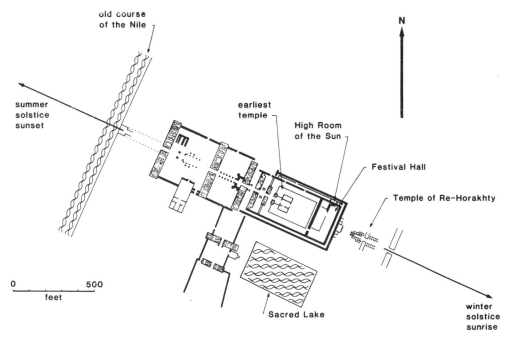

According to Sir Norman Lockyer, the main axis of the Great Temple of Amun-Re at Karnak, near Luxor, Egypt, pointed to the setting summer solstice sun. Lockyer also claimed that a smaller temple, southeast of the main structure, opened to the winter solstice sunrise. *(Griffith Observatory)*

Their general plan is based on a longitudinal axis—the line of symmetry—which starts at a high wall, or pylon, with an entrance on the axis, continues through a court and several halls, and finally reaches an inner sanctuary. The design introduces a tunneling effect, and Lockyer thought the temples might have been designed to allow the light from the sun or a star to beam all the way down the long passage on the axis and fall upon the statue of the god lodged in the sanctuary. In principle, the idea is sound, but Lockyer's analyses, particularly those that involved stellar alignments, are contradicted in most cases by the known dates and identities of the temples he considered.

Lockyer had concluded the Great Temple of Amun-Re at Karnak was designed to catch the light of the setting sun on the summer solstice because its entrance faces northwest. With a sightline of 1,996 feet, baffles (to narrow the central passage) could have allowed the Egyptians to use the temple as an astronomical observatory that would determine the exact day of winter solstice and the exact length of the solar year. At least, Lockyer thought so.

Gerald Hawkins recognized that the Temple of Amun-Re was directed as Lockyer

From inside the sanctuary of the Great Temple of Amun-Re, one's gaze is guided through massive pylons, columns, and gates to the northwest horizon across the Nile. *(Robin Rector Krupp)*

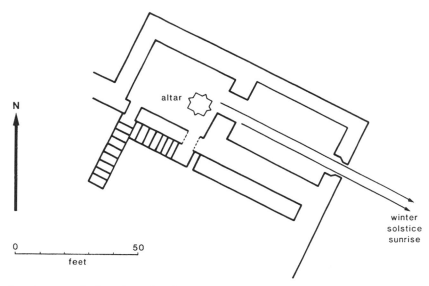

A window in a small chapel in the Great Temple of Amun Re opened to the southwest and framed the rising disk of the sun at the winter solstice. *(Griffith Observatory, after Gerald S. Hawkins)*

had claimed, but he felt the Theban hills, across the Nile, were high enough to displace the sun from true alignment. In searching for an explanation of the temple's orientation Hawkins decided the southeast and the winter solstice sunrise made more sense. Although the main body of the temple opens in the other direction, a small, two-roomed chapel, far to the rear of the Amun-Re assemblage and part of the Festival Hall built by Tuthmosis III, offered a possibility. Its window looked out to the southeast. Inscriptions in the temple were studied by the French Egyptologist Paul Barguet and found to contain astronomical references, particularly to the rising sun. In the chapel, which Hawkins has called the High Room of the Sun, a relief shows the pharaoh facing the window, and an accompanying text says "Make acclamation to your beautiful face, master of the gods, O Amun-Re, primordial god of the two lands . . ."

A large stone altar, carved in the shape of the hieroglyph *hetep,* or "offering," strengthens the idea that the ritual observation of the sun, Amun-Re, occurred here. Another inscription in the lower rooms of the Festival Hall offers more information.

One goes toward the Hall, horizon of the sky, one climbs to the *aha,* lonesome place of the majestic soul, high room of the ram who sails
across the sky; one there opens the doors of the horizon of the primordial god of the two countries to see the mystery of Horus shining.

We don't know what *aha* means, but this appears to be an invitation to watch the sunrise through the window of the High Room of the Sun.

In November, 1976, I received a call from Mr. and Mrs. R. L. Mikesell, who

The massive stone block in the foreground is the damaged "*hetep* glyph" altar in the High Room of the Sun at Karnak. To the right is what remains—the lower frame—of the southeast window. *(Robin Rector Krupp)*

planned to be in Egypt in December and wondered if there were anything astronomical they might see. I immediately thought of Gerald Hawkins's High Room of the Sun and asked if they could be in Karnak for the winter solstice. They said they could, and I "invited" them to watch the sunrise. This turned out to be more complicated than I'd thought.

The Mikesells were careful to survey the situation at Karnak before the morning of the solstice, and they discovered that a palm tree had grown high enough to completely block the view of sunrise through the window of the High Room of the Sun. Looking back now at the photographs Gerald Hawkins took of the window several years before, one can detect the upper tips of the interfering fronds just beginning to intrude above the window sill. By 1976 the tree was in full view.

Undeterred by the temple's locked gate and the offending palm, the Mikesells, in the company of a Canadian schoolteacher who was also traveling in Egypt, made their way through the darkness to the High Room of the Sun on the morning of the winter solstice. Just as the sun began to rise, the schoolteacher jumped from the outer wall into the palm tree, and, spread-eagled, held its fronds down on either side to permit the sun to rise in the *V* between them; the sunrise alignment was confirmed.

Other New Kingdom temples were also aligned on the winter solstice sunrise. Lockyer had indicated that the Temple of Re-Horakhty and the temple of Amenophis III both faced this direction. Hawkins confirmed these claims 80 years later and also found another solar chapel like the High Room of the Sun at Karnak. This is the small side temple at Abu Simbel, and Hawkins's analysis shows that it, too, is oriented on the winter solstice sunrise.

In some respects, then, Lockyer was right. Some New Kingdom temples were astronomically aligned in accordance with their meaning and their use. The other things we know about these temples, however, make them improbable observatories.

Not all Egyptian temples were astronomically aligned, but the inscriptions and pictorial reliefs in some of them detail other ceremonies in which celestial events played a part. The walls and the ceilings of the Temple of Hathor at Dendera, for example, are covered with astronomical imagery and portrayals of rituals focused on heaven. The Temple of Hathor is rather late as Egyptian monuments go, belonging to Egypt's Ptolemaic period. Its foundations were laid late in the second century B.C., and work on the temple continued into the first century after Christ. The site was probably occupied by New Kingdom temples, and possibly even Old Kingdom ones. Because it is so late, its astronomical reliefs include influences from beyond Egypt's borders, but some of the traditions preserved at Dendera are thousands of years old.

With its main axis oriented about 18½ degrees east of north, there seems to be no obvious astronomical significance in the temple's alignment. Lockyer proposed that Dendera's axis may have been pointed at a star: Dubhe in the Big Dipper, or Eltanin in Draco the Dragon. By Lockyer's own calculations, however, these alignments would have worked only for a Temple of Hathor 3,000 to 4,000 years older than the one that exists. Whatever the reason behind the orientation of the temple, we do know it was consciously and carefully selected.

One of the walls inside the temple carries a description of the ceremony of the Stretching of the Cord. In this ritual a loop of rope was drawn tight between two poles, or stakes, to establish the reference line from which the temple's foundation would be measured and oriented.

Dendera was the site of an important festival: the New Year Ceremony. A long stairway that leads to the temple's roof is decorated with the figures in the New Year procession. On the roof itself is their destination: a kiosk open to the sky and the sun. The texts tell us what takes place there.

> Radiant rises the golden one (Hathor-Isis-Sirius) above the forehead of her father (near but in advance of the sun), and her mysterious form is at the head of his solar boat ... As her fellow-divinities (the other stars) unite with her father's rays and as they merge with the glittering of his disk, Dendera is joyful ... There is a festive mood as they behold the great one, the firmly striding creator of feasts in the holy city, on that beautiful day of the new year.

This and other texts describe metaphorically the heliacal rising of Sirius. Hathor, by Ptolemaic times, was a very ancient goddess with complex associations. She was con-

An open kiosk on the roof of the Temple of Hathor at Dendera was used to observe the predawn reappearance of Sirius in the New Year ritual. *(E. C. Krupp)*

nected with the idea of motherhood, but on a cosmic scale. She was the source of all creation—the Great Mother—and the one who nourishes the world order. These attributes are evident in her mythology and symbols. Her child was the newly risen sun, and so she was identified with the sky and the goddess Nut. Isis as Sirius was the mother of the New Year. Isis in her way, then, also reestablished the world order by creating the new year as she rose heliacally in the form of Sirius. Another text tells us that Hathor is

> ... the beautiful one who appears in heaven, the truth which regulates the world at the head of the sun barge, the queen and mistress of awe, the ruler (of gods and) goddesses, Isis the great, the mother of gods.

Hathor's identity accordingly extended to Isis and Sirius. In any case, the New Year procession, after ample ceremony, sacrifice, and offering, carried the temple's statues of Re (the sun) and Hathor-Isis (the star Sirius) to the roof kiosk. There the light of Sirius and then the sun's would shine on the unveiled statues.

The heliacal rising of Sirius involves only a brief appearance of the star before it is lost in the light of the sun. The event is a union, or marriage, which, when consummated, re-creates the world order by celebrating the sun's "birthday," the New Year. Certainly this astronomical event was watched from the roof of Dendera temple, but the inscriptions tell us the observation was carried out in a ritual environment, an environment made sacred and given meaning by a revelation in the eastern sky.

II

The Cities We Plan

The cosmos itself is what mattered to our ancestors. Their lives, their beliefs, their destinies—all were part of this bigger pageant. Just as the environment of their temples was made sacred by the metaphors of cosmic order, entire cities and great ritual centers were also astronomically aligned and organized. Each sacred capital restated the theme of cosmic order in terms of its builders' own perception of the universe. Principles which the society considered its own—which ordered its life and gave it its character—were borrowed from the sky and built into the plans of the cities. In this way the ancient capital represented and reaffirmed the identity of the people who built it. It was the seat of government and the center of religion and so served as the source of law, order, and sovereignty.

The Forbidden City

Beijing is the only world capital still laid out according to a sacred cosmological plan. Its primary axis is north-south. Although we encounter cardinal orientation in the street grids of many other cities, these arrangements are not consciously linked to the idea of cosmic order. In Beijing, however, the cosmological motive behind the city's layout is known and preserved. Even today, the monuments of the secular government of the People's Republic of China adhere to the ancient sacred plan. The flagpole in Tian an men Square, the Monument to the People's Heroes, and Mao Ze dong's Mausoleum all occupy stations on the city's main axis, between the Tian an men and the Qian men, two great gates of old Imperial Beijing. Tian an men Square is itself bisected by this meridian of Beijing, and it is flanked on the west by the Great Hall of the People (National People's Congress Building) and on the east by the

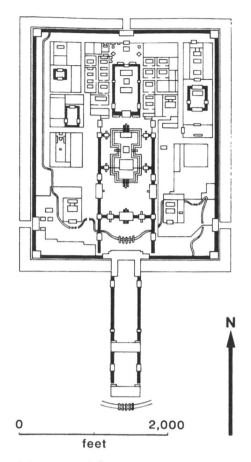

Beijing's Imperial Palace dates from China's Ming Dynasty and reiterates the relationship between the sacred capital and cosmic orientation. Its main axis is the meridian, a north-south line oriented by the north celestial pole. The emperor, who was likened to the pole, held ceremonial audiences in the pavilion in the center. *(Griffith Observatory)*

N

0   2,000

feet

Museums of Chinese History and of the Chinese Revolution. Modern Beijing is still organized by the cosmic axis of the old "Imperial Road."

Thirty-three hundred feet of Imperial Road reach due north from the Gate of Heavenly Peace (Tian an men) to the Hall of Supreme Harmony and the Emperor's Golden Throne within, at the heart of the Forbidden City. Here the emperor held his public audiences and made official announcements. This spot, inside the deep red walls of the Imperial Palace, was the world's symbolic center as far as the Ming (A.D. 1368–1643) and Qing (A.D. 1644–1911) Dynasties of China were concerned, and their capital was based on traditions of city planning that were already centuries old. The *Li Chou,* an ancient account of early customs and rituals compiled in Confucian times (sixth and fifth centuries B.C.), included detailed instructions for a proper city layout. The place of the emperor was the center. He was the pivot of the world. Fixed like the pole of the sky, he steadied the world, and all its affairs revolved about him. His Imperial Palace was known as *Zi jin cheng,* the "Purple Forbidden City," just as the realm around the north celestial pole was called the "Forbidden Purple Palace."

Beijing's Imperial Palace is rectangular and oriented on a north-south axis marked by the Imperial Road. Every gate, wall, court, and hall is bisected with a line oriented

A meridian of imperial marble sidewalk stretches north along the cosmic axis of the Forbidden City to the Tai he dian, or Hall of Supreme Harmony, the hub of the empire and the terrestrial echo of the circumpolar realm of the sky. A sundial is mounted on the far right of the upper platform, and a small shrine shelters a grain measure on the far left. *(Robin Rector Krupp)*

by the pole of the sky. In the words of the ancient texts, the emperor's palace was where "earth and sky meet," and from the Qian men one must cross five majestic gates, four courts, and two ceremonial rivers to get there. Stations of the route to the Golden Throne read like a litany to the congruence of heaven and earth:

> Straight-Toward-the-Sun Gate*
> Gate of Heavenly Peace
> Gate of Correct Deportment
> Meridian Gate
> Gate of Supreme Harmony
> Hall of Supreme Harmony

And beyond the throne, the names of the halls reecho the theme of celestial order brought to earth—"Central Harmony," "Preserving Harmony," "Heavenly Purity," "Union," and "Earthly Tranquility."

---

* This was the official name of the Front Gate, or Qian men.

Chinese concepts of order defined the way in which ideas were associated and thought was organized. For example, the number 5 was symbolically important because it was related to the traditional Chinese sense of terrestrial order. For the Chinese, the earth was encompassed by Five Directions, which are the four cardinal directions and the center. A square or rectangle symbolized the earth, and for that reason the Forbidden City has four sides, four corner towers, and four cardinal gates in its 35-foot-high walls. The fifth direction, the center, belonged to the emperor.

There are, therefore, in Chinese thought Five Activities, or principles of movement, action, and change: Water, Fire, Metal, Wood, and Earth. Something like the Four Elements of Classical antiquity, they have a more complex function and dynamic character. Associated with them are the Five Atmospheric Phenomena, the Five Sacred Animals, the Five Colors, the Five Internal Organs, the Five Metals, the Five Planets, and more quintets. Five arches tunnel through the Gate of Heavenly Peace. Five marble bridges cross the Golden Water River in the court beyond the Meridian Gate (each is said to represent one of the Five Virtues). And, as we have already seen, the gates that lead to the Hall of Supreme Harmony total five.

Yong Le, the third emperor of the Ming Dynasty, built the Forbidden City. He seized the throne from his young nephew, moved the capital from Nanjing, and in 1404 started construction in the center of Beijing. Legend has it that an astrologer and geomancer, Liu Po-wen, designed the Forbidden City according to interactions of the Five Activities and the anatomy—including the Five Internal Organs—of No Cha, a hero in Chinese mythology. Liu Po-wen is supposed to have delivered the plans in a sealed envelope to Yong Le, who followed the instructions implicitly. Fourteen years later, the Forbidden City was complete.

Cities, houses, and tombs were sited and oriented according to a system of symbolic—as well as practical—rules. The process is known as geomancy, a kind of divination that may involve astrology, celestial alignments, terrestrial magnetism, dowsing, topographical symbolism, and other magical techniques. The Chinese form of geomancy is known as *feng-shui,* which literally means "wind and water," and it was intended to produce a plan that in all respects would harmonize the human presence with the natural environment. The Forbidden City of Beijing advanced this principle to the grandest scale and, in Chinese terms, harmonized heaven and earth.

As the "Son of Heaven" the emperor received his mandate from the sky. It was his responsibility to preserve the world's order, and his capital was designed to facilitate the harmonious convergence of all influences.

On the terrace of the Hall of Supreme Harmony stands a sundial and a miniature temple that housed a grain measure. Useless under cloudy skies, the sundial stood for ethical government and the Emperor's virtue. The grain measure symbolized the proper distribution of justice. It is a container of standard size, and the little square temple, by its shape, like the Forbidden City, represented the whole empire. Both instruments also imply other fundamental duties of the emperor; he announced the

The sundial on the platform of the Hall of Supreme Harmony invokes the order of celestial time and symbolizes ethical, orderly government. *(Robin Rector Krupp)*

new calendar at the Meridian Gate each year, and he authorized the system of weights and measures.

China was first unified in 221 B.C. by Qin shi Huang di, whose astronomically oriented tomb is described in Chapter 5. He, too, built a capital, on the south bank of the Wei River, in central Shan Xi province, west of the modern city of Xi an. A copy of the palace of each prince he defeated was added with each victory, and for his own public audiences he erected a huge, cardinally oriented hall. From this palace a covered avenue crossed the river. It seems this symbolized an astronomical bridge across the Milky Way that reached from the "Apex of Heaven" to the "Royal Chamber."

Qin's capital is completely gone, swept away by the armies of Liu Pang, the founder of the Han Dynasty. Qin's "spirit city," or tomb, still survives, however, and replicates the cosmological design of a capital in its own cardinal orientation. Although we do not know exactly where Qin was buried, we might guess his body resides in the center of the "spirit city" as he occupied the "center of the world" in life.

The Solar-Efficient City

Cardinal directions spring from the sky to orient the earth. We have already learned that another people, American Indian platform mound builders, also laid out their city (Cahokia) along cardinal direction axes. Most of the towns of the Pueblo Indians in the Southwest and all Hopi villages save one also are accurately oriented with the cardinal directions. Similar interest in cardinality is also evident in the prehistoric Pueblo centers.

The central plaza of Pueblo Bonito (a D-shaped, 800-room "apartment house" in Chaco Canyon, New Mexico) is neatly split by a low wall on a north-south line. The wall is skewed from an exact north-south orientation by only 45 arcminutes to the east, or about ¾ degree. The great kiva that adjoins this feature is also well aligned: only 15 arcminutes—¼ degree—west of north. These alignments were established in the field by Dr. Ray Williamson, who also showed that about half of the south wall—

The cliff face on the north side of Chaco Canyon acts as a monumental backdrop for the ruins of Pueblo Bonito, seen here looking along the axis created by the low, north-south wall that bisects the plaza. *(Robin Rector Krupp)*

the west half of the straight leg of the "D"—is oriented east-west with an error of only 8 arcminutes south of east (all of this masonry belongs to the later period of Pueblo Bonito, about A.D. 1000 to 1100).

Pueblo Bonito's north-south wall actually creates two separate and distinct plazas, each with its own great kiva. Here the architecture suggests division of the population of Pueblo Bonito into two groups, perhaps based upon kinship or affiliation with different religious fraternities. Dr. Travis Hudson has assembled circumstantial evidence that favors the use of two different units of measure, one used to build rooms in the east zone, the other for the west. The fact that among some of the present-day Rio Grande Pueblos responsibility for the year's ceremonials is split between two groups supports the theory of duality. Besides being the city's centers for public gatherings and ceremonies, the plazas were the headquarters for the community's daily work. And new evidence has been uncovered that shows the Indians were adept in using architecture, environment, and the sun itself to make Pueblo Bonito comfortable and energy-efficient. Ralph Knowles, a researcher interested in architecture and energy use, noticed in 1974 that Pueblo Bonito's overall design is a successful exercise in passive use of solar energy. Much light and heat from the low winter sun was reflected back into the central plazas by the curved north wall. Williamson and his collaborators, physicist Howard Fisher and architect Peter Paul, observed the rooms to be

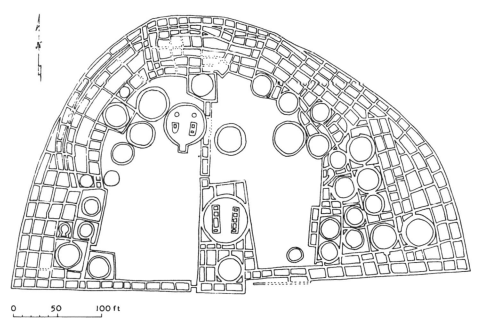

0   50   100 ft

Pueblo Bonito in Chaco Canyon, New Mexico, is neatly split by a wall that runs true north-south. Each half appears to have its own great kiva, clan kivas, plaza, and apartments. *(National Park Service)*

better shaded by the canyon walls in summer, when it is. hot, than in winter, when the sun's heat is attenuated. By measuring the temperature in several rooms, they proved Pueblo Bonito is energy-efficient. Thick walls and reflection from the face of the cliff keep the room temperature constant despite the extreme fluctuations in desert temperatures from day to night.

The City That Nourished the Sun

Tenochtitlán, the island capital of the Aztec Empire, was more populous, more vast, and more organized than any European city its Spanish conquerors had ever seen. The metropolis was quartered by four great avenues that all began at the central, sacred precinct around Templo Mayor, the main pyramid. Three of these roads, joined with bridges—on the north, south, and west—crossed Lake Texcoco, and connected Tenochtitlán with the mainland. This plan is still preserved in the present-day capital. The main north-south axis of Mexico City coincides with Republica de Argentina and Seminario Street. Calle de Tacuba heads west in line with the old west avenue and causeway. Most of the old capital is still buried beneath the streets and buildings of later times, and the exact placement of some of the temples, pyramids, and palaces described in the post-Conquest chronicles and included on early maps is still in dispute. Even so, the main structures in the center of Tenochtitlán all seem to have helped mark the center's main axes or duplicated their orientation.

Although the cardinal directions inspired the city's four-part division, Tenochtitlán's sacred buildings, thoroughfares, and causeways were skewed from true cardinal orientation. Anthony Aveni's 1976 report of measurements of an exposed foundation of Templo Mayor indicates the pyramid's main, east-west axis was actually shifted about 7 degrees south of east. We can't dismiss this deviation from a true east-west line as the product of half-hearted surveying by Tenochtitlán's planners and architects, for a 1532 chronicle, *Historia de los Mexicanos por sus Pinturas,* described the concern of Moctezuma II, the Aztec king, for proper alignments. Referring to Templo Mayor, it remarks:

> The festival called Tlacaxipehualiztli took place when the sun stood in the middle of the Temple of Huitzilopochti, which was at the equinox, and because it was a little out of the straight, Moctezuma wished to pull it down and set it right.

Tlacaxipehualiztli was one of the eighteen 20-day "months," or *veintenas,* of the Aztec 365-day year. It fell within the period we allot to the month of March and included the vernal equinox: a time of fertility and renewal, the time of spring. From the account we learn the sun "stood in the middle" of Templo Mayor. This means the sun was visible in between the two temples at the summit of the twin pyramid, Templo Mayor. An early (1524) map of Tenochtitlán, said to have been drawn by

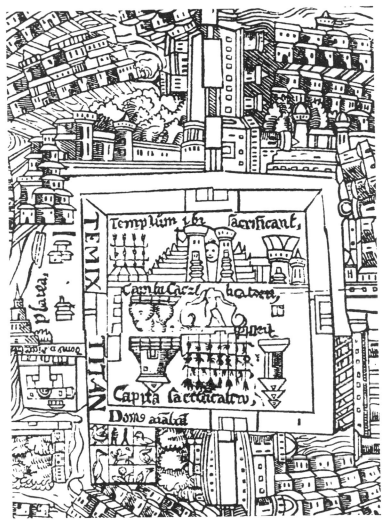

Four avenues extend from the four sides of the central zone of Tenochtitlán. This detail from a map drawn in 1524 also depicts Templo Mayor—the huge twin pyramid that dominated this sacred precinct—as a stepped structure with two towers. The face of the sun peers through the gap between them. *(from Lucien Biart,* The Aztecs: Their History, Manners, and Customs, *1886)*

Cortés, includes a stylized Templo Mayor with the sun's face caught between its two towers.

Everything implies Templo Mayor was intended as a stage for ritual events and reenactment of myths. Apparently, the equinox sunrise played a part in one of these ceremonies, but Aveni's measurements tell us the line of sight through the gap between the two temples was 7½ degrees too far south of the point where the equinox sun appears on the horizon. More than 270 feet wide at its base and with its upper platform perched perhaps 90 feet above the ground, the huge bulk of Templo Mayor would have blocked the view of sunrise. By taking the height of the pyramid into account, however, Aveni and his collaborator, Dr. Sharon L. Gibbs, made some sense out of the misalignment. As the equinox sun climbs higher, it follows a slanted trajectory, a path angled almost 70½ degrees to the horizon. This carries it toward the south. From a point in the plaza west of Templo Mayor and on its axis, the sun would eventually be seen between the twin temples, or, as is now thought more likely, over the temple on the south, dedicated to Huitzilopochtli. It seems likely that the east-west axis of Tenochtitlán was established by the equinox sun, not on the horizon, but in a direction staged by the architecture for dramatic appearance in conjunction with ritual and sacred drama.

At Tenochtitlán, then, we have all the elements of sacred space. Through its symbolic identity with the primeval mountain, the heart of Tenochtitlán is the heart of the world. It was regarded as the site of the world's creation, and this identity was forged anew each year when the birth of Huitzilopochtli, the patron god of the Aztec, was ritually reenacted there. The east and the vernal equinox were associated with birth and renewal. Templo Mayor links the creation of the cosmic order of the sun with

Although it is possible the skew of Templo Mayor, 7½ degrees south of east, was intended to accommodate the sun after it had risen high enough—on the equinox—to appear between the temples on top, as shown here, recent research suggests alignment was altered to permit the sun to rise over the Temple of Huitzilopochtli, on the right. The monumental stone relief that depicts the goddess Coyolxauhqui was found at the base of the right-hand (south) stairway.   (Griffith Observatory)

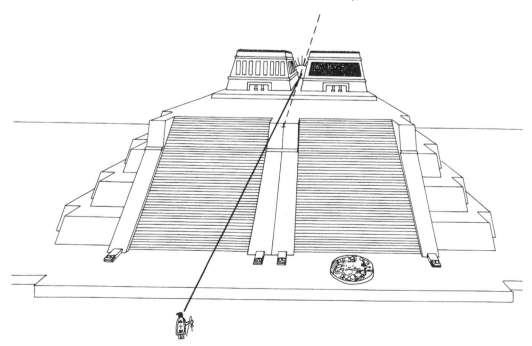

A page in the Aztec *Codex Mendoza* illustrates the legend of the foundation of Tenochtitlán in 1325. In this detail, the famous eagle is perched upon the cactus, about to grasp the
fruit of the plant in its talon. The fruit is symbolically equivalent to the human hearts the
Aztec procured through ritual sacrifice for the sun, represented by the eagle. Tenochtitlán's name derives from the cactus and its fruit, and its sovereignty was sustained by the
Aztec vision of their celestial mandate. This celestial power was transmitted to earth at
the center of the Aztec world, the place where their empire began. This spot, where they
came upon the eagle consuming the cactus fruit, is also marked by the intersecting waters
of Lake Texcoco. These channels quarter the Aztec realm symbolically and echo the four
world directions. *(The Bodleian Library, Oxford, MS. Arch. Selden A.1, folio 2)*

the foundation of Tenochtitlán itself. According to legend, the city was founded when the early Aztec, or Mexica, saw an eagle perched upon a cactus. The cactus was growing out of a rock in the middle of the lake, and its fruit was being devoured by the eagle. This symbol is the emblem of Tenochtitlán, which means "Place of the Fruit of the Cactus." Here, the eagle represents the sun, and the cactus fruits are human hearts, which they resemble in shape and color. On the spot where they found the cactus, the Aztec built their temple, and at the heart of this city, they sustained the sun with sacrifices and, in their minds, kept the cosmos alive. In the very name of their capital—Tenochtitlán—the Aztec stated the purpose of the world's center. It was the place of the sun's nourishment. In the symbolism of its architecture, the metaphors of its myths, the cycle of its festivals, and the layout of its temples and streets, Tenochtitlán expressed the Aztec's sense of their own cosmic destiny.

The City at the Navel of the World

In Peru, the Inca empire was centered on its sacred capital and administrative hub— Cuzco. In Quechua, the language still heard in the Andes today, *Cuzco* means "navel of the earth." As such, Cuzco nourished the world with the order it imposed— through its military strength and effective government—upon the lands in its domain. At the pinnacle of Inca social order, the Sapa Inca, or Supreme Emperor, held complete and absolute power. His authority originated in the sky, for he was the son of the Sun. Through him, the city itself was the point of articulation between the earth and heaven. Cuzco was regarded as the source of the power that organized and mobilized the lands and peoples of the earth. As a center of sacred sovereignty, Cuzco conducted its affairs, organized its people, arranged its temples, and oriented itself according to the pattern of the sky. The earth's responsibility to heaven was fulfilled in its rituals, festivals, and sacrifices. Here then, like Beijing or Tenochtitlán, is another holy city. Its layout not only clearly acknowledges the sky; Cuzco's debt to cosmic order may be greater and more complex than that of any other sacred capital.

The Inca called their empire *Tahuantinsuyu,* the "Land of Four Quarters." Its four geographic divisions were reflected in the empire's social and political organization, for the next level below the Sapa Inca in the rigorous social hierarchy was occupied by four *apus,* or prefects. They were the Sapa Inca's chief advisors, and each represented one of the four provinces. These regions were named for their bearings from Cuzco, but intercardinal directions—and not cardinal—oriented the empire:

Antisuyu	"northeast quarter"	Cuntisuyu	"southwest quarter"
Chinchaysuyu	"northwest quarter"	Collasuyu	"southeast quarter"

This four-part division extended right into the center of Cuzco, and each of the city's four districts was known by the same name. The city's quarters were defined by

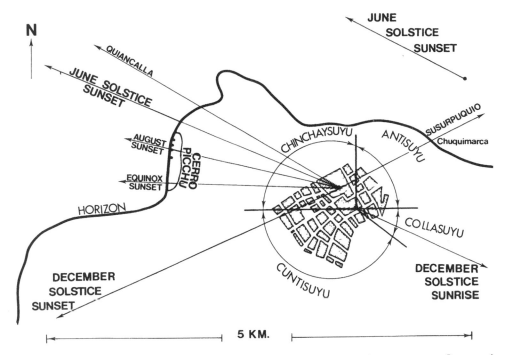

N

JUNE SOLSTICE SUNSET

QUIANCALLA

JUNE SOLSTICE SUNSET

SUSURPUQUIO

Chuquimarca

CHINCHAYSUYU

ANTISUYU

AUGUST SUNSET

CERRO PICCHU

EQUINOX SUNSET

HORIZON

COLLASUYU

CUNTISUYU

DECEMBER SOLSTICE SUNSET

DECEMBER SOLSTICE SUNRISE

5 KM.

Politically and geographically, the Inca empire was divided into four quarters. Cuzco, the sacred capital at the center of the Inca world, reflected this same organization in its four divisions. In some cases, astronomical directions determined the orientation of buildings. The Coricancha, or Temple of the Sun, for example, was oriented with the June-solstice sunrise/December-solstice sunset line. *(Griffith Observatory, after A. F. Aveni)*

the details of Inca administrative organization, social hierarchy, and kinship. With *ceques* and *huacas,* the whole plan was incorporated into a system of lines of shrines. Huacas, as we have seen, were places or features with sacred meaning, and the line formed by several huacas was a ceque. Almost anything could be a huaca—a spring, an ancestral tomb, a distinctive rock or hill, even a deposit of ore.

Father Bernabé Cobo provided a detailed description of the Cuzco huacas in his lengthy and comprehensive commentary on the Inca empire, *Historia del Nuevo Mundo.* Although he wrote from the point of view of an early sixteenth-century Spanish priest, Cobo spent most of his life in Peru. His familiarity with the country and its people provides us with a reasonably reliable account of the ceque system and the organization of Cuzco.

Although each municipal quarter was associated with an intercardinal direction, these zones were not split equally or symmetrically. The southwest "quadrant" enclosed 37 more degrees than a real quarter of 90 degrees, and the southeast zone was shorted by that amount. Three of the city's *suyus,* or quarters, had 12 ceques each within their jurisdictions, but the Cuntisuyu zone had five more. These deviations from symmetry probably were the result of other conditions the ceque system had to meet. Clan lineage and the intricate rules of the Inca social system perhaps were responsible for the uneven split, but there could have been several reasons for it, given the multiple symbolic character of the ceques.

By noting the remarks of the sixteenth-century chroniclers Cristóbal de Molina and Juan Polo de Ondegardo, Dr. R. T. Zuidema, an anthropologist at the University of Illinois, Champaign-Urbana, began, in 1953, to realize the ceque system was even more complicated. According to the old accounts, each of Cuzco's 328 huacas stood for one day. Together, Zuidema suggested, they represented the time it takes the moon to complete 12 circuits with respect to the background stars. He already had concluded, from the evidence in post-Conquest reports, that the offerings, services, and maintenance of each huaca were the responsibility of a specific group of people, its members determined in some way through kinship and perhaps in some cases through position in the bureaucracy or through administrative regulation.

Royal lineage established its own bloodlines, each of which traced its roots to an Inca emperor. Society as a whole was broken into three groups by status: the royal aristocracy, or *Collana;* the servants and assistants, or *Payan;* and the general population, or *Cayao.* Some of the ceques in Cobo's survey are named in terms of these groups. The general pattern of ceques in one of the quarters of Cuzco involved, then, three sets of three ceques each. In each set, one ceque belonged to the Collana line of the group, one to their Payan, and one to the Cayao affiliated with them.

The entire ceque system has been likened to the *quipu*—or sets of knotted strings—that were the Inca substitute for writing. Each ceque was like a string, and its huacas were like the knots tied in sequence on the string. Seen in this light, the ceque scheme preserved the family identities, the kinship, and the social hierarchy, in a set of sacred places, all organized into straight lines that radiated through the quarters of Cuzco.

The center from which all the ceques extended was the Coricancha or Temple of the Sun. As Cuzco was the center of the world, the Coricancha was the center of Cuzco. According to Inca myth, the temple's site was determined by Manco Capac, the first Son of the Sun. He was sent to earth to bring civilization to its barbaric inhabitants. In the company of his wife, the Daughter of the Moon, Manco Capac traveled in search of the place to establish his realm. When he reached the valley of Cuzco, Manco Capac followed his Sun-father's instructions and drove a rod of gold into the earth. At every other place he had tried this test, the staff simply stood upright in the ground. But here, where two rivers joined, the golden staff slipped right into the earth and vanished. At this spot Manco Capac built the Coricancha. In myth, the sacred Temple of the Sun marks the place where Cuzco was founded. This is also where the empire was centered, and that makes it, as well, the place where cosmic order was brought to earth. The Coricancha's alignment with the solstice was an umbilical to the sky. Through this conduit the principles of cosmic order flowed into the Coricancha and circulated through the ceques to permeate all aspects of the highly organized life of the Inca. All of this was apparently related to the calendar, and although not all ceques were straight lines, some of the ceques were astronomically oriented.

Ceques with astronomical significance are clear expressions of the celestial order. Cobo gave us a description of one in the southwest quarter (Cuntisuyu) that turns out to be oriented to the December solstice sunset. Cobo describes the line in terms of planting time, but the details he provided demonstrate his error in that respect. In particular, he wrote that the third huaca of the thirteenth ceque is called Chinchincalla, and "It is a great hill where there are situated two pillars, and when the sun arrives there it is time to sow."

The "pillars" are also called *mojones*, and these stone towers were described by other chroniclers and even illustrated by Poma. What is interesting here is Cobo's description of what seems to have been a working "observatory." Other descriptions of ceques imply that the huaca on a line could be such a tower. Anthony F.

The First New Chronicle and Good Government, a post-Conquest commentary by Felipe Huamán Poma de Ayala, illustrates the *mojones,* or stone towers, the Inca built. Some of these marked the terminations of *ceques,* and others acted as foresights for astronomical observations. Here, the Inspector of Roads fulfills his duties while three *mojones* stand in the background. *(Det-Kongelige Bibliotek, Copenhagen)*

Aveni and R. T. Zuidema, collaborating in a study of astronomical alignments of ceques, knew they had an astronomical line, but its exact direction was unknown. They had to figure out where it started and where it stopped. Painstaking detective work—crosschecks against Cobo's descriptions of adjacent ceques and field searches for the places he mentioned—made it possible to locate the place where the towers stood and the place from which they were observed. Aveni and Zuidema discovered that the December solstice sunset line does fall between ceques 12 and 14 and passes south of a feature that fits the description of Ravaypampa. Cobo's reference to sowing time may in some way relate to the fact that the last time to sow occurred in December.

Other astronomical alignments involving huacas also fell into place. These lines were not part of the ceque system, for they originated in places other than the Coricancha. Cobo's comments about specific huacas, however, alerted Aveni and Zuidema to further astronomical possibilities. Astronomy turned out to be threaded through and through the fabric of Cuzco's organizational features.

The place Cobo called Sucanca was the seventh huaca on the district's eighth ceque, and it was

> ... a hill where the irrigation canal of Chinchero comes. On it were two towers or monuments to signal that when the sun arrived there it was time to begin to plant maize. The sacrifice made there was addressed to the sun, and they asked that it would arrive on time so that they would have time to plant ...

Another sixteenth-century writer, whose name is not known, also seems to have described these pillars, and he provided more detail. After explaining that there were four of them, he described their use: "When the sun passed the first pillar this began the time when they were warned about the planting of vegetables at the highest altitude." The sun is headed south in August, and so the first *mojón* it reached was the one farthest north. Because crops grown at higher elevations take longer to ripen, their planting must start earlier.

> When the sun entered the space between the two pillars in the middle it became the general time to plant in Cuzco; always the month of August.
>
> And when the sun stood fitting between the two inner pillars they had another pillar in the middle of the plaza, a pillar of well-worked stone about one estado high, called the Ushnu, from which they viewed it. This was the general time to plant in the valleys adjoining Cuzco.

Anthropologist Gary Urton's meticulous study—*At the Crossroads of the Earth and Sky: an Andean Cosmology*—of contemporary skylore and cosmology among the Andean people in a small community near Cuzco sheds considerable light on other aspects of Cuzco's symbolic design. Mismanay is a settlement of 350 Quechua-speaking people, who make their living by farming. Two principal pathways give the village its structure: Chaupin Calle ("middle road") runs northwest-southeast, and Hatun

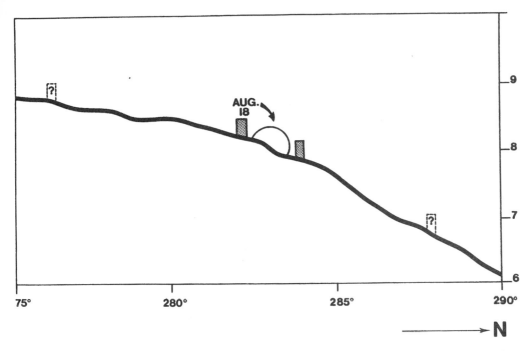

From the *ushnu*, the sun set between the two middle towers on the flank of Cerro Picchu in mid-August. This was the time to plant in Cuzco, and the event probably coincided with the date on which the sun passed through the nadir at midnight. *(Griffith Observatory, after A. F. Aveni)*

Raki Calle ("the great road of the four-part division") cuts northeast-southwest. These two routes cross at a chapel known as *Crucero,* or "cross," and they divide and order the landscape into four intercardinal quarters, repeating the partition of Cuzco and the entire Inca Empire. The two primary irrigation arteries are similarly oriented and also cross at the chapel.

In Mismanay, the people see the sky in similar intercardinal terms. The Milky Way establishes "intercardinal" horizon directions; it rolls up each night from the eastern horizon like a cresting wave and eventually arcs through the zenith. One end of this band of diffuse starlight intersects the southeastern horizon, while the other drops out of sight in the northwest. Because the axis of the Milky Way is skewed with respect to the earth's rotation, its orientation changes. About 12 hours later, its other half transits overhead, and then the band runs from northeast to southwest. All four intersections with the horizon provide two more intersecting cosmic axes—like the main routes and canals of Mismanay. On the ground everything intersects at the Crucero; everything crosses at "the Cross." In the sky the corresponding point is the zenith.

The zenith is one of the organizing principles of the Andean world, and it estab-lishes the character of sacred space. The little chapel of Mismanay is such a spot, and

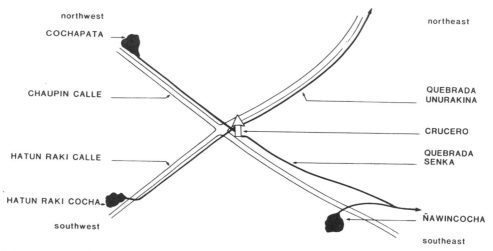

northwest
COCHAPATA

northeast

CHAUPIN CALLE

QUEBRADA
UNURAKINA

CRUCERO

QUEBRADA
SENKA

HATUN RAKI CALLE

HATUN RAKI COCHA

ÑAWINCOCHA

southwest

southeast

Inhabitants of the small Andean village of Mismanay think of their world as oriented by the intersection of intercardinal cosmic axes. These are associated also with sunrises and sunsets on the zenith and antizenith passage dates and on the June and December solstice dates. The Milky Way's intersections with the horizon when it crosses the zenith also fit into the scheme. Accordingly, the main footpaths (Chaupin Calle and Hatun Raki Calle) of the village run in roughly intercardinal directions. As it turns out, the primary irrigation canals (Quebrada Unurakina and Quebrada Senka) flow from the reservoirs, or *cochas,* and cross near the center of Mismanay, where a chapel known as *Crucero* ("cross") was built. All of these earthbound crosses, converging as they do, mirror the celestial crosses the Peruvian Indians perceive in the sky.   *(Griffith Observatory, after Gary Urton)*

it is located at the hub of a landscape partitioned intercardinally. It mimics all of the "intercardinal" axes of the sky, including the intersection of heaven's river with the zenith.

In the Andes the Milky Way is a river. Its Quechua name, *Mayu,* means "river." Urton applied the present Andean concepts of the "cross" at the zenith and the "river" in the sky to explain the location of the Coricancha—Cuzco's heart. Cuzco's most sacred spot was the intersection of its two rivers. The Inca built the Coricancha at the confluence because that place represented terrestrially the organizing pivot of heaven. Urton believes the unsymmetric division of Cuzco's southern quarters also can be understood in celestial terms. The southeast boundary is set by a ceque that could have been oriented on Alpha Crucis, the brightest star in the Southern Cross. Urton showed this was associated with the "center" of the Milky Way, the place where the oppositely flowing "waters" of its two halves "collide." The Milky Way is bright there, and its light is the "foam." Urton believes the Inca expanded the southwest quarter in order to keep the ceques that were aimed at the rising and setting points of Alpha Crucis in the same *suyu.* If so, we have yet one more intricate link between the organization of the Inca world and the pattern they perceived overhead.

The City Where the Gods Began

Long before the Aztec built their huge twin pyramid at the heart of Tenochtitlán or the Inca strung huacas on the ceques of Cuzco, a people—whose own name for themselves has disappeared with their empire—planned and built an urban metropolis and ceremonial center that others, centuries later, called Teotihuacán—"the Place of the Creation of Gods."

It was still a site of worship in Aztec times, for the Aztec inherited many traditions of Teotihuacán through the Toltec and other peoples of central Mexico. The first major structures of Teotihuacán were erected around the time of the birth of Christ, but the city was burned and abandoned around 750 A.D.—five and a half centuries before the Aztec entered the Valley of Mexico. In the interim, the Toltec state—centered at Tula—reconsolidated central Mexico and later fell to barbarian invaders. Despite the continuity of some cultural traditions, much was lost. We therefore have to rely on archaeology for most of our knowledge of Teotihuacán.

Two monumental pyramids dominate the Teotihuacán ruins. Called the Pyramid of the Sun and the Pyramid of the Moon, they certainly sound as if they have something to do with the sky. The names were provided, however, by the Aztec, whose myths marked Teotihuacán as the sacred place where the Fifth Sun the sun of the

The Pyramid of the Sun dominates the landscape of Teotihuacán. Here it is viewed from the Viking Group, the site of one of the pecked crosses that defined the city's axes of orientation.   *(Robin Rector Krupp)*

present cosmic age—was created through the gods' self-sacrifice. In this myth, Nana-huatzin, an impoverished god with a decaying, ulcerated body, threw himself into the ceremonial fire when the four Creator gods—one for each cardinal direction—could not bring themselves to jump in. Consumed by the flames, Nanahuatzin became the new sun. His companion in sacrifice, the wealthy Tecciztécatl, overcame his fear and followed Nanahuatzin into the fire and was transformed into the moon. More con-flicts and sacrifices took place before the Fifth Sun agreed to move in his proper course, orient the world by his risings, and organize the passage of time, but these details—most of them with celestial symbolism—need not concern us here. The myth embodies the familiar theme of the cycle of cosmic order. Decay and death lead to purification and resurrection. The whole world is re-created through celestial meta-phor, and the Aztec saw in Teotihuacán's ruins the end of one age and the monu-ments to the next. The two large pyramids were built, they said, for the two gods who became the sun and the moon.

It is clear that Teotihuacán was regarded by the Aztec as the original source of civilization and government, and the place where cosmic order was established. Whether the people of Teotihuacán told themselves the same myths and actually dedicated the Pyramids of the Sun and Moon to those two celestial gods is unknown, but archaeology tells us the city was all the Aztec thought it to be and more. Teoti-huacán was a sprawling urban center where people lived and worked, and it was the principal focus of ceremony and religion throughout the large area of central Mexico under its control.

Under the direction of Dr. René Millon, the University of Rochester's Teotihuacán Mapping Project established that the whole city was laid out by its planners on an orderly grid. Its buildings and blocks were organized into neighborhoods where vari-ous crafts, industries, activities, "immigrants," and social or family groups were con-centrated. At its height nearly 100,000 people lived there. More came on pilgrimages and to trade. It was a real city, and despite its urban complexity, its grid plan imposed visible order upon all of it. The rectilinear plan extended several miles from city center, out into the "suburbs" and on the surrounding hillsides. Even today the rows of agave plants in cultivation on the land once occupied by outlying *barrios* west of the main city seem to preserve the old Teotihuacán city plan in sight of the two great pyramids.

The main axis of Teotihuacán is the "Street of the Dead." It is not a street at all, however, but a series of plazas, flanked by platforms and multiroom structures. They open onto the Street of the Dead and seem suited for ceremonial use. At the Pyramid of the Sun, situated just east of the Street of the Dead and with its sides in confor-mance with the city's rectilinear plan, the central axis turns into something like a processional way which ends at the Pyramid of the Moon.

Streets and structures of Teotihuacán were oriented according to its main axis. In its time Teotihuacán was unique. No European city could claim comparable foresight and elegance in its overall plan. Within the well-laid scheme is a mystery, however. The central axis of Teotihuacán—the Street of the Dead—is oriented 15½ degrees east of north. The direction seems arbitrary. We have seen cardinal and solstitial orientation of cities and ceremonial centers, but there is nothing obviously significant in Teotihuacán's orientation. Practical considerations for the local topography seem unlikely, for the inhabitants of Teotihuacán modified the landscape to suit them. Left to its natural course, the San Juan River would have run diagonally through the plazas and pyramids on the south end of the Street of the Dead. Instead, the builders canalized the river and made it conform to the city's axes.

The full extent of the Teotihuacán city planners' commitment to the city's main axis and grid became clearer as Millon's heroic mapping effort continued. Altogether, Teotihuacán extended over 12½ square miles, and here and there in that area— sometimes in enigmatic locations—Millon and his team found a symbol pecked into floor pavements and rocky outcrops. The design has come to be called a "pecked cross." It is actually a set of concentric circles, usually two, quartered by two axes that cut through the circles' center. Neither the circles nor the axes are drawn as continuous lines. Instead, the design is executed in sequences of shallow holes chipped into the rock or punctured into the floor. In a way, the effect is like speed limit signs on which the numbers are formed of reflective buttons. In "downtown" Teotihuacán,

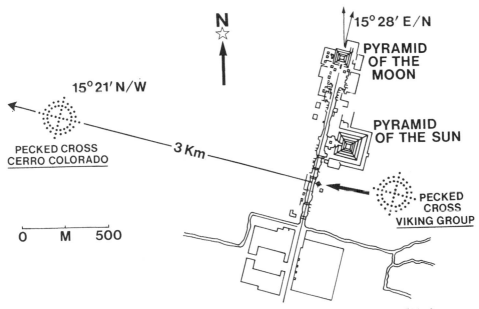

The overall plan of Teotihuacán is a very orderly grid, but it is skewed 15½ degrees east of north. The Street of the Dead is one of the grid's principal axes, and the pecked crosses in the Viking Group and on Cerro Colorado establish a line perpendicular to this primary feature of the city. *(Griffith Observatory)*

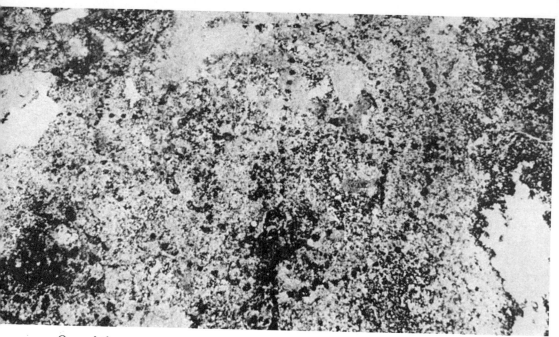

One of the quartered circle, or pecked cross, designs that establishes the orientation of Teotihuacán was inscribed in the plaster floor of a room in the Viking Group, located on the east side of the "Street of the Dead." *(Robin Rector Krupp)*

one of these markers (cross number 1) is in the middle of the floor in a room that opens onto the Street of the Dead. The room is in a complex known as the Viking Group—named after the foundation that funded its restoration and not a reference to transatlantic contact with Scandinavia.

A total of 233 holes complete the Viking group pecked cross. A very similar design (cross number 5) with 201 holes was found on the low hillside of Cerro Colorado, a little less than 2 miles to the west of the Viking Group. Unlike the Viking Group cross, Cerro Colorado's marker is carved into the sloping surface of a small outcrop of rock—one of few, in fact, on the hillside. The slope is empty today, except for the scrub and cactus—including the cholla, which seems to jump out of nowhere and lodge its hooked spines into shoelaces, pants cuffs, and ankles. A few small houses are situated near the base of the hill. Cerro Colorado was just about the edge of the city in this direction during prehistoric times.

Both pecked crosses have two rings. The arms of the cross in both cases are identical, with 18 holes in each (or 20 if we count the holes in the circles where they are intersected by the arms). Also the pattern of the arm holes is the same: ten inside the inner circle, four between the circles, and four beyond the outer circle. Both designs are about 3 feet in diameter. They are rather small details in the entire Teotihuacán environment, but three things about them suggest they might be important. First, they are similar. Second, their locations are in sight of each other. Third, the line they

The Cerro Colorado pecked cross is carved into a rocky outcrop on the steeply sloped side of the hill. *(Robin Rector Krupp)*

form between them is almost exactly perpendicular to the Street of the Dead—within 7 arcminutes, in fact. Their relationship to the main axis of Teotihuacán moved Dr. James W. Dow, an archaeologist, to propose that the crosses were used to establish a baseline for the layout of a perpendicular line: the main axis. The quartered circles could be surveyor's benchmarks, preserved because of their symbolic importance.

Actually, the grid pattern of Teotihuacán is more complicated than this. Most of the streets and alleys don't actually run through the city from one end to the other, and only one main east-west thoroughfare has been identified. It is really two avenues. One extends eastward from the Ciudadela, a large ritual plaza at the south end of the Street of the Dead. Inside the walls of the Ciudadela is the famous pyramid known as the Temple of Quetzalcóatl (featured on the back of Mexico's twenty-peso note). The west avenue departs from a symmetrically placed plaza known as the Great Compound. It was probably the site of the city's major market. No doubt pilgrims, merchants, craftsmen, farmers, and traveling traders made their way into the city by its east and west avenues to participate in the city's religious life and to shop its great markets. Teotihuacán was the political, religious, and economic hub of Mesoamerica.

Although the Street of the Dead is twisted 15°28′ east of north, the "east-west" line so clearly defined by the east and west avenues is not perpendicular to it. Were it truly perpendicular, it would point 15°28′ south of east. Instead, it forms a second orientation axis—16°30′ south of east. This is only a small difference, just one degree,

From a point near the Viking Group pecked cross at Teotihuacán, the slope of Cerro Colorado, about 2 miles away, may be seen. The position of the Cerro Colorado pecked cross is indicated. *(E. C. Krupp)*

but the care with which the whole city was planned implies it was intentional. A marker something like a pecked cross near the peak of Cerro Gordo, the large mountain north of the city, establishes a line with the Viking Group cross that is nearly perpendicular (17° east of north) to the "east-west" axis. What all this means is very intriguing. The "east-west" orientations of Teotihuacán avenues parallel the main "east-west" axis. The "north-south" orientations of streets parallel the main "north-south" axis. These axes are not exactly perpendicular to each other, and the streets and avenues aren't either. And yet the evidence from the quartered circles and the streets themselves confirm that this is how the people of Teotihuacán wanted it. Why?

Local topography offers few answers. Nothing is really aligned on prominent mountains. If anything, Teotihuacán ignores the terrain. James Dow said, "The city gives the appearance of a colossus erected in defiance of natural topography and certainly not in conformance with it."

If Teotihuacán's alignments are not terrestrial, perhaps they are celestial. One astronomical explanation of the city's plan is often repeated. Guidebooks and other accounts of Teotihuacán invariably state the city was oriented in a direction perpendicular to the position of sunset on the day of zenith passage. This would make the Pyramid of the Sun face that sunset, but the claim is untrue.

Dow considered another astronomical interpretation and concluded that the rising point of Sirius in A.D. 150 could explain the east-west line. This was the time when the ancient surveyors of Teotihuacán were laying out the Street of the Dead and planning their city. They might, Dow thought, also have been interested in the direction where the Pleiades set, for that, too, looked like it fell on the "east-west" axis.

Aveni verified the true orientation of the Viking Group–Cerro Colorado line in the field by surveying it and found it agreed well with the setting of the Pleiades. Dow had noted that the Pleiades, in A.D. 150, rose heliacally on the day of spring zenith passage (May 18), and for the next several centuries the cluster could have heralded this solar event. This may have given the Pleiades added importance and provided a motive for the city's layout.

Further field study by Anthony Aveni, Horst Hartung, and Beth Buckingham demonstrates, however, that it is much too soon to draw detailed conclusions about the orientation of the Teotihuacán grid and the role of the pecked crosses. In 1978, they catalogued 29 known examples of the design, found either by them or by others. Since then the total has exceeded 70. A cross very similar to those at Teotihuacán was marked on the floor of a building at Uaxactún, a Maya center (ca. A.D 300–900) more than 600 miles to the southeast in the Guatemalan Petén jungle. A large pecked cross, 13 feet in diameter, was reported—with a drawing—in 1889; said to be near the northern frontier and U.S. border but without a detailed account of its location, it has not been seen again. What may be related pecked designs are also known in New Mexico and California.

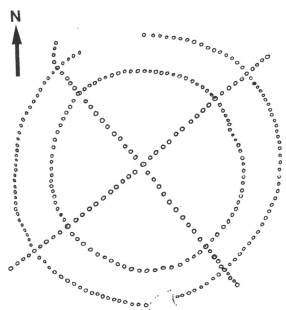

Several dozen crosses have been found throughout Mesoamerica. This one, #5 from Tepeapulco, displays the canonical 10-4-4 count in each of the four arms. *(Griffith Observatory, after A. F. Aveni)*

Among the most interesting pecked crosses outside of Teotihuacán is a pair carved into the rocky shelf at the east edge of a small mesa known as Cerro El Chapín. Now Chapín is just about 4½ miles south of Alta Vista, the apparent Teotihuacán outpost near the Tropic of Cancer. And the crosses at Chapín are very similar. Two hundred and sixty-one holes make up one of them (Chapín #1), and 265 holes the other (Chapín #2). In the former the inner and outer rings have 80 and 100 holes respectively. The split is 83 and 101 in the latter. All arms on both symbols have the usual 10-4-4 pattern of dots. The "east-west" axes of both crosses actually point toward the northeast, and from them Picacho Montoso—Alta Vista's equinox indicator—is conspicuous. But the exact placement of the Chapín crosses is, perhaps, more important than anything else about them. From these stations the summer solstice sun rises precisely behind Picacho Montoso. Recalling that Alta Vista and Cerro El Chapín are practically on top of the Tropic of Cancer, Aveni, Hartung, and archaeologist J. C. Kelley realized the crosses also mark the one day the sun passes through the zenith. Here, at the Tropic, zenith passage and summer solstice combine. It is hard to believe the ancient peoples of this area were unaware of this, and the Chapín markers verify that pecked crosses have something to do with astronomy.

In 1978, two pecked crosses were found near Tepeapulco, about 20 miles northeast of Teotihuacán. Tepeapulco was an important node on the obsidian trade route. Obsidian tools, crafted by Teotihuacán's workers, were a major commodity in the city's market. Teotihuacán architecture exists at Tepeapulco. More than three dozen more pecked crosses have been found since, and several of these point to the summit of Cerro Gordo or to Teotihuacán. Long-distance orientation on landmarks and important monuments may imply a complex system of sacred places in the landscape of central Mexico. Proper placement and alignment of structures and markers within the rules of such a system—like the Chinese feng shui—reflect an ordered, cosmic vision of the world.

Recent discoveries at Teotihuacán further extend our appreciation of that ordered vision. It appears now that the centers of the Pyramid of the Sun and the Ciudadela—two of Teotihuacán's most significant architectural features—fall on the line that connects the summit of Cerro Gordo on the north with the flank of Cerro Patlachique on the south. Another line due west from the center of the Pyramid of the Sun intersects a mountain known as Cerro Maravillas, 4½ miles away.

Other considerations also may have played a part in the planning of Teotihuacán. Cerro Gordo is the source of the San Juan River, the main water supply for the city. A multichambered cave with a long, twisting passage that opens to the east is underneath the Pyramid of the Sun, and superficially it resembles the "labyrinth" at the Alta Vista's Hall of Columns. Archaeologist Stephen Tobriner believes Cerro Gordo's association with water may have inspired the people of Teotihuacán to orient—albeit roughly—their city with the mountain. Doris Heyden, also an archaeologist, suggested the orientation of the cave and the ceremonies associated with it set the

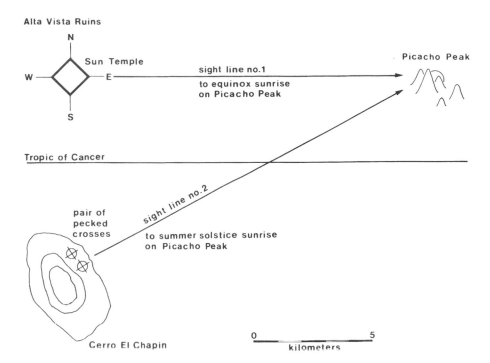

Alta Vista Ruins

N

Sun Temple

W — E ———————— sight line no.1 ——————————→

to equinox sunrise
on Picacho Peak

S

Picacho Peak

Tropic of Cancer

pair of
pecked
crosses

sight line no.2

to summer solstice sunrise
on Picacho Peak

Cerro El Chapin

0   5
kilometers

The Hall of Columns, or "Sun Temple," at Alta Vista and the pair of pecked crosses on Cerro El Chapín are both very near the Tropic of Cancer. At this latitude the sun passes through the zenith on the summer solstice. From the pecked crosses, the summer solstice sun rises behind Picacho Montoso, and the same peak is the foresight for the equinox sunrise observed from the Hall of Columns. *(Griffith Observatory, after A. F. Aveni)*

direction of Teotihuacán's grid. Both of these ideas could be valid, and neither of them diminishes the importance of the various astronomical and geomantic alignments already catalogued. Teotihuacán had, from the first, a conscious and carefully contrived order. Millon called it "an initial vision that is both audacious and confident." Locked within the city's monuments, avenues, and relationships to outlying sites seems to be a complex system of associations. The order is there. The quartered circles are part of it. And these symbols have something to do with sky. They tell us that somewhere in the planning of Teotihuacán its builders' view of the cosmos took root.

12

The Symbols We Draw

There is no question that the pecked crosses of Mesoamerica stand for something. They are symbols that "spoke" in a compressed, visual language to those who used them. Their design emerged from the key elements of the message they contained, and anyone sharing the belief system that produced them could understand what they meant. Our culture has symbols of its own, and when we see them, we recognize their meaning without necessarily dealing with it in detail. One major television network puts its identity in a stylized peacock whose fanning tail contains a feather for each color in the rainbow. An ancient Teotihuacano would be hard-pressed to explain the periodic appearances of this fantastic "mythological" bird on the screen of ever-changing images. Numerological significance would be sensed with the realization that the bird only appears when a movable dial indicates a certain number. Of course, we who interact with the environment television creates realize the peacock represents the network's pride in its programming, and the multihued tail feathers are an allusion to color broadcasting. Another network, by contrast, confronts us with an "eye" abstracted into a symbol that implies the act of viewing, which, of course, is what television is all about.

These are commercial emblems from a secularized age, of course, but they do their job. Through symbolic recognition they create identity. We know what we are dealing with when we see them. It is equally possible to condense the principles of cosmic order into sacred symbols. These become familiar images which orient the viewer and integrate him or her into the world structure they represent.

Crosses, Calendars, and Cosmic Gameboards

Let's take a closer look, then, at ancient Mexico's pecked crosses, for they may be able to tell us something more about themselves. Even if we did not find them associated with the orientation of architecture and alignments on celestial events, we would guess that their axes have something to do with direction. The idea of world quarters is widespread and is expressed by cultures throughout the world in architecture, ritual, and myth. Anthropologist Claire R. Farrer asked her key consultant among the Mescalero Apache of New Mexico to draw a picture of the universe, and he provided her a circle quartered by a cross. This symbol is, as Dr. Farrer terms it, a "template" that structures all aspects of Mescalero life. Metaphorical transformations of the principle of four permeate the way they look at the world and, accordingly, how they behave.

Even formal speeches at Mescalero community gatherings are organized into four parts. The speaker begins with a prologue that adheres to a standard formula, and in it he explains that he doesn't really know much about the topic at hand. This is followed by the speaker's thoughts on the subject. He then offers a brief summary of what he has said. Finally, he finishes with a standard closing in which he acknowledges his ignorance but says that his words have expressed how the situation looks to him. In structure, the speech is a model for effective communication. Bordered by the speaker's admissions of his own limitations, his argument slips into the public discussion with tact and grace. This form is intended to harmonize such exchanges in an environment in which conflicts and sincere differences of opinion are bound to arise. By adopting a principle of cosmic order that originates in the sky, the Mescalero inject the world's "rightness," or congruence, into the public life of their community.

In Mescalero tradition, the four directions are the four "Grandfathers" who hold up the universe, and this metaphor penetrates Mescalero architecture and ritual. The four primary poles that support the sacred tipi constructed for the girls' puberty ceremony are equated with the "Grandfathers," and the ritual itself is designed to last four days. As a principle of natural order, the number four is also recognized by the Mescalero Apache in the number of the seasons, and they see the pattern of their own lives in terms of four stages, or "seasons," as well: infancy, childhood, adult life, and old age. The girls' puberty ceremony celebrates a fundamental transformation in life—the transition to womanhood—as certainly as the winter solstice marks the turning of the year. In most traditional cultures, puberty rites embody the idea of renewal or rebirth and so rely on the imagery of the cycle of cosmic order. Mountain God dancers participate in the Apache girls' puberty ceremony, and each of their bodies is decorated with a four-pointed star. The star is said to be the "same thing" as the Four Grandfathers.

The number four's significance is rooted in the sky. It encodes a perception of space and time with symbolism that is economical, immediate, and accessible. It is like a button that fastens Mescalero life to the sky and secures it to its proper place in the cosmos. Through the symbols of their belief system the Mescalero Apache—in Farrer's words—"live the sky."

Crosses with astronomical connotations are known from other parts of Mesoamerica. The Maya *kin* glyph looks like a St. Andrew's—or "intercardinal"—cross and is known to symbolize "sun," "day," or "time." Associations between all three of these ideas and the concept of world quarters evolved naturally out of the celestial phenomena that put structure into space and time. The sun that calibrates the calendar and deals out the days also travels through the territories that define direction. It is, in fact, the sun's movements in these realms that reveal and mark the passage of time. Our ancestors not only saw this happening around them, they also measured it and systematized it.

The *kin* glyph may not allude just to directions but to the technique—the instrument—that established the prime directions and the positions of celestial timekeepers. A pair of crossed sticks—something that looks quite a bit like the *kin* "cross"—most

The chest of this Arizona Apache Mountain Spirit dancer is painted with a four-pointed star similar to those worn by the Mescalero Apache of New Mexico. *(E. C. Krupp)*

A temple, a pair of crossed sticks, and, perhaps, an astronomer comprise this glyph from the *Codex Bodley*, a pre-Conquest Mixtec manuscript. The circular ornaments that look like half-closed eyes are star symbols. *(Horst Hartung)*

likely established astronomical sightlines from Mesoamerican temple platforms. Numerous examples of these ancient observatories are illustrated in several of the codices, and Zelia Nuttall, an amateur archaeologist who often did professional quality work, inventoried the astronomical symbolism in the codices as early as 1906.

Horst Hartung has taken a fresh look at the *Codex Bodley* and the *Codex Selden I* and summarized the crossed-stick symbolism they contain. Both of these documents come from Mixtec territory, in Oaxaca, and seem to record dynastic history. Crossed sticks frequently show up in the doorways of temples, between seated figures, and even as a headdress. Often a star glyph sits in the V created by the sticks. These stars are not the multipointed star symbols we use but look more like round eyes with the lid half-closed to the pupil. The same star symbols stud some of the little temple drawings themselves as though to designate them as observatories. In one explicit case an ancient Oaxacan skywatcher has extended his head out the temple door and looks across the V.

In the codices, these symbols are not meant to chronicle everything that went on at a Mixtec observatory. Instead, at least some of the temple glyphs with stars and crossed sticks are place names. They refer to a settlement identified by the observatory that once stood there. In fact, the present-day city of Tlaxiaco, in the Mixteca Alta, is thought to have been the "place of the observatory." A modern parallel can be imagined. Griffith Observatory in Los Angeles is located in Griffith Park, and the Observatory is a prominent landmark. Some future history of Los Angeles, written centuries from now when details of the past are lost and confused, may recall the "ancient observatory" so closely associated with this park and refer to the land as

Griffith Observatory Park. The reference might not tell us much about the observatories of old, twentieth-century Los Angeles, but it would verify that an observatory once existed there. This is what the Mixtec codices tell us about astronomy in ancient Oaxaca. Fortunately, the picture writing also gives us some clues about the observatories themselves. Similar crossed sticks are shown in a temple glyph carved on an exterior slab of Monte Albán's Structure J. There, too, it probably has nothing to do with astronomy but is an emblem of conquest of a town known by its observatory.

Whether the quartering of the pecked circles refers to the world quarters, crossed sticks, or both is not something we can prove; the people of Teotihuacán didn't tell us how their symbols came to be. However, detailed information may even be contained in the peck marks that form the circle-cross petroglyphs. Aveni noted that the 10-4-4 pattern in each of the arms provides the number 18 when the marks are added, and 18 was an important number in the Mesoamerican calendar. The 365-day year had 18 *veintenas* with 5 extra days that had no "month" of their own. Inclusion of the marks where the arm intersects the circles bumps the number to 20, another important number in the ancient calendrics. Although it is true that the total number of peck marks varies considerably from cross to cross, three that seem well drawn (the two at Cerro El Chapín and the lost marker in the north of Mexico) have totals that come close to 260, the number of days in the Sacred Count. At Chapín the totals are 261 and 265, and the third apparently also had a total of 261. The Uaxactún cross in Maya territory was made with care, and the total in just its two rings is 256.

Suggestive tallies of the dots in these crosses imply some kind of connection with the calendar, and yet there is too much variation from one cross to the next to let us conclude the crosses were actually meant to serve as calendars. Rather they are symbols related to the calendar and to the passage of time.

This rather elaborate design, certainly related to the quartered circle symbols, is pecked into the floor directly across the Street of the Dead from the Viking Group cross. The dots in the outer perimeter total 260. The calendric significance in this emblem is verified by similar patterns of dots in central Mexican and Maya codices. *(Griffith Observatory, after A. F. Aveni)*

N

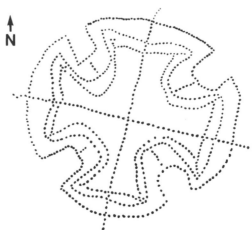

The border of this Aztec design in the *Codex Fejérváry-Mayer* is marked by 260 dots. Directional, calendrical, and mythological symbolism make it a cosmogram, a picture of cosmic order. *(Bellerophon Books, Santa Barbara, CA 93101)*

Page one of the *Codex Fejérváry-Mayer* screenfold combines cosmology and calendrics into a design based on forms similar to a pair of intersecting Maltese and St. Andrew's crosses. Direction symbolism orients the design with east on top. An eight-pointed sun disk above a pyramid corresponds to sunrise, and the color of the border is red, a color associated with east in central Mexico. West, then, is the blue branch of the cross at the bottom. North, in yellow, is on the left, and south, in green, is on the right. Each zone contains a different kind of tree, growing from a different source, with a different bird at its summit and a different pair of attending gods on either side.

Four directions imply horizontal orientation, but some of the symbols cue us to vertical organization as well. In the north the bowl that harbors the tree has a star or sky symbol in it and signifies the realm above. On the right, in the south, the tree grows from the reptilian jaws of an earth monster. It is the portal to the realm below. Vertical structure is, of course, also a component of the native Mexican cosmology.

Xiuhtecuhtli, the old Fire God and Year Lord, stands in the center square. He is the first of the 13 Lords of the Day and also the first of the nine Lords of the Night. If we imagine the realm of day as a seven-level pyramid, the steps up one side and down the other total 13 and correspond to the Lords of the Day. An inverted, five-stage pyramid provides nine steps for the Lords of the Night. These ideas are related to the 13 "heavens" and the nine "hells" of the layered cosmology of Mesoamerica, and they also symbolize the passage of time in a day and night. Xiuhtecuhtli, then, sponsors the first "hour" of night and is a master of time. He is energized, perhaps, by the blood that flows diagonally into his realm from each corner of the page. Four parts of the body are the sources of this blood: a hand, a head, ribs, and a foot. The foot is severed to the bone and may belong, with the other parts, to Tezcatlipoca, who had this attribute.

Clear calendrical references also appear in the diagonal zones, next to the "St. Andrew's Cross" arms. Each domain contains five of the 20 day-name glyphs from the *tonalpohualli*, or 260-day Sacred Count. At the very end of each diagonal branch is a different type of bird, each with a different glyph upon its back. The glyphs are easy to recognize: rabbit, house, flint knife, and reed. These are, of course, the four year-names we have already encountered, and the four birds are the four year-bearers. They are the same *voladores* we encountered at El Tajín.

A 260-day calendar cycle was also used by the Maya, and the perimeter of 260 dots around this drawing from the *Codex Madrid* echoes the shape and significance of the design in the Aztec *Codex Fejérváry-Mayer*.   *(Stephen D. Peet,* Myths and Symbols or Aboriginal Religions in America, *1905)*

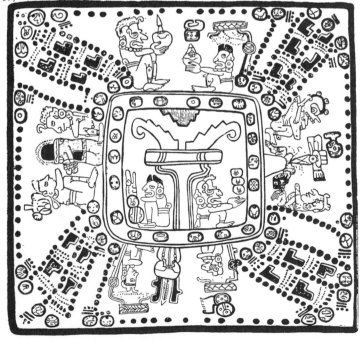

By now it is evident that this opening page from a pre-Conquest "book," most probably produced in the Mixteca-Puebla or Gulf Coast regions, has cosmological, astronomical, and calendrical meaning. Its relevance to the pecked crosses, and especially to #2 at Teotihuacán, goes beyond their similarity in shape. A careful examination of the entire outer perimeter of the Fejérváry-Mayer diagram reveals a continuous sequence of dots that resemble the peck marks in the pecked crosses. Every thirteenth dot is replaced by one of the 20 day-name glyphs, and the total around the entire border is 260, exactly the same number as in the rim of cross #2 at Teotihuacán.

Fray Diego Durán, a Dominican friar, came to Mexico as a child and later wrote an extensive account of the history, traditions, religion, and calendar of the Aztec. In it, the 52-year cycle is symbolized in a kind of quartered circle. Four spokes radiate from a central hub, occupied by the sun. Each spoke bends into a quarter-arc of a circle at the same radius, but the arcs do not touch. These spoke-arc combinations are divided, in turn, into 13 little boxes, and in the boxes are the number-name designations for the 52 years. All the boxes in a single arm have the same day-name—rabbit, say, or house. The numbers are not sequential but fall in a pattern determined by the actual sequence of year designations. Cardinal direction names orient the four sides of the

Another Mesoamerican calendar cycle, the 52-year xiuhmolpilli, is symbolized in this drawing included with the writings of the sixteenth-century chronicler Durán. Each square contains the name of a different year, given by a number from 1 to 13 (indicated by the number of dots) and by one of four day-names: rabbit (upper right quarter), reed (upper left), flint knife (lower left), and house (lower right). *(Fray Diego Durán,* Book of the Gods and Rites *and* The Ancient Calendar, *trans. by Fernando Horcasitas and Doris Heyden. Copyright 1971 by The University of Oklahoma Press)*

page, and winds, depicted in conventional European style, blow in from the corners.

Both of these calendar wheels look like gameboards, a notion that isn't so far-fetched. Mesoamericans played a game called *patolli,* which has been likened to par-cheesi. Durán tells us that the gameboard was sometimes prepared by carving out small cavities in the floor stucco. Bean markers were moved from cavity to cavity (or from square to square in gameboards painted on reed mats) in accordance with a throw of bean "dice." The course of play was laid out as a cross. Intense play and high stakes were part of the game, and losers could be brought to complete ruin.

Post-Conquest accounts also indicate patolli had religious meaning, and calendrical, numerological, cosmological, and astronomical symbolism determined the rules of play. For example, each player's tokens migrate through 52 spaces by the time they complete their circuit of the board. Francisco López de Gomara, the household chaplain to Cortés, wrote, "Montezuma showed the greatest pleasure . . . at the Spaniards' game of cards and dice." (With 52 cards in a pack, four suits, 13 cards to a suit, and plenty of numbers and royal imagery, I should think Montezuma would have been pleased.)

Two Indian tribes in a remote region of Puebla were still playing the game in 1925, and from their accounts and post-Conquest descriptions, Timothy Kendall, of the Museum of Fine Arts in Boston, reconstructed the original game. His study accompanies a patolli game manufactured by a U.S. game company for today's market.

When the ancient Mexicans played patolli, they put their fortunes and fates in the throw of the dice and the hands of the gods. Organized like the cosmos, the game and its board symbolized the order of time and space. Patolli's connection with 52 associates it with the 260-day Sacred Count ($52 \times 5 = 260$) and the 52-year Calendar Round, and it is likely the game was employed in divination. Through the calendrical diagrams, the cosmological designs in the codices, and the pecked crosses we can trace its meaning all the way back to the ordered layout of the "place of the creation of the gods," Teotihuacán.

A Celestial Seal of Approval

One of ancient Mexico's most familiar symbols is the complex, circular design known as the Aztec "Calendar Stone." Pottery, purses, and pendants decorated with the design can be found in the tourist markets of Mexico; it is tooled into the leather of imported wallets; wall calendars and ashtrays ornamented with this picture from Mexico's past find their way into homes throughout the world.

The Aztec Calendar Stone is not really a calendar at all, but a huge symbolic carving—53,000 pounds of olivine basalt—that once occupied a prominent place in the sacred central precinct of the Aztec capital, Tenochtitlán. Like most of the city, the Calendar Stone lay buried in rubble and ruin below the streets and buildings of Colonial Mexico City. But on December 17, 1790, while the Zócalo, or main plaza, was excavated and the Cathedral repaired, the massive sculpture, 13 feet in diameter, was brought, again, into the sun. Another name for it is the "Stone of the Suns," and the central face has been long identified as a portrait of Tonatiuh, the present Sun.

Tonatiuh, in Aztec tradition, was the fifth in a line of Suns, each of which ruled a previous age that fell in cataclysm and was replaced by the next. According to the *Leyenda de los Soles* ("Legend of the Suns"), a post-Conquest native manuscript written in Nahuatl in 1558, the first age was known as 4-Jaguar. It was ruled by

The ring of day-name glyphs that surrounds the central design of this monumental Aztec sculpture is responsible for its popular but incorrect name, the Calendar Stone. Among the symbols identified here are the emblems of the four cosmic ages that preceded in the minds of the Aztec their own time—the age of the Fifth Sun. *(Bellerphon Books, Santa Barbara, CA 93101)*

Another Aztec sculpture also carries the emblems of the four previous Suns. On the left side here is the symbol for 4 Water (the goddess Chalchiutlicue), and the right side has the jaguar's head that stands for Tezcatlipoca and 4 Jaguar. *(E. C. Krupp)*

The other two sides of the "Stone of the Four Suns" portray 4 Wind (the god Éhecatl) on the left and 4 Rain (the goggle-eyed god Tlaloc) on the right. *(E. C. Krupp)*

Tezcatlipoca, and wild jaguars devoured everyone living in this era. In the end, the jaguars themselves were consumed, and the second great age, 4-Wind began. This Sun was swept away by hurricane winds, and followed by the third world age, 4-Rain. Fiery rain—perhaps a metaphor for volcanic eruption or possibly for the lightning of a cataclysmic thunderstorm—destroyed the third Sun, and the fourth age, 4-Water, began. It, too, eventually fell, and those who lived then perished in a great flood. With the death of the fourth Sun, Tonatiuh ascended the throne of cosmic time to rule the fifth age, 4-Movement.

Emblems of the four previous Suns occupy the four squares that surround the central face on the Stone of the Suns. Cosmological time begins in the upper right square—which contains a picture of a jaguar's head—and circulates counterclockwise. Age number two is symbolized in the upper left square by the head of Éhecatl, a

The combined outline of the face in the center of the "Calendar Stone," the four squares around it, and the claws on each side of it is in the shape of the *ollin* glyph, which means "earthquake" or "movement." Although the central face traditionally has been identified as that of Tonatiuh, the Fifth Sun, some interpreters now regard it as a portrait of the earth-lord Tlaltecuhtli. In part, these new conclusions are based upon other representations of Tlaltecuhtli recovered during the recent excavations of Templo Mayor. In either case, the cosmological, directional, and calendrical symbolism of the rest of the design authenticated Aztec sovereignty. *(Robin Rector Krupp)*

form of Quetzalcóatl associated with the wind. Below it, the square on the lower left frames the face of the god Tlaloc. He is most often connected with rain and stands for the downpour of celestial fire. Finally, the goddess of rain and water, Chalchiutli-cue, completes the design by representing the fourth age in the last square, on the lower right.

Taken together, the four squares, the round face, and the two claws that protrude to the left and right outline the shape of the glyph *ollin,* or "movement." This alludes to the present age, the Fifth Sun, which is to end in earthquake, according to the myth.

Surrounding the central design is a ring of 20 glyphs. These represent the names of the days in the 260-day calendar cycle, and because of them the huge disk became known as the "Calendar Stone." The signs start with *Cipactli* ("alligator") at the top and continue, counterclockwise, around the ring to end at *Xóchitl* ("flower").

Many other symbolic images are incorporated into the disk. What look like eight triangular rays are arranged in the form of a compass rose around the day-name ring. A circle of dots surrounds the very edge of the disk and stands for the starry nighttime sky. Just inside this ring, two *xiuhcóatls,* or "fire serpents," form another ring, their tails at the top and their mouths, each enclosing a human head, faced off against each other at the bottom. These serpents symbolize the path and movement of the sun across the daytime sky and are like the feather and paper serpent carried, east to west, down the steps of Templo Mayor at the festival of Panquetzaliztli.

A fire serpent, or something like one, descends a stairway of the main pyramid at Chichén Itzá in Yucatán. Originally Chichén Itzá belonged to the Maya, but Toltec from Tula in central Mexico or, more likely, Toltec-influenced Maya from the Gulf region, took control of this ceremonial center in the tenth century and created a hybrid architectural style with the original Maya inhabitants. The big pyramid, called the Castillo, reflects the Toltec influence at Chichén Itzá. It appears to combine considerable celestial numerology into its various components (four sides and stair-ways, 52 sunken panels on each side, 91 steps to a stairway plus the top level for a total of 365, and nine levels) and may be oriented toward zenith passage sunset. Most recently, however, J. Rivard noticed that in the hour before equinox sunset a pattern of light and shadow forms on the west-facing balustrade of the north stairway. Actual-ly, this phenomenon was well known to the local people.

In recent years the vernal equinox has been transformed practically into a national holiday in Mexico, for 10 to 12 thousand people turn up then at Chichén Itzá to watch the serpent's descent. Most of this interest has been stimulated by Luis E. Arochi, an enthusiastic amateur and the author of a successful book on the phenomenon.

The pyramid was carefully oriented and proportioned to let the profile of its north-west corner create first one inverted triangle of light and then another below it in a descending image of a diamondback serpent. At the bottom are serpent heads. The

A symbol in sunlight, this undulating sequence of triangles transforms the north balustrade of the main pyramid, or Castillo, at Chichén Itzá into the body of a serpent during the last hour before equinox sunset. *(Robin Rector Krupp)*

serpent heads argue well that the alignment and effect were intended. It seems reasonable that the display played a dramatic part in a ceremony timed by the equinox. This serpent of sunlight matches the markings of the indigenous rattlesnake of Yucatán, and the many sculptured feathered serpents of Chichén Itzá can be identified, by their rattles, as rattlesnakes, too. This links the equinox serpent to rattlesnake symbolism that involves the year, the passage of time, and the idea of renewal. These same associations were embedded in the ceremonies of the Aztec god Xipe Totec, whose main festival was celebrated in Tenochtitlán at the vernal equinox. We can't be sure that the rituals of the Toltec-Maya at Chichén Itzá paralleled what the Aztec inherited in central Mexico, but the Castillo draws a cosmic symbol in celestial light that looks like something the Aztec would have understood.

Richard Fraser Townsend has studied the cosmological symbolism of ancient Mexico and concludes the east-to-west motion of the sun makes the top of the Aztec Calendar Stone—and the serpent's tails—east, and the bottom—by their heads—west. He offers a good case for the proposition that the Calendar Stone never occupied a wall or was even displayed upright. Instead, he argues convincingly that the great disk lay flat with its "top" pointing east.

East was the primary sacred direction for the Aztec, and an arrow, formed by another triangle just above the central face and a stylized feather tassel below it, points that way. East is the direction of the sun's daily creation, and the Fifth Sun's Creation was, by tradition, 13 Ácatl ("13 Reed"). This date appears in a box at the very top—or east rim—of the disk: 13 dots surround the glyph "reed." Equally important to the Aztec, 13 Reed opened the year in which the Aztec king Itzcóatl began his reign. He allied the Aztec with the forces of Texcoco to defeat the Tepanecs and so started the Aztec on their march to empire. Aztec rule derived its authority from a cosmic source, the Fifth Sun, whose creation paralleled in myth the creation of the Aztec state.

If the "Calendar Stone" really were set flat into a pavement on the floor, another of Townsend's interpretations is strengthened. The central face, he argues, represents not the sun at all but the earth. It is the face of an earth-lord, or *Tlaltecuhtli,* and other Aztec sculptures of this god portray him with a face upturned to the sky, a knife-blade tongue that consumes the dead, and jaguar claws. It makes even more sense, then, that the central design forms the "earthquake" or "movement" sign. The eight symmetrically spaced triangles point to the cardinal and intercardinal directions—the directions that organize the earth—and make what looks like a compass emblem exactly that.

Altogether, the design on the "Aztec Calendar Stone" is a cosmogram, a compressed symbolic "picture" of the universe. A cosmogram is not really a portrait or map of the cosmos, but a picture of cosmic order. Through orientation in space and commemoration of time, the Aztec Calendar Stone unites the Aztec concepts of terrestrial territory, political supremacy, and sacred stewardship of the cosmos. It is a cosmological-astronomical-calendrical-mythological-religious symbol. It invoked the very structure of the universe to legitimize the Aztec's right to rule.

Marking the Moon

An intricate network of meaning is woven into the "Aztec Calendar Stone," and its symbolic treatment of time, space, and their interrelationship, is complex. The imagery is the product of an imperial vision laid upon an ancient and evolved foundation of central Mexican myth and cosmology. A much simpler treatment of celestial time—though perhaps no less important to the stone age artists who carved it—may be what is symbolized on a stone embedded in the side of a prehistoric passage grave.

Many mysterious symbols and designs were cut into the megalithic slabs of the tombs built by the neolithic farmers of Ireland. Newgrange, for example, is adorned with spirals, lozenges, zigzags, and complex, geometric compositions. Nearby, a curbstone on the east side of the passage grave at Dowth carries at least seven images that look like rayed sun disks. What seems to be a representation of the moon's monthly cycle on another curbstone, at a different monument, the Knowth passage grave, has been pointed out by Martin Brennan, an artist and designer interested in the rock art of prehistoric Ireland and the author of *The Boyne Valley Vision*.

Although excessive speculation and inadequate evidence mar Brennan's book, some of his remarks about the carving at Knowth deserve attention. Knowth is a little over a half mile northwest of Newgrange, just upstream on the River Boyne. Although it differs in detail from Newgrange, Knowth is roughly the same size. Many of its stones are carved with the symbols and in the style we see at Newgrange, but the stone noticed by Brennan is not duplicated there. Brennan calls it "The Calendar Stone," but in the literature it is Kerbstone 69, located on the southwest side of the mound.

The stone's central design is a spiral. Flanking it—in fact, crossing right over it—is a sequence of 17 "half-circles," or crescents, all of them open to the left. Brennan guessed that the first crescent left of the spiral represents the first appearance of the waxing crescent moon, which is seen in the west in the early evening after new moon. If each crescent represents a day, on the eighth day the crescent is reversed and opens to the right. This could stand for the first quarter, but it does not, of course, look like a "half-illuminated" moon.

Continuing clockwise around the top of the stone, we encounter ovals or circles. These roughly resemble the waxing gibbous phase. Those at positions 14, 15, and 16

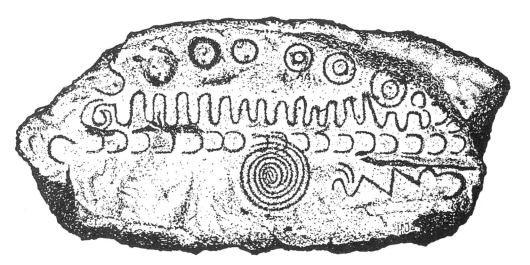

More than a hundred large and massive stones were laid into place as a kind of curb surrounding the entire base of Knowth, a prehistoric passage grave in Ireland. Enigmatic symbols and designs were carved on most of these megaliths, and although this ornamentation has not been interpreted systematically, this stone, #69, may carry a symbolic representation of the changing phase of the moon. The crescents—most of them lined up across the center—and the circles that arc along the top total 29. They could stand for days in one lunation, or month, that is, one cycle of lunar phases. *(Griffith Observatory)*

are distinctive double circles, and they correspond to the days of full moon. The rest of the positions are marked by crescents. Altogether, the "moon symbols" total 29, a good approximation of the lunar month. The opening of the first crescent after full moon period is relatively small and faces down. This could be interpreted as a reference to the waning half of the cycle. The moon starts it as an oval, passes through another half-illuminated stage, and ends as a crescent. Superposition of the last two crescents with the spiral could have meant something about the time of new moon, the moon's invisibility.

We cannot be sure the design is really a daily representation of the lunar cycle, but the symbols, their sequential changes, and their total number are remarkably consistent with a lunar interpretation. Still, there is always a hazard in an analysis like this, for it isolates the design and removes it from its context. There may be, however, a way to test the interpretation. Already archaeologist George Eogan has unearthed many other decorated stones in his masterful excavation at Knowth. Some of these are carved with similar motifs and designs. When the inventory is complete, comparison and analysis between stones may permit us to determine if time was symbolized by the moon at Knowth.

Barley, Bees, and the Passage of Time

Symbolic presentation of another type of calendar count may explain details of a 9,000-year-old mural in a neolithic shrine at Çatal Hüyük, in south central Turkey. Dr. Mary Kilbourne Matossian, an historian at the University of Maryland, has interpreted the symbols in a honeycomb pattern on the east wall of Room VI.B.8 in terms of the growing period of barley and the annual swarming of bees.

James Mellaart, a distinguished archaeologist and the primary excavator of Çatal Hüyük, suggested two decades ago that the insect forms that appeared in some of the cells of the mural's lattice referred to the stages in the life cycle of bees. At the latitude of Çatal Hüyük, swarming occurs 48 days after the vernal equinox, in early May, and this seems to be a significant number in the mural's pattern. Clearly defined columns of honeycomb cells—but no well-defined rows—prompted Matossian to count the cells from bottom to top. Elements of the mural seemed to increase in complexity from left to right, and Mellaart thought it should be "read" in that sense. Accordingly, Professor Matossian continued her tally by that rule.

There are 72 cells visible in the mural and 62 of them are filled with one symbol or another. Counting from the first filled cell, the compartment with the first bee turns out to be number 48 (bottom to top) or 49 (top to bottom). Based upon the behavior of bees, the first filled cell on the left was assigned the date of the vernal equinox. This is, to some extent, an arbitrary choice, but the last filled cell, number 62, then corresponds to 22 May. This is consistent with the time it takes barley to "ear," or flower, about 60 days.

Much of this prehistoric mural from a shrine at Çatal Hüyük, in Turkey, is actually a set of "compartments" or "cells," some filled with symbols and others empty. Altogether, 72 cells create the general appearance of a honeycomb, and some of the symbols resemble blossoms and bees. The sequential placement of these symbols corresponds to intervals of time related to the vernal equinox, the swarming of honeybees, and the flowering of barley. If the natural calendar implied by these events really is represented in the mural, this neolithic shrine is another example where the cyclical and ordered passage of time is incorporated into the environment of sacred space. *(Griffith Observatory, after James Mellaart)*

Matossian speculates that the circle-and-dot symbols on the left side of the mural, near the vernal equinox and time of sowing, stand for seeds. "Flower" symbols at cells 54 and 60 may indicate the ripening of the grain. Some of the other symbols are crescents and may in some way relate to a lunar count. The spacing between most of the cells loosely fits a quarterly breakdown of the lunar month.

Evidence for calendric symbolism in the mural at Çatal Hüyük is sketchy and circumstantial, but the symbols in the shrine do seem to reflect springtime phenomena significant to the neolithic farmers, who probably used honey as a natural sweetener and grew barley as a principal crop. Unfortunately, there is little chance we can prove that calendric information was applied to the wall of a religious shrine. However, other imagery in the room and in many of the 40 other shrines at Çatal Hüyük seems to be concerned with fertility, pregnancy, and birth. Astronomical information—embodied in a calendar count and encoded in the behavior of bees and barley—would not be out of place in the sanctuaries of neolithic Turkey.

The Star of Ishtar

Because some astronomical objects move through the sky in repeated and known intervals of time, the behavior of the celestial gods associated with them can be symbolized numerically. Ishtar, as the planet Venus, perhaps was handled this way in the eight-pointed star that usually stands for her on Babylonian boundary stones.

References to Venus as early as 3000 B.C. are known from evidence at Uruk, an important early Sumerian city in southern Iraq. One clay tablet found at the site says "star Inanna," and another contains symbols for the words "star, setting sun, Inanna." Inanna is Venus, known later as Ishtar, and the Uruk tablets specify her celestial identity with the symbol for "star": an eight-pointed star. At this early stage the symbol seems to carry no more meaning than that, though it eventually evolves, in cuneiform writing, into a sign that means "god" and is placed before the actual names of deities. If the relationship between gods and the sky were not already explicit enough, this development in Mesopotamian writing would confirm it.

By the Kassite Dynasty, roughly 1600–1150 B.C., the eight-pointed star had acquired a more specific meaning. It belonged to Ishtar, as Venus, and shows up on numerous *kudurru*, or boundary stones, which were an innovation of the Kassite kings. Such stones were set up to mark field boundaries. The earliest of them record and confirm royal grants of land and therefore establish title to the territory they .

Symbols of three celestial gods top this *kudurru*, or Babylonian boundary stone. From left to right, they are the eight-pointed star of Ishtar, or Venus; the crescent of Sin, the moon; and the rayed disk of Shamash, the sun. Most of these stones belong to the second millennium B.C., and this one records a gift of land in 1120 B.C. from Eanna-sum-iddina to Gula-ereš.   *(E. C. Krupp)*

represent. Most of them are 2 to 3 feet high. Elaborately carved with the emblems of sky gods and a detailed text, they verify celestial approval of the transaction and warn others to watch their step.

After an appropriate description of the land in question and a list of those involved in effecting the transaction, the boundary stone of King Marduk-ahe-erba forcefully counsels,

> Whenever . . . any one
> shall arise and against
> that field shall raise a claim
> or cause a claim to be raised,
> shall say the field
> is not the gift of the king
> and shall order
> a thoughtless man, a fool, a deaf man
> to approach that inscribed stone
> and shall throw it into the water,
> burn it with fire,
> hide it in a field where it cannot be seen—
> May the great gods, as many as on this stone
> by their names are mentioned
> with an evil curse, that is without escape,
> curse him.
> May Anu, Enlil, and Ea
> in anger look upon him and destroy
> his life, [and] the children, his seed.
> May Marduk, the lord of constructions (?),
> stop up his rivers, and
> Zarpanitum, the great mistress,
> spoil his plans.
> May Ninib and Gula, the lords of the boundary
> and of this boundary stone,
> cause a destructive sickness to be
> in his body, so that, as long as he lives,
> he may pass dark and bright red blood as water.
> May Sin, the eye of heaven and earth, cause
> leprosy to be in his body, so that
> in the enclosure of his city he may not lie.
> May the gods, all of them, as many as are mentioned
> by their names, not grant him life for a single day.

It was not a good idea to overrule the gods of the sky.

Not all of the identities of the gods named and symbolized on *kudurru* are known, but most (and perhaps all) of them are celestial. Three prominent symbols included on most stones refer unambiguously to Shamash, the sun; Sin, the moon; and Ishtar, the planet Venus. The emblem of Shamash is a four-pointed disk with undulating lines radiating intercardinally, and this is a standard Mesopotamian symbol for the sun. The wavy lines could be radiating sunlight, the "net" of Shamash. For Sin, the stones have an obvious crescent moon, and the other large star—almost always with eight points—is Venus.

Very direct symbolism in the signs for the sun and the moon and in several of the other symbols whose meaning is understood tempt a guess that the symbolism in the Star of Ishtar is in some way equally direct. Perhaps the number eight is itself symbolic, for Venus experiences an eight-year cycle. During that time it passes through its complete evening star/morning star/evening star pattern five times. This means that a configuration of Venus recurs on the same solar calendar date after eight years, which is how long five complete back-and-forth passes to either side of the sun take.

To establish the importance of this cycle we must verify that the Mesopotamians were familiar with it and made something special of it. In fact, we know they were well aware of it. Omen texts from the First Babylonian Dynasty (ca. 1900–1660 B.C.) confirm that the old Mesopotamian skywatchers understood that Venus as the morning star and as the evening star were the same thing. By the Seleucid period (ca. 301–164 B.C.), we have a number of late goal-year texts in which the eight-year period was used to predict the appearances of Venus. These goal-year texts are clay tablets that list astronomical data for a given year and also for years specified by adding an appropriate number to the starting year. For Venus, the number to be added is eight. Accordingly, the pattern in the table for Venus will work for every eighth year from the year for which the table was prepared. For example, Professor Otto Neugebauer, one of the foremost historians of ancient science, described one of the Venus goal-year texts and showed that it provides dates and positions for Venus at last visibility as a morning star in steps of eight years. Another lists the planet's reappearance as an evening star over three eight-year intervals.

Although the eight-year, five-cycle Venus period is close, it is not exact. After eight years, Venus is actually a little ahead of schedule, about 2.4 days. One text from the Neo-Babylonian period (626–539 B.C.), referring to Venus as Dilbat, records "Dilbat 8 years behind thee come back . . . 4 days thou shalt subtract." Here, the Mesopotamian planetwatcher is instructed to subtract four days to get the right date for Venus. This may appear to be an error, but it isn't. The 2.4-day correction applies to a solar calendar and the Mesopotamians kept their calendar by the moon. Because the moon arrived 1.6 days late, Venus configurations recurred four days early, and the Neo-Babylonian astronomers adjusted their predictions.

Unfortunately, the goal-year texts are rather late and do not confirm that the eight-year cycle of Venus was known in Kassite times. We have, however, copies of a much earlier set of astronomical texts, the so-called tablets of Ammizaduga. Ammizaduga (or Ammi-saduqa) was the next to the last king of the First Babylonian Dynasty and probably ruled between 1650 and 1550 B.C. The exact dates are somewhat uncertain. Three decades after the end of his reign the Hittites deposed his successor, and somewhere in that period the Kassite Dynasty began.

The original tablets of Ammizaduga probably were inscribed around 1700–1600 B.C., but they are long gone. Copies survived, however, in the library of the Assyrian king Ashurbanipal (668–626 B.C) at Nineveh and are in the British Museum today. In them, 21 years of Venus data are given—dates of first and last appearances as a morning star and as an evening star and durations of invisibility—along with appropriate omens.

> If on the 25th of Tammuz Venus disappeared in the west, for 7 days remaining absent in the sky, and on the 2nd of Ab Venus was seen in the east, there will be rains in the land; desolation will be wrought. (year 8)

Despite scribal errors, the texts clearly exhibit the eight-year cycle and indicate Mesopotamians in the middle of the second millennium B.C. were aware of it.

Apart from a few exceptions, an eight-pointed star is used exclusively for Venus on the Kassite boundary stones. Other stars are usually represented by dots, and Sebitti, a group of stars, is illustrated as a cluster of seven dots and appears on many of the *kudurru* with the Star of Ishtar. In later times the Ishtar symbol may have fallen into more general use, but during the time of the celestial boundary stones, the eight-pointed star meant Venus.

The Sun Takes Wing

During the Assyrian period many of the same Old Babylonian symbols for celestial objects persist on commemorative stelae, on temple walls, in cylinder seal impressions, and in other formal contexts. A tablet that marks the restoration and refoundation of the temple of Shamash at Sippar displays the three main symbols—sun, moon, and Venus—as a celestial stamp of approval upon the enterprise. Shamash is seated inside on a throne, and a large version of his wavy-lined, four-pointed sun disk rests upon a table. In this period, however, the sun's emblem sometimes took a different shape. A winged disk replaced the Shamash emblem, and often the primary god of the Assyrians, Assur, was ensconced in the flaming disk. When the Assyrians ruled Mesopotamia, their national deity assumed most of the characteristics of Marduk and occupied the same role as creator and sustainer of order. Similarly, Assur was associated with the sun, and so his appearance in the flying disk of the sun was altogether natural.

The winged sun symbol was common in late Mesopotamian art. After the Assyrian and Neo-Babylonian periods, the Achaemenians, a Persian dynasty (558–330 B.C.), ruled Babylonia and Assyria. Identical winged disks "fly" upon the walls of the great Achaemenid ceremonial center at Persepolis. Of course, the winged sun disk also appears on temples throughout upper Egypt. The form is slightly different, for the Mesopotamian version often sports a feathered tail in addition to the outspread wings.

The Assyrian god Assur crosses heaven in the winged disk of the sun. *(E. C. Krupp)*

It looks like a bird—as was intended—to suggest the idea of flight through the sky. No less prepared for its daily journey, the Egyptian sun managed with just the plumage of two wings. Sometimes the disk was framed by two *uraei*—divine cobras that spit fire, protected the righteous, and were as deadly as Huitzilopochtli's fire serpents.

Lodged above the entrance on the pylon, or front wall, of New Kingdom and Ptolemaic temples, the Egyptian winged disk not only symbolizes the sun but may allude to the way its position was once measured. Egyptian temple pylons have a very distinctive form. Face on, a pylon looks like a wall surmounted on each side by a tower. The archway entrance is centered in the pylon wall, beneath the gap between the two "towers." Winged sun disks on temple façades are nested, then, just below the pylon gap, in the "notch" created by the towers. Inside the temple, winged sun disks may hover from the undersides of the cornices of successive entryways, an architectural equivalent of the pylon. There is no doubt about the symbolic meaning of the pylon; the Egyptians explained it in their word for the structure. In texts, the hieroglyph for pylon looks like a pair of mountain peaks, between which the sun rises, and the name means "Luminous Mountain Horizon of Heaven." This is where the sun's yearly progress could be measured: among the features on the horizon.

This idea of the horizon-lodged sun appears very early in Egypt. A piece of predynastic pottery, noticed by Egyptologist N. Rambova, has a pair of joined, triangular peaks at its center and images of the eastern and western sun on either side. A

In painted relief, the Egyptian winged sun hangs above the entrance to the first hypostyle hall of the Temple of Ramesses III (1198–1166 B.C., Twentieth Dynasty) at Medinet Habu.   *(Robin Rector Krupp)*

Two distinctive towers combined into a single wall, or *pylon,* front the Egyptian New Kingdom Temple of Khonsu at Karnak. The pylon is a stylized horizon, and a winged sun, in relief above the central doorway, occupies the notch between the two "mountain peaks."
(E. C. Krupp)

product of the neolithic Amratian (or Naqada I) culture of upper Egypt, this design was drawn around 3800 B.C. By the time of the Old Kingdom, hieroglyphic writing was in use, and the concept had developed into the word for "horizon." The symbol for this includes a sun disk which rests in what looks like a low and very soft cushion. With the "padding" pushed up on each side of it, the sun disk nestles between two stylized and softly rounded mountains. In the *Pyramid Texts* the mountain element of the symbol is used to treat the sunrise metaphorically. The event is equated with a god coming into being when the "two mountains divide." Related versions of the mountain horizon remain in Egyptian art and retain the same meaning throughout its long existence.

Earlier we encountered the same idea in the Akkadian cylinder seal (ca. 2360–2180 B.C.) that illustrated the emergence of Shamash from the mountain in the company of the goddess Ishtar. The same wavy lines that undulate between the four points of the Shamash sun symbol issue from the god's shoulders, as he makes his first appearance in an unmistakable gap between two peaks. In another seal Shamash stands fully erect, one foot on one of the peaks and ready to climb the sky. A post-Akkadian seal seems to portray a solar eagle taking wing in the V between two conventionalized mountain forms.

We know that natural horizon features have been used by various ancient skywatchers to pinpoint the position of the sun, and perhaps other celestial objects. The Hopi horizon calendar at Walpi is well documented, and the pecked crosses and

The Egyptian hieroglyph for "horizon" consists of a sun disk cradled by the rounded peaks of two mountains. Embedded in the symbols that comprise this word is the idea of monitoring the passage of time in terms of the shifting position of sunrise among the details of the mountainous horizon. *(Robin Rector Krupp)*

Another reference—this time from ancient Mexico—to horizon measurements may be shown in the *Codex Bodley.* The sun is perched in the gap between two triangular mountains. Also, a star symbol rests in the V created by the crossed-stick observing instrument on the platform in the foreground. *(Griffith Observatory)*

"Camino del Sol" at Cerro El Chapín and Alta Vista imply a similar technique. California Indians also seem to have monitored the sun this way, for one informant said,

> The sun would pass the middle peak on the way south, pass the valley, remain two days, and on the third day would come up again over the middle peak on its way north. He would notify the other Indians of the New Year.

Residing in North Ossetia, in the Soviet Caucasus, is an Indo-European group related to the ancient Sarmatians. These people, known as the Ossetes, continued to use a well-defined and highly detailed horizon calendar as recently as the first years of this century. All that is needed for this kind of astronomy is a place to stand and a place to look. The place to stand in the Ossetian communities was carefully preserved in many villages. Next to the church–assembly hall, a bench marked the spot. From there, sunset was observed on the distinctive mountainous profile to the west. The village's designated sunwatcher determined the passage of the year by observing the sun's return to the solstice point. Dates for all of the year's holidays were established in the same way. Between 40 and 50 of the year's 365 days were specially marked by natural horizon features, and the intervals in days between them were part of the ancient calendar lore transmitted orally from one generation to the next.

Evidence for horizon-feature astronomy is widespread and was a natural way to monitor and master celestial cycles. The horizon transforms the passage of time into a visible pattern in the landscape. Time is converted into space. Through celestial phenomena certain directions take on special meaning, and because they are places where events occur, they order time as well as space. The relationship between time and space is reciprocal, and obviously of considerable importance, for the symbols that embody the idea of the horizon-nested sun are much more than cue cards that alert the ancient astronomer what to do next. In Egypt the idea was turned into a huge temple façade, one of the principal elements of the era's religious architecture. Wrapped up in the winged sun about to ascend from the mountain is a sense of an integrated, ordered cosmos—renewed and recognized in each sunrise at its proper place.

Shining Disks and Other Worlds

Speculation about flying saucers and ancient astronauts has taken full advantage of the winged-disk motif. Because they correspond symbolically with the idea of a flying disk, Assyrian and Achaemenid sun disks and Egyptian winged suns have been promoted as ancient eyewitness reports of unidentified flying objects. Mythology—in the minds of some entrepreneurs and flying saucer enthusiasts—is the legacy of our alleged contacts with visitors from outer space. Mythology certainly is involved in this matter, but in a different way, and by turning the argument around we can learn something about the infatuation we have with flying saucers and aliens from other worlds.

Most UFO reports are misidentifications. Any major observatory that provides information for the general public soon reaches that unambiguous conclusion. There are, now and then, however, very strange stories told by very sincere people who are profoundly affected by what they have experienced. Often the setting and events described are very dreamlike. Peculiar things happen: not extraterrestrial things, but things that seem out of place or nonsensical. This aspect of the UFO experience aroused the interest of Dr. Jacques Vallee, an information specialist, and he has adopted a rather unique approach toward the phenomenon. Spending relatively little time on attempts to rationalize the physical reality of UFOs, he is much more concerned with the belief systems that sustain public interest in them.

Vallee points out that some of the "truest" UFO experiences are also the most absurd. That is, very often the persons who seem to be most sincere, honest, and affected by what they believe they have seen report circumstances and events that are contradictory, disconnected, and even silly. He mentions one case in which a witness was asked the time by a UFO occupant. When the witness replied "Two-thirty," the correct time, the alien minced no words and said, "You lie—it is four o'clock." The UFO pilot then asked, "Am I in Italy or Germany?" Here, again, the conversation had taken an absurd turn because the country was France. This is altogether a very odd exchange to take place between a resident of earth and a visitor from beyond. In content, it is restricted to a dispute over time and place and seems to be a conversation neither party needed.

Such accounts seem bizarre when we hear them secondhand, but they are absolutely real to those who feel they have experienced them. We are not, however, talking about sightings of unidentified lights in the night or daytime sky. The reports that interest us here are those that seem accompanied by a transcendental change in the personality and attitude of the witness. Vallee reports that their "lives are often deeply changed, and they develop unusual talents with which they may find it difficult to cope." The experience has a mystical or a religious quality to it. An element of the experience may involve an odd twist of time or space. The witness may be unable to account for his whereabouts for a certain period of time or may have a sense that the environment in which the encounter took place was unfamiliar or permitted things impossible in ordinary, three-dimensional space to happen. Later the witness may be preoccupied with the concepts of time and space, and the story of the two-thirty meeting in France reflects this kind of emphasis.

Vallee has shown that many parallels exist between the experiences described as contacts with UFO folk and the lore that surrounds the fairy folk. The fairies come from another unworldly realm—out of normal space and time. Encounters with the fairies often involve a supernatural passage of time. The experience is considered dangerous, and at times the perilous otherworld of the fairies is associated with chaos. To mingle with the fairies is to risk death. In a similar way, other UFO encounters parallel what other people—survivors of serious physical trauma—describe as the "after death" experience. In these, too, space and time take on different meaning. The person who has such an experience feels he or she has dealt with something fundamental, profound, and uncanny. Usually a major change in personality accompanies these feelings.

While all of these mystical and supernatural elements of UFO experiences continue to be reported with little fanfare, governments, media, scientific institutions, and most individuals continue to deal with the UFO phenomenon in very different terms. Whether one believes in UFOs or not, they are treated as machines—spacecraft—whose technology exceeds our own. This is the public side of the UFO myth, and its images are congruent with our public belief system, which is secular and not mystical.

More than 20 years ago the psychologist C. G. Jung offered some insight into the UFO experience by drawing parallels between its imagery and that encountered in dreams and visions. In his book *Flying Saucers (A Modern Myth of Things Seen in the Sky)*, he concluded that the round, shining objects seen in the sky are best understood as symbols. They are bridges between what a traditional society would regard as the realm of the sacred and the world of everyday life. Jung suggested that tension, anxiety, or a life-threatening crisis may trigger the vision. The vision itself is a hierophany. If the mind, through such an experience, is able to reintegrate itself, the witness reorders his or her life and achieves an equilibrium not present before the crisis.

Jung identifies the disk, or circle, as an archetype—an image that conforms to the structure of thought and so communicates its meaning directly and unconsciously. It is a symbol in the complete sense of the word, and it is related to order, orientation, equilibrium, and wholeness.

By this point we should be able to identify the source of the archetype. The shining, flying disks that bring order to the psyche are visions that originate with the celestial lights that bring order to the world. The natural order is challenged by the intrusion of death, the fairy realm, or the aliens from other worlds. In that conflict, however, is the seed of its own resolution. Just as the threat of chaos passes in the myth of cosmic order, the UFO witness returns from his or her encounter with the uncanny. Juxtapositions and discontinuities of time and space are part of this "genuine" UFO experience, because time and space are the fabric of the natural order. The flying disk is the primary agent. It brings on the "peril," and after a close encounter, it departs, only its image lingering as a symbol in the mind of the witness.

Certainly not all UFO reports can be understood in terms of this myth, but at its heart the UFO experience is an exercise in the perception of cosmic order. Quite naturally, it is staged in the sky. In another age we might have placed winged suns above the entrances to our sacred spaces or fixed them to the walls of ceremonial palaces where the authority behind the social order was symbolically centered. But in our time these symbols no longer speak to us. Today, our metaphors of transcendence are "other worlds" and "outer space." And even though the language of our belief system has changed, we are still dealing with the sky. Like our ancestors, we still think in symbols. We depend on them to communicate the ordinary and the sublime. They can distill the forces that move our lives and the phenomena that fill our world to their essential principles. Cosmological and celestial symbols provide instant—even unconscious—recognition of the structure our minds impose upon the universe. Our brains must function as they always have. In one form or another, then—even in the imagery of flying saucers and astronaut gods—the myth of sacred order will emerge.

13

The Universes We Design

Cosmos means the "ordered whole," and a sense of the cosmos—of the ordered whole—puts us on the trail of the patterns and principles our brains impose upon the events we experience. Pattern, cycle, and order stand out most clearly overhead, and, invariably, the cosmos echoes the sky. The language of the metaphor may range from mathematics to myth, but our models of the ordered whole are celestial. The arena of cosmology is the sky.

Energized by the Sky

The universes we design reflect our perceptions of order. For the present-day Desana Indians of northwest Colombia, the principles of cosmic order are visible overhead. The same energy that animates life in the cyclical pattern of reproduction moves the stars and sets the seasons in sequence. Nearly everything of consequence follows the celestial cycles—the weather, the growth of plants, the availability of fish, the abundance of game—and so the sky is the key to the state of the world at any given time.

The Desana are interested in the way the universe behaves as much as in what it looks like. They reside in a rain forest environment, and the survival and the continuity of their way of life depend on their ability to preserve a delicate balance between their needs and the resources available to them. Although they grow some food and obtain an adequate supply of fish from the rivers, the Desana must cope with alternating periods of scarcity and abundance. This is especially true of the availability of game.

Hunting is the focus of the men's lives, and so the pursuit of game animals and their preparation into food define, to a great extent, what it means to be a Desana.

The Indians are well aware they could—with little difficulty—deplete the forest of its game animals, and so strict rules of behavior, based upon the Desana concept of the world, help preserve the environmental balance.

As the Desana see it, the animals and the human beings continue to survive by participating in a closed circuit of vitalizing energy. The source of this creative energy is the Sun Father, a supernatural being who brought the world into existence. His agent is the visible sun, and its light and heat are the conduits of Sun Father's power. Only a finite amount of this energy is available in the circuit, however, and so every time it is used, it must be replenished. By hunting or fishing, the Desana take some of this energy, and it is transformed into new life through human procreation. Were this to continue without some kind of reciprocation, the Desana soon would consume themselves out of a home. Instead, they magically enhance the fertility and abundance of the game through a symbolic injection of energy into the creatures they kill. Hunting is equated with the sexual act, and sexual abstinence is required as preparation for the hunt. By following this rule and other restrictions that define when the hunt may occur, under what circumstances, and the amount of game that may be taken, the Desana—in their minds—return some of the procreative energy to the circuit. The other side of sexual abstinence is a part of the Desana birth control system that insures that the population will not outstrip the availability of food.

To adapt successfully to the environmental niche they occupy, the Desana must know their place in the world and to learn this must know the world itself. Thus they devise elaborate systems of classification of animals, plants, and food that give the abstract concepts of energy and fertility concrete meaning. These systems define what may and may not be done, and by association they extend to other aspects of Desana life, ordering human behavior. For example, the same principle of reciprocity involved in the preservation of the world's energy applies also to marriage. Separate groups exchange women and by doing so establish a system of kinship and social cohesion. These same groups trade goods with each other. Each has its own distinctive products, and the exchange not only keeps the economic cycle in balance, it reinforces the distinctive identity of each group. In the same kind of reciprocation, men and women divide the labor involved in food production. This joint activity is part of their strategy for survival, just as their reciprocal roles in reproduction also guarantee the continuity of life. The Desana world view clearly modulates the way they deal with the world.

There is a link between the sense the Desana have of the organization of their territory and the organization of the sky. To see this aspect of their cosmology, we have to look at their creation myth, as reported by Gerardo Reichel-Dolmatoff. The center of their homeland was determined, they say, by Sun Father, who, when time began, picked a place where his upright staff could cast no shadow. There, at a whirlpool entrance to the womb of the earth, Sun Father impregnated the earth, and from that spot the Desana and their neighbors emerged, transported by living anaconda canoes to the places they settled along the river. Figuratively, the shaft of light from the zenith sun fertilized the earth with the procreative energy. The place where this happened defined the center of the Indians' world. The boundaries of the Indians' territory were established by six mythically huge anacondas, each outstretched to form one side of a hexagon.

Until we encountered the hexagonal border of six giant anacondas, the elements of this story were consistent with the idea that the world's structure is derived from the sky. But how does a sense of six-sided bounded space enter the picture? The answer is the shaman; the answer is quartz. In Gerardo Reichel-Dolmatoff's words, the shaman is an "ecological broker." Rituals designed to promote the fertility of the game and successful hunts are his responsibility, and he acts as an environmental protection agent for the Desana to make sure they don't consume themselves out of their habitat. Quartz crystals are essential tools in the shaman's kit of magical paraphernalia. Quartz, along with hallucinogenic drugs, permits him to communicate with the unseen world. Associated with Sun Father's creative energy, it sometimes is called crystallized semen. With it the Desana shaman diagnoses illnesses and effects cures.

Recognizing an abiding principle of order in the hexagonal shape of a rock crystal, the Desana integrate its geometric regularity and symmetry into their system of cosmic order. For that reason, the hexagon defines the shape of their sacred space. All

The geometric structure of quartz crystal has made its hexagonal shape a metaphor of cosmic order for the Desana Indians of Colombia. They see a similar shape among the stars and relate it to their picture of the world. Radiance and an ability to manipulate light extend the celestial associations of quartz to the sun, another source of cosmic order. *(E. C. Krupp)*

Petroglyphs on the Rock of Nyí, a sacred site of Colombia's Barasana Indians, depict the myth of creation. The tradition is the same as that held by their Desana neighbors to the north. The large human figure portrays Sun Father at the moment he erected a perfectly vertical rod. It cast no shadow. The world was fertilized by Sun Father at this spot, and from it the first people emerged. In fact, the sacred Rock of Nyí is almost exactly on the equator. The sun, therefore, rises due east and passes through the zenith on the equinoxes. When, at equinox noon, the sun is straight overhead, an upright staff loses its shadow. In a sense, the zenith sun's rays penetrate directly into the earth, and when this happens the world is seasonally renewed. *(G. Reichel-Dolmatoff, U.C.L.A. Latin American Center Publications)*

hexagonal shapes in nature have a special significance for them: the honeycomb, the spider's web, even the shell of a particular land tortoise. Each cell in the shell's pattern of hexagons symbolizes a character in the creation myth or an organizational principle of society—the family, for example, or marriage into another family. Desana rules for marriage exchange are visualized, in fact, in terms of a hexagon.

Because of its importance as an organizing principle of thought, the hexagon metaphor reappears in one aspect of Desana tradition after another. A giant hexagon of stars, centered on the belt of Orion, is equated with a hexagon of landmarks on the earth that establishes the limits of the tribal territory. The corners of this terrestrial hexagon are marked by six waterfalls, each a place where the head of one of the six original giant anacondas meets another's tail. Each of these snakes stands for one of the six rivers that frame the traditional homelands. The center of the celestial hexagon corresponds to the intersection between the Pira-Paraná River and the earth's equator. Here, where the sky is said to cohabit with the earth, is the place where Sun Father erected his shadowless staff and fertilized the world.

The Desana time the arrivals and departures of the seasons by the stars. Orion's belt is one of their seasonal indicators, and as the center of the earth's celestial template, it occupies a key spot in the cosmic structure. It is nearly on the celestial equator, and so it can pass through the zenith just as the equinox sun does. The center of the world is associated with the vertical axis to the zenith. This explains why Orion's belt is the center of the celestial hexagon.

These same three stars are also identified with an important figure in Desana mythology—the Master of Animals. He is the supernatural gamekeeper, and he helps maintain equilibrium in the flow of procreative energy by controlling the availability of the rain forest's game and by fathering the river's fishes. Both game and fish adhere to seasonal behavior, and the function of the belt of Orion as an announcer of seasons is consistent with its association with the Master of Animals. Also, because these stars are located very close to the celestial equator, they rise nearly due east and set nearly due west. Accordingly, they reinforce the importance of the east-west, or equinoctial line.

In the Amazon, the equinoxes are important, for they signal the start of each rainy season. One begins in March, the other in September. At the equinoxes, when the rivers rise, fish head upstream to spawn and so become relatively scarce. Likewise, game is less available. The rainy seasons are regarded as periods of gestation.

The Desana associate a pair of intertwined copulating anacondas with the celestial equator's intersections on the ecliptic, or annual path of the sun. These two intersections are, of course, the places occupied by the sun at the equinoxes, and at these times of the year, with the onset of the rains, anacondas swim upstream to mate. At night, when these great reptiles make their way upriver, they can be heard far from the banks. They lift two-thirds of their dark bodies high out of the current like poles and slap themselves down upon the water's surface with a thunderous crack. The Desana say the anacondas are lifting their heads from the water to watch the stars.

Oriented by the Zenith

We can argue that the way we treat the geometry of the sky—imagining it as a sphere that surrounds the earth—is very sensible because the part of the sky we see looks like a hemisphere. Now in reality, the sky looks no less spherical in the country of the Desana, and yet they impose another pattern upon it, the hexagon, because that shape reinforces their perception of order. It is quite possible, therefore, to come up with a whole sky—and a cosmos as well—that is not spherical at all. The Warao Indians of the Orinoco delta in Venezuela do just that and visualize the sky, for example, as a sort of bell-shaped canopy. They live at a latitude between 8 and 10 degrees north. This is fairly close to the equator, and the paths of celestial objects as seen from this zone make the zenith—straight overhead—seem more significant than the celestial pole, which is close to the horizon. Accordingly, the Warao universe is structured by the references that order the Warao sky, and they give it the look of a bell.

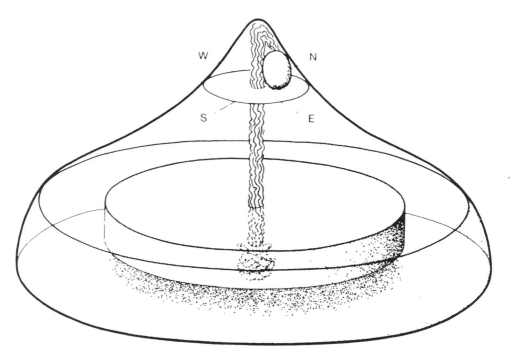

To the Warao Indians of Venezuela, the earth is flat—a disk that floats in the world sea. Overhead, the sky canopies the world, like a bell-shaped tent, supported at the zenith by the world axis. The knotted snake shown at the base of the world axis has four cardinally-facing heads and is the "Goddess of the Nadir." Another great serpent, "the Snake of Being," resides in the sea and encircles the earth, but it is not shown here. Up where the bell narrows, at the height that corresponds to the highest angle of the sun on the solstices, there is another level in the Warao heaven. In its northeast quadrant rests the House of Tobacco Smoke, an egg-shaped place of shamanic power. Warao shamans travel there supernaturally by ascending the tobacco smoke ropes of the world axis to the zenith and crossing the bridge of tobacco flowers and smoke from there to the western doorway of the shaman's celestial egg. *(Patrick Finnerty, U.C.L.A. Latin American Center Publications)*

Dr. Johannes Wilbert, the U.C.L.A. anthropologist who has lived with the Warao and studied their cosmology, reports that they see themselves as occupying the center of a flat disk—the earth. They live at the base of the universe's principal axis, a line that reaches straight up to the zenith, the highest point of the sky. The earth, they say, is surrounded by water that extends in every direction to meet the circular horizon at the rim of the cosmos. Below the surface of this encircling ocean lives a giant serpent, the Snake of Being. The Warao myths treat this great snake as the source of all life, the womb of the world. Her body surrounds the earth, and her tail meets her head in the east, the direction they associate with life and creation.

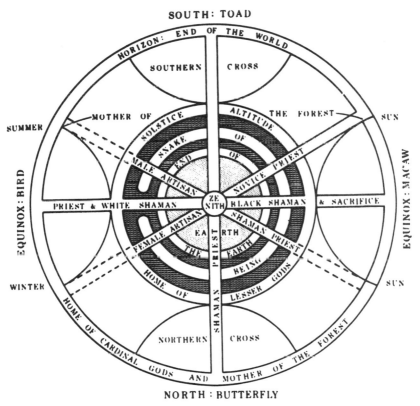

The rim of the Warao cosmos is the horizon—the outer circle in this plan. Here we are looking down upon the zenith and the bell-shaped Warao sky. The next circle inside the horizon indicates the altitude of the solstice sun and the level where the sky nipples more sharply toward the zenith. Far below, in the world sea, the Snake of Being encircles the disk-shaped earth. This serpent's head and tail nearly meet in the east (left side). Various lines cross the zenith from one side of the horizon to the other. These are the paths that certain shamans and the souls of the dead may follow on their transcendental journeys.
(Reprinted by permission from Death and the Afterlife in Pre-Columbian America*)*

Another huge snake, the Goddess of the Nadir, lives in an inaccessible, dark realm below the earth's center. She has four heads, each fitted with a set of deer horns, and each head faces one of the cardinal directions. Out on the horizon, at the edge of the world, the spirits of the cardinal directions make their homes in great petrified trees. These fulfill the same function as the world mountains of other cosmologies; they hold up the sky. Four more stone trees—residences for the "intercardinal" spirits— also support the heavens where the sun rises and sets at each solstice.

In the Warao territory the daily paths of the winter and summer solstice suns are nearly symmetric on opposite sides of the zenith of the due east–due west vertical circle that passes through it. The maximum height of the sun (reached at noon) on either solstice also corresponds to the minimum noontime height reached throughout the year. At the solstice, the noon sun establishes that special zone in the Warao cosmos where the dome of the sky spreads out into a wider bell. The bell rests, in turn, upon the petrified trees of the world directions. At the equinoxes the noon sun appears at its highest—at the zenith—and emphasizes the significance of the world axis. Because the sun is higher at the zenith, the sky is higher, and the Warao trans- late ideas that relate the positions of the sun, the orientation of the world, and the passage of time into the bell-shaped canopy—held up in the center, like a circus tent, by the zenith-nadir axis. And the whole structure—sky, earth, ocean, and horizon—is shouldered by the solid, featureless subterranean realm of the four-headed Serpent Goddess of the Nadir.

By now we see that a relatively unknown people who make much of their living foraging the swamps of the Orinoco delta have a very complex vision of the cosmos. But this is only half the story. They use their cosmology to integrate themselves into their environment. In a direct, concrete way a metaphysical concept assists the Warao to maintain a stable population despite a 50 percent infant mortality rate. The shaman possesses this special cosmological knowledge and through his power exer- cises control over the social and economic structure of Warao society. This makes cosmology a tool for survival.

Oriented by their cosmology architecture, the Warao organize all of the other experiences of their lives into a complex set of interlocking relationships with numerous gods and spirits, their supernatural dominions, and the natural phenomena in their environment. They see themselves and their gods as participants in a world system that dependes on all of its components for its continued existence. Like other peoples who practice shamanism, the Warao depend on their shamans to facilitate communi- cation and exchange with the spirits and gods. Through this process of reciprocation, the world's balance and harmony are sustained. The shamans depend, in turn, on their cosmological imagery and hallucinogenic drugs for the insight they require to perform their duties.

Tobacco is the only mind-altering substance consumed by Warao shamans, but the tobacco in their 2-foot-long cigars is very powerful. Our havanas and rum-soaked crooks pale by comparison. Warao shamans smoke to "feed" the gods and then "ascend" in nicotine trances to the supernatural realms. The shaman's route carries him to the zenith via the world axis, which is taken to be a bundle of energy paths made, of course, of rising tobacco smoke, the substance that lets the shaman travel upward.

There are three types of Warao shamans, and one—the "light shaman"—negotiates the interaction between the Warao and the Land of Light. This realm is the east, and it is associated with the Creator Bird of the Dawn and with the perpetuation of life. In particular, the light shaman is a curer, and his usual patients are women. By treating them successfully they become obligated to him and are, in a sense, "daughters." The shaman influences the marital stability of the society because the women's obligations are discharged by their husbands, who must devote some of their time as members of the shaman's work team. This structure stabilizes the group, and the shaman is even able to renegotiate pairings that do not produce children.

It is the shaman's job to make sure—through his magic—every element of nature is integrated into a balanced, functioning system. In effect, the shaman preserves cosmic order by making sure every component participates. To do this, the shamans play a mental game whose symbols reflect the intricate balance and order they see in the world around them.

The game involves speaking aloud the name of one of four social insects—the black bee, the wasp, the termite, and the honeybee—between intervals of silence. Each insect carries many symbolic connotations, and the shamans keep mental track of them, much as they would monitor tokens on a gameboard. In this case, however, the "gameboard" is a transcendental realm in the Warao cosmos—the House of Tobacco Smoke. Actually, the House of Tobacco Smoke is the residence of the primordial light-shaman and patron of light-shamanism, and it was brought into being by the Creator Bird of the Dawn. This bird is an incarnation of the sun and comes complete with all the imagery we expect to see associated with creation and new life: the dawn, the east, light, and the rising sun. The House itself is white and egg-shaped, and these attributes suggest that other familiar symbol of life's continuity through renewal, the egg. It is located near the zenith—that special place in the Warao heavens—and the four insect-spirits reside there in four separate chambers on the east side.

To reach the House of Tobacco Smoke the shaman first must ascend to the zenith, where he meets an invisible guide who will conduct the shaman—after a rest and another cigar—across a bridge of tobacco smoke ropes. On the other side, the shaman reaches his destination. The insects play the game there, their moves controlled by the tobacco smoke the shamans blow upon them. At some point, a snake enters the room by extending itself vertically, right through the floor. Its plumed head hooks over, a luminous white globe appears from its mouth, and the game ends.

All of this is very mysterious, but it is closely related to the Warao cosmology and their sense of cosmic order. Through this game the shaman experiences a vision that provides him the knowledge and power of the winning insect to cure his patients.

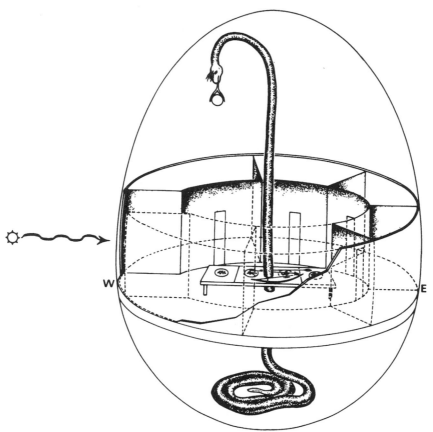

Four pairs of four different supernatural insects reside in their respective chambers in the House of Tobacco Smoke. The two remaining rooms, near the doorway on the west, are occupied by the ancestral shaman and his wife and by the Creator Bird of the Dawn. Under the influence of powerful tobacco smoke, the shamans play a mental game—with the insects—on the table in the circular room in the center. This game determines, as far as they are concerned, the fate of life on earth and is terminated when a supernatural Thread Snake ascends vertically through the floor of the room, hooks over, and produces a luminous white sphere from its mouth. *(Noel Diaz, U.C.L.A. Latin American Center Publications)*

The erect snake and the shining sphere are reminiscent of the sun at the zenith, the cynosure of the Warao sky. The globe-bearing snake is also a font of shamanic wisdom. When, for example, it is encountered during initiation, the novice shaman acquires instantaneous understanding of the entire shamanic tradition. He is oriented in the cosmic order by the experience.

In a remarkable way, the serpent also figures in the shaman's effort to preserve the balance of life. Professor Wilbert's many years of attentive work with the Warao allowed him to penetrate the complex metaphors of Warao cosmology and reveal them as an elegant, interlocking, and logical system of thought that reflects the Warao's most basic concerns. The supernatural plumed serpent, for example, with the glowing white sphere at the tip of its forked tongue, has a real counterpart—the thread snake (genus *Leptotyphlops*)—in the Orinoco basin. Although the thread snake produces no white globe from its mouth, it does erect itself in much the same way the supernatural serpent emerges—as a "world axis"—through the floor of the House of Tobacco Smoke. The purpose of the real snake's activity is orientation. It is blind, and when upright, its senses pheromones and uses chemistry to locate its food. This snake is an insectivore, and its sense of smell guides it to globular termite nests. After breaking into the nest at the base, it burrows in and feeds on the termites. In a sense, it does produce a white globe, for its eggs, laid in the nest, hatch there, and they can represent both new life and the sun.

This snake is part of a food chain that begins with the flowers of the tobacco plant. The bee takes nectar from these flowers, but its young are consumed by the wasp. The blind snake eats the wasps and bees, as well as the termites, and is in turn eaten by the avatar of the Creator Bird, the swallow-tail kite. The Warao are alert to their environment and know this food chain well. It represents the balance and order of nature to them. The game the shamans play is a metaphysical manipulation of these natural relationships.

It is not at all surprising the Warao incorporate the metaphor of a snake orienting itself into their cosmology. Cosmic orientation is the purpose of the Warao shaman's mystical journey to the House of Tobacco Smoke. From this visionary ascent he acquires his power in the community, and this power enables him to orient and stabilize his people and optimize the reproductive capacity of the group. This very practical advantage is, therefore, a consequence of cosmology—perception of the world order. The shaman sees it. He participates in it. He uses it. It is a weapon in the arsenal of survival.

Moved by the Spheres

How people perceive the world has a lot to do with how they live. This is as true for us as it is for the Desana or the Warao. Our first photographs of the earth from space made us change our minds and feelings about the planet. It no longer looked like a world of unclaimed territories, wide open spaces, and unexplored frontiers. Instead, we saw a world that was no more than a speck compared to the stars, the galaxies, and the vast extent of the known universe. The earth—wrapped in ocean and scarfed with air—looked like a fragile, self-contained bubble of life and color in the cold indifference of space. This vision of Spaceship Earth coincided with a new appreciation for the environmental balance that makes life possible. We now find ourselves caught up in a mixture of priorities yet to be reordered, and we are still trying to figure out how to harmonize our needs with our resources.

Just as we began to see the world as a planet, our moon landings and probes to other planets transformed those places into worlds—real landscapes rather than images trapped within telescopes. The unexplored frontiers have not evaporated. They simply have been moved—and multiplied—beyond the bounds of the earth's surface, and even space itself is being turned into a habitable environment by the Space Shuttle.

Moving into space brings us to grips with our concept of the world, and that, in turn, makes us think about the whole cosmos. Our Greek ancestors invented scientific cosmology, and by grappling with space, time, and the heavenly spheres, they set us on the course that has lifted us off the earth and carried us to worlds beyond our own. In the sixth century B.C., Greek philosophers tried to understand the shape and place of the earth and the nature of the sun, moon, planets, and stars in a new way. They analyzed their ideas about the structure and behavior of the world critically and embraced—or discarded—them through rational thought.

Eventually, the Greeks began comparing detailed models of the motions of celestial objects, and one clear theme was shared by all: uniform circular motion. The fundamental celestial motion—the daily rotation of the sky—looked uniform and circular, and this idea became the Greeks' fundamental principle of cosmic order.

Now when we put together a cosmology, we are not drawing a picture of the cosmos; we are drawing a picture of cosmic order. And the Greek cosmologies naturally called upon uniform circular motion to explain what they saw happening in the sky. They evaluated their theories in terms of their ability to match observed phenomena. Although it would take more than two millennia, Greek science evolved into modern scientific method and made it possible for us to see firsthand the earth from space and reach—with robot explorers—the planets whose motions kept the Greek cosmologists so busy.

Before the sixth century B.C. and the time of the Ionian Greek philosphers, cosmology in ancient Greece was costumed in myth and in the language of human procreation. Homer's two epics, *The Iliad* and *The Odyssey*, give us some feeling for the early Greek ideas, and the poems of Hesiod are even more explicit. Both writers probably belong to the eighth century B.C.

A flat circular disk surrounded by a "river" of oceanic waters seems to be the best description of the earth portrayed in the Homeric poems. Its extent was limited by Greek knowledge of the earth's geography. An important realm—Erebus, or the land of the dead—was reached by traveling west through the Pillars of Hercules. Heaven vaulted high above the earth and an unseen counterpart, Tartarus, completed the cosmos an equal distance below.

Hesiod's *Theogony* relies on the metaphors of sexual intercourse and birth to provide an account of the events that brought the cosmos into being. In the very beginning, there was only the empty space known as Chaos, but Gaia, or Earth, then spontaneously appeared. She was joined by Tartarus, the deepest realm of the underworld, and Eros, or Love. Eros represented the creative force that brings things together, and so embodied the ideas of desire and attraction associated with love. In a sense, Eros is like Sun Father's energy that charges the Desana world with the power of sexual procreation. With the arrival of Eros, Gaia—and even Chaos—start adding components to the cosmos. From Chaos came Night and Erebus. Erebus in this sense seems to be a subterranean mirror of the night, a darkness under the earth that separates the earth from Tartarus. With Erebus, Night conceived Day and Space.

To the Greeks, Space meant the pure upper atmosphere, the ether. Heaven itself—the Sky—was born of Earth, who then also gave birth to the mountains and the sea. Up to this point, the only cosmic mating to have occurred was the conjunction between Erebus and Night. The rest of the "children" were products of self-procreation. Now, however, the stage was set for some serious moving and shaking.

Earth coupled with the Sky to create the Titans. These twelve children were the original stewards of the world, although what exactly they represented is not that clear. Some personified natural phenomena—the earth-girdling oceanic river, for example—and others were associated with concepts like memory and cosmic law. The Sky was castrated by one of his sons, the Titan Kronos, but the procreation of the world continued as all of the others continued to take part in the enterprise, bringing forth Doom, Destiny, Death, Sleep, the "whole tribe of Dreams," the Rivers, the Sun, the Moon, the Dawn and many, many more. Kronos sired the Olympian gods upon Rhea, his sister, and one of them, Zeus, overthrew his father, as Kronos had deposed *his* father, the Sky. The Titans were cast into Tartarus and confined there, and the Olympians inherited the cosmos.

Hesiod's poetic account of cosmogony attempts no detailed, physical explanation for the world's phenomena, but describes the cosmos and gives it order and unity in terms of relationships defined by myth. Thales of Miletus (624–547 B.C.), the first Greek philosopher of whom we have any real knowledge, abandoned the metaphor of myth and attempted to describe the world physically. His picture of the cosmos was not so different from Homer's. Thales imagined the earth to be a flat disk that floated, like wood, on the cosmic ocean and was surrounded by its waters. These waters were the source of all creation, for Thales maintained that the world's fundamental element—the first substance underlying all things—was water. Far above the ocean waters and the earth, the sky bounded the cosmos. Thales said nothing about the actual dimensions of the earth, ocean, and sky, but the earth, at least, he judged to be finite.

Other Greek philosphers also tried to describe the universe physically and match the appearance and behavior of the world with a true portrait of reality. Anaximander (611–546 B.C.), a contemporary of Thales and also from Miletus, offered a more detailed picture of the universe. Although Anaximander's earth was still flat and circular, its proportions were more like a cylinder than a disk. He specified the earth's height to be one-third its diameter. Imagining the sun, the moon, and the stars to be

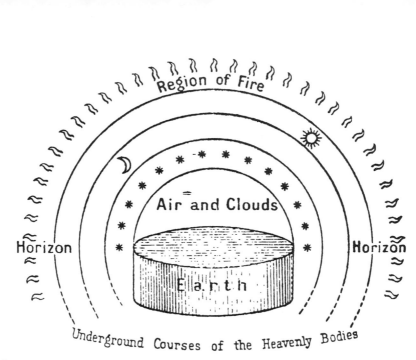

According to Anaximander, the earth was a disk with a diameter three times its thickness. The sun, moon, and stars were not thought to be objects but holes in moving rings that light from more distant fiery realms shined through and down upon the world. *(Yerkes Observatory)*

holes that permitted the light from three fiery rings to penetrate the sky, Anaximander estimated their dimensions and tried to match their movements with the rotations of wheel-like rings at the appropriate distances and with the appropriate rates. In Anaximander's cosmos the sun was the most distant celestial object; the stars were the nearest. Of course, he was wrong, but his ideas were based on rational argument and observation.

Anaximander also thought a single principle—rather than the gods—governed the form of the world, but it was an intangible substance he called *apeiron*, and not water, as Thales had favored. It is not exactly clear what Anaximander had in mind with his *apeiron*, but the idea seems to have carried connotations of the infinite—the undefined and all-pervasive. His universe was eternal—infinite in time—and boundless. This does not mean infinite in extent, however. Just as the surface of a bubble has no boundary or end but still embraces only a finite area, the cosmos, as Anaximander saw it, was without bound but finite. Uniform circular motion powered this cosmos.

Many more ideas about the cosmos were contrived by the Greeks. Anaximenes of Miletus (585–526 B.C.) said the elemental substance was air. An infinite ocean of it supported the flat disk of the earth, while overhead the sun, moon, and planets—all imagined as disks—and the stars moved in their cycles. The stars were thought to be attached to a rotating, crystal sphere. Pythagoras (580–500 B.C.), who founded a religious and philosophical brotherhood in a Greek colony in southern Italy, saw in pure number the underlying principles of the cosmos. Numerical and geometric relationships acquired considerable symbolic significance as part of a secret lore shared by the Pythagoreans. These ideas, applied to the distances and the movements of celestial objects, were described in terms of cosmic order and symmetry—the "harmony of the spheres," an actual music created as the spherical heavenly bodies moved uniformly on their circular paths around a spherical, immobile earth. Philolaus of Croton (the same Italian city, Croton, where Pythagoras had founded his school of mysticism and philosophy) was also a Pythagorean, but he moved the earth from the center of the cosmos and placed the "Central Fire" there instead.

This Central Fire was also the source of the central force that moved the celestial

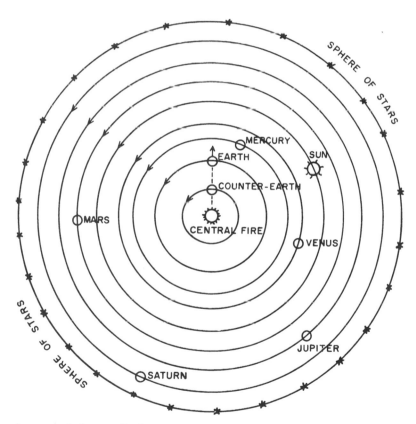

Philolaus incorporated some Pythagorean ideas into his picture of the universe but centered it on the Central Fire. A Counter-Earth was added to the scheme. The moon also was part of this system, but it is not shown here. *(Griffith Observatory)*

objects, including the earth, on their circular paths. In addition to the earth, the sun, the moon, the five planets, and the Central Fire, Philolaus added a tenth object, the counterearth, to the system. Geometry and motion kept the counterearth out of sight of those on earth. What evidence motivated Philolaus to include such an invisible object is not clear. He could have been trying to account for the invisibility of the Central Fire, presumably eclipsed by the counterearth, or, perhaps, the earth's counterpart was added to bring the number of celestial objects up to ten, a most sacred number for the Pythagoreans.

Moving the earth from the center of the cosmos was rather a radical step. Another Ionian, Aristarchus of Samos (310–230 B.C.), tried the same idea and attempted to estimate the sizes and distances of the sun and the moon. Aristarchus concluded the moon was about one-third the size of the earth and 9½ times the diameter of the earth away. The sun's distance he calculated at 180 earth diameters, and the sun itself nearly seven times larger than the earth. In fact, the moon is about one-fourth the diameter of the earth, and the sun's diameter is about 109 times larger than the earth's. The sun is about 65 times farther away than Aristarchus guessed, and the moon is a little more than three times farther than his estimate. His answers were incorrect, but his method, based on geometry, was essentially valid. The instruments of his day simply were incapable of the high precision his technique required.

Aristarchus was right about the sun, of course. It does belong at the center of the solar system. His ideas never caught on, however, perhaps because the idea of an earth-centered cosmos was too well entrenched or, more likely, because Aristarchus provided no evidence to support his revolutionary idea.

There was, in fact, evidence against a sun-centered cosmos. From a moving earth, a shift in the positions of stars should be seen as the earth occupies first one side of its orbit and then the opposite. No such shifts were seen, and Aristotle used this argument in support of the earth's centricity. Neither Aristotle nor Aristarchus knew that shifts in the star's positions actually do occur. The stars, however, are far more distant than the Greeks ever imagined, and detection of their displacement due to the movement of the earth was well out of the reach of their instruments.

Greece and Turkey's Ionian coast, which harbored numerous and successful Greek settlements, became places where philosophical thought prospered, and what the early Greek philosophers were best at was telling each other what the universe was really like and why each other's ideas were wrong. There was no universal belief system to confine them to a single concept of the cosmos. Their speculations were based more upon deductive reasoning than on detailed observations, but their goal still required them to confront the real phenomena anyone could see and measure in the sky. They believed there was a fundamental unity and order to the universe. In trying to generalize about it—to describe and account for how the cosmos is regulated—they had to encounter the observed departures of celestial objects from regularity and symmetry. This prompted more detailed explanations. The Greeks began to apply mathematics in the form of geometric models to the problem of celestial motions. When Eudoxus of Cnidus (408–355 B.C.) tried to duplicate the real celestial motions in terms of geometry, he invented scientific cosmology.

Eudoxus, like the other Greek philosophers, believed that uniform circular motion explained the behavior of the universe. This is not as arbitrary an assumption as we might guess, for, as we have seen, the fundamental motion of the earth—its daily rotation—creates the impression of uniform circular motion in the paths of the stars. Even the motions of the sun and moon—if not subjected to intense scrutiny—suggest the same underlying principle in the cosmos. Anaxagoras (500–428 B.C.), who ascribed the creation of cosmic order to an intelligence—although not to a divinity—within the physical universe, pictured the process as starting with uniform rotation in the primordial chaos. From that single motion—still visible in the sky—the regular and systematically arrayed components of the cosmos emerged to form the "ordered whole."

Eudoxus arranged the "ordered whole" into a set of 27 nested spheres, all centered on the earth. This cosmology was designed to account for deviations in uniform circular motion—especially evident in the behavior of the planets. By combining motions of several spheres, Eudoxus could approximate what was observed in the sky. The scheme was not perfect, however, and one student of Eudoxus added 7 more spheres in an effort to replicate reality more accurately. This clearly shows scientific thinking. The theory, when matched against nature and found wanting, was modified into a better mirror. Later Aristotle (384–322 B.C.) added even more spheres. He brought the total to 55 and made the scheme much more complicated, but no more correct.

Aristotle's theoretical picture of the cosmic order was replaced by the system of Claudius Ptolemy, an Alexandrian astronomer of the second century after Christ. Ptolemy and his predecessors worked out a series of geometric inventions in order to match the theory of planetary motions to the observations and at the same time preserve uniform circular motion. Instead of letting the planet go around the earth on a circular orbit, they put the planet on an epicycle. This was a second circle, and its center traced out uniform circular motion around the earth on a path called the *deferent.* At the same time, the planet moved in the opposite sense around the center of the epicycle. In later versions, the epicycle was allowed to move with constant speed around an off-center point, known as the *equant.* The earth was still at the center of the epicycle's orbit, but with respect to the earth the epicycle's motion no longer looked uniform. It was possible to fit combinations of these motions to the observed data, but it took numerous deferents, epicycles, and equants in a highly convoluted system to do it. The elegance and simplicity of uniform circular motion was obscured by the labyrinth of celestial gear trains.

Greek science, unfettered by any mythic premise, generated its own assumption

COELVM EMPYREVM IMMOBILE.

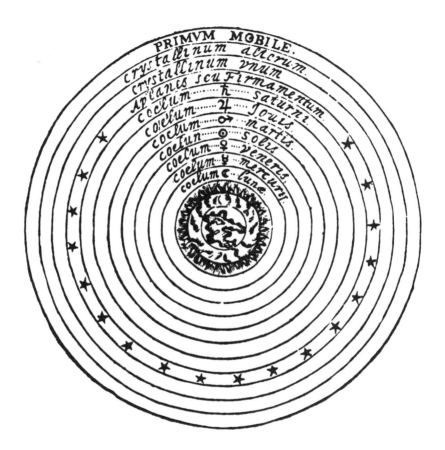

The world system of Ptolemy was centered on the earth, the heaviest of the four "elements." Surrounding it, in turn, were the waters, the air, and a sphere of fire. Higher still in heaven revolved the spheres of seven "planets": the moon, Mercury, Venus, the sun, Mars, Jupiter, and Saturn, all nested inside the sphere of fixed stars. This version of the Ptolemaic system was included in a 1647 work by the French philosopher-astronomer Pierre Gassendi. He shows Ptolemy's crystalline sphere beyond the stars divided into two that deal with "trepidation," a combination of precession and other long-term cycles. Beyond these is the sphere of the Prime Mover, the sphere that set all the others in motion, and beyond that is the "immobile empyreal heaven." *(Reprinted with permission from S. K. Heninger, Jr.,* The Cosmographical Glass: Renaissance Diagrams of the Universe, *The Huntington Library, 1977)*

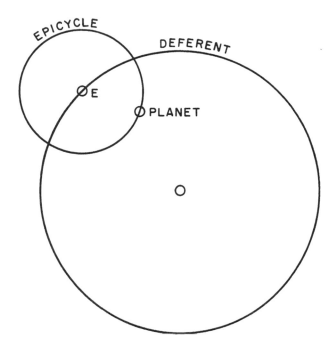

Better approximations to the planets' motions motivated the invention of the epicycle. It was a circle of planetary motion that itself circled in orbit. The center (*E*) of the epicycle moves on the deferent while the planet itself moves on the epicycle. The object at the center is the earth. *(Griffith Observatory)*

about the same cosmic order it was intended to probe. The Greeks stalked uniform circular motion in the planet's deviations from it and never really deserted the notion. It would take a millennium and a half, the Copernican revolution, better observations of the planets, and Johannes Kepler, a German astronomer and mathematician, to demonstrate that the celestial order revealed by the data had nothing to do with uniform circular motion.

Organized by Heaven

In some respects, astronomy's development in China during the Han Dynasty (206 B.C.–A.D. 220)—especially in the first centuries after Christ—paralleled the pattern that had begun in Greece a few centuries earlier. Like the Greeks, the Chinese also tried to understand the architecture of the universe and express the structure of cosmic order in cosmological theories that had to mimic the sky. Without deductive geometry, however, the Chinese never immersed themselves in the detailed modeling of planetary motion that led to the complexities of epicycles, deferents, and equants. In fact, even though the Chinese recognized the ecliptic—the path of the sun and the planets—their system of astronomical measurement and concept of the sky remained essentially equatorial, while movement along the ecliptic itself was the organizing principle of planetary motion in European tradition.

By the first century after Christ, Chinese astronomers were measuring the positions of celestial objects through bronze sighting tubes attached to finely engraved armillary spheres. These instruments were relatively sophisticated and incorporated rings for the meridian, horizon, the celestial equator, and the ecliptic. With them, any celestial object could be observed, charted, and considered in the light of any of the three major cosmological theories that had evolved by this time. The oldest of these was known at times by the name of the mathematical-astronomical treatise that described the instruments and concepts around which the theory was formulated. This work was called *Chou Pei,* a name which means "the gnomon and the circular paths of heaven." An alternate name for the theory, *kai thien,* means "heavenly canopy" and refers to the hemispherical celestial cocoon that was thought to enclose the earth.

To imagine the *kai thien* cosmos, you must picture the earth as an inverted bowl whose rounded surface culminated in a square rim, in keeping with the Chinese tradition that the earth is square. The sky's round contours completely enveloped the earth and joined its rim to form a trench that encircled the world. There, rains collected and created an oceanic rim analogous to that imagined by the Greeks. The difference here, however, is the earth's shape. It is not a flat disk but a kind of half-bubble blown inside another half-bubble, so that where they touch, their surfaces merge.

As the celestial vault rotated, the sun, moon, and planets, although attached, were free to move through their own cycles. Of course, the sky rotated around the celestial pole, construed by the Chinese as the center of the sky. Originally, the point directly below it was the center of the earth. For the Chinese, the center of the earth had to be in China, but anyone could see the celestial pole was no longer overhead but about 36 degrees or so above the northern horizon. They explained this as the result of a battle between a monster known as Kung Kung and the fourth mythical emperor, Yao. During the fight, Kung Kung ran into one of the supports, or pillars, of heaven. It broke, and the whole sky tilted over.

Although we don't really know how old the *kai thien* cosmology is, tradition attributes it to the Bronze Age Shang Dynasty of the second millennium B.C. Certainly this is possible, for all of its characteristics reflect the basic observations obtainable with an instrument the Shang used, the gnomon. The meridian, the length of the day, noontime, the length of the year, the solstices and equinoxes, the cardinal directions, and the duration of the seasons: all are embodied in the double hemisphere formed by heaven and earth.

Attempts were made to estimate the "height of heaven"—the distances to celestial objects—through gnomon shadow measurements. These were no more successful than the Greek efforts, and the height obtained was considerably less—about three times less—than the radius of the earth. An entirely independent school of cosmology

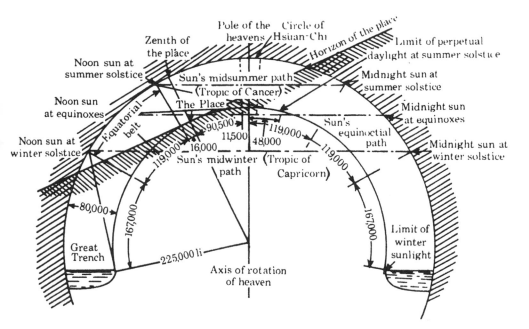

An earth tented by a hemispherical sky, with its bottom edge tucked under and attached to the earth's own square footing: This was the picture of the universe provided by the *kai thien,* or "heavenly canopy," cosmology. Part of the effort of some Chinese astronomers was directed to measuring the "height of heaven," shown here as 80,000 *li,* or perhaps about 6,700 miles. *(Joseph Needham,* Science & Civilization in China, *Vol. III,* Cambridge University Press)

maintained, by contrast, that the heavens were immensely high, far away, and without bound. This theory was known as *hsuan yeh,* words which mean "brightness and darkness" or "all-pervading night" and refer to the principle of infinite empty space on which the theory rests.

According to the *hsuan yeh* ideas, the heavens are empty, a void. They look blue in the day, but that is because the eye distorts reality. The sky looks blue for the same reason distant mountains take on a blue or violet cast. All of the celestial objects—the sun, moon, planets, and stars—float without restrictions in space. In many ways, these ideas are accurate (even the explanation for the blue sky has a grain of truth in it) and modern. Yet, by A.D. 180, one commentator was moved to write that no teacher of the *hsuan yeh* theory still lived to preserve and promote its vision of the cosmos. The same writer declared the *kai thien* theory to be "proved incorrect and lacking in many ways when tested in explaining the structure of the heavens." Only the remaining theory—the *hun thien* cosmology—approximated what was really going on in the sky.

Hun thien means "celestial sphere." It was, to some extent, similar to Greek concepts of a spherical earth situated in the center of a spherical sky. In his *Commentary on the Armillary Sphere,* Chang Heng, the first-century Chinese astronomer and seis-

The armillary sphere is actually a model of the Chinese *hun thien* idea of the universe. The sphere created by the rings mimics the sky, with the earth visualized at the center. By sighting along the diameter through a movable lensless tube, Chinese astronomers measured the angular positions of celestial objects against precise scales engraved upon the bronze rings. This armillary sphere was fabricated in A.D. 1437 and is an exact copy of an early instrument used in 1279 by the astronomer Guo Shou jing. It is now on display on the grounds of the Purple Mountain Observatory near Nanjing. *(Robin Rector Krupp)*

mologist, described the heavens as like "... a hen's egg and as round as a crossbow bullet; the earth is like the yolk of the egg, and lies alone in the center." The analogy of the egg yolk is crucial; it implies a spherical earth. Chang Heng continued with further details on the relationship between earth and sky. "Heaven is large and earth small. Inside the lower part of the heavens there is water. The heavens are supported by vapor, and the earth floats on the waters."

After explaining why only half of the sky may be seen at any one time (the invisible half is below the earth), Chang Heng delineated the positions of the north celestial pole and the unseen south celestial pole and correctly described why stars near it are never seen. There was a bit of the *hsuan yeh* cosmology and philosophical perspective in Chang Heng as well, for when he considered what might be beyond the celestial sphere he wrote, "What is beyond no one knows, and it is called the cosmos. This has no end and no bounds."

By A.D. 520 only the *hun thien* cosmology was considered to be true. Although it never dispatched the Chinese on the quest for planetary theory or into the trap of uniform circular motion, it did furnish an image of cosmic order consistent with Chinese measurements of the sky. Whereas cosmology in Greece was an intellectual exercise, in China it was also linked with astronomy's official role in the order of the empire. It was a flexible scheme, laced here and there with a notion of infinite empty space that suited the mystical component of Chinese philosophy. At the same time, Chinese cosmology coordinated the celestial cycles and references that oriented Chinese life.

Chinese cosmology did not culminate in a revolution in thought spearheaded by a "Chinese Copernicus." Why not? There is no single reason that explains why cosmology in China took a different turn than it did in Europe. The connections between new technology, scientific observation, and innovative thought are real enough, but they are also complex and not fully understood. With a vision of the sky that emphasized the celestial pole and equator and had no use for precise geometric imitations of celestial motions, the Chinese astronomers persisted in their sacred view of the cosmos and delayed the secularization of cosmology until long after the European Renaissance.

Framed by Space and Time

Anyone intent on drawing a picture of cosmic order has two things to manipulate: space and time. Now, the cosmologies of the ancients tell us they realized the structure of the universe was evident in the organization of space and in the passage of time. And even though concepts of the universe changed, the scaffolding of space and time was not abandoned, just rearranged. From the earth-centered systems of the Greeks, with their deferents, epicycles, and equants, we shifted to the sun-centered universe of the Polish astronomer Nicolas Copernicus (1473–1543). It took 14 centuries to do it, however, and uniform circular motion, the fundamental principle of the Greek cosmologies, did not get lost in the shuffle. Copernicus believed in uniform circular motion and represented the planets' orbits as circles with epicycles—and epicycles on epicycles. This allowed him to discard Ptolemy's equants, but it took more epicycles than Ptolemy used to do so.

No one felt moved to move the earth until Copernicus did it in the sixteenth century. His model was based on ancient Greek observations, however, and could not survive long after the Danish astronomer Tycho Brahe (1546–1601) made the most accurate and precise observations of the positions of celestial objects yet obtained. Each observation represented a combination of space and time; it told where something was and when it was there. From these reliable couplings of space and time, Johannes Kepler (1571–1630) fabricated an entirely new pattern of cosmic order and

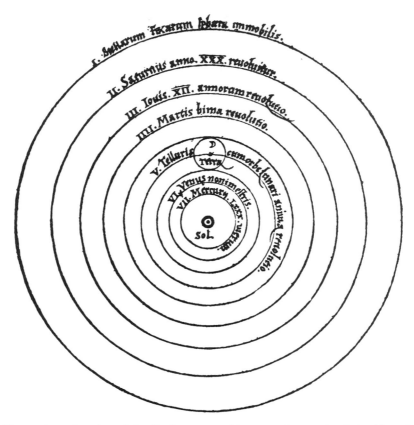

When Copernicus abandoned the Ptolemaic world system, he retained the idea of uniform circular motion but placed the sun at the center of the universe. Continuing outward, the order of the planets became Mercury, Venus, earth (with the moon in orbit about it), Mars, Jupiter, and Saturn. Beyond the planets, as before, revolved the sphere of fixed stars. This drawing of the Copernican system appears in the original manuscript of *De Revolutionibus Orbium Coelestium* by Copernicus. *(Reprinted with permission from S. K. Heninger, Jr.,* The Cosmographical Glass: Renaissance Diagrams of the Universe, *The Huntington Library, 1977)*

dispensed—once and for all—with uniform circular motion. The planets' orbits, he showed, were ellipses, not circles, and a planet's speed along its path around the sun was not uniform, but varied in accordance with a simple geometric law. Time and space were still at the heart of the cosmic harmony in Kepler's laws of planetary motion, however, for he demonstrated that the time it took a planet to complete one circuit of its orbit is related in a simple way to the orbit's size. In fact, he called the book in which he announced this relationship *The Harmony of the World*. With Kepler, the movements of celestial objects through time and space—as seen upon the sky—were represented by laws that explained the relationships among them.

Eventually, Sir Isaac Newton (1642–1727), an English mathematician and physicist

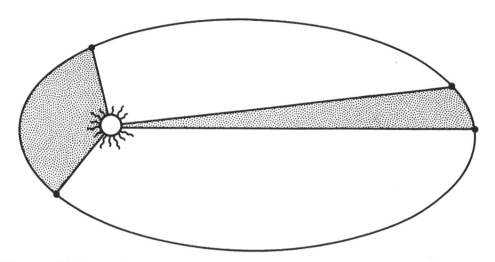

The actual behavior of the planets, as deduced from the precise observations of Tycho Brahe, led Johannes Kepler to forsake the idea of uniform circular motion. Instead, he concluded the orbits are elliptical. The shape of an orbit is greatly exaggerated here, for the real orbits of the planets deviate only slightly from true circles. This drawing, however, is able to illustrate clearly Kepler's first law of planetary motion: the elliptical orbit with the sun at one focus. The shaded areas demonstrate the second law: A planet sweeps out equal areas in equal time. Because the distance traveled on the orbit for the area left of the sun is greater than that on the right—even though the two *areas* are equal—the planet moves faster when nearer the sun. The planets' motions are not only not circular, they are not uniform as well. *(Griffith Observatory)*

and one of the most influential scientists of all time, was able to show that Kepler's cosmic harmony could be explained by a universal force—gravity. Newton described the mutual attraction everything in the universe experiences, in mathematical terms, and his formulas allowed him to calculate any object's trajectory in space and time, whether an apple falling from a tree in his Woolsthorpe garden or a planet falling around the sun. With Newton came the realization that whatever the universe may be like, gravity was its underlying principle.

By the eighteenth century, our sense of the landscape of the universe had changed considerably. Although an accurate measurement of the distance to even one star would have to wait until nearly the middle of the nineteenth century, by 1785 the English astronomer William Herschel (1738–1822) was able to draw a map of what he called "the construction of the heavens." Herschel's cosmos was a vast flattened system of millions of stars. Branches and arms of stars extended somewhat irregularly from the disk. Herschel placed the sun and its planets near the center, not for philosophical reasons but because his observations from inside this galaxy, or stellar system, were consistent with that interpretation. We now know we are near the rim of the Milky Way galaxy. Herschel managed to give us a sense of our galaxy, but its true dimensions defied measure until the third decade of the present century. Herschel speculated that the Milky Way galaxy might be an island floating in empty space— one, in fact, of many such islands of stars, but he recognized his observations were incapable of proving this.

Newton's law of universal gravitation accounted for the universe very well, and it was appreciated that all of the stars in the Milky Way galaxy—as the planets in the solar system—must march to gravity's orders. But the subtleties and assumptions of a Newtonian universe gradually came under closer scrutiny, and in 1916 Albert Einstein (1879–1955), a German mathematician and physicist, proposed a new way to look at gravity in his "general theory of relativity." By recognizing time and space as a single fabric, its threads woven intimately and inextricably together into the web of events we call the universe, we can no longer let gravity be a force in the Einsteinian cosmos; it is instead, the curvature—the contour—of space and time. By describing the curvature accurately and mathematically, we describe neither space nor time but the events that occur—the way the universe behaves.

Einstein's way of looking at the universe came as a real surprise to the world, but it does provide the most accurate representation we have of the way things are. A sense of the intimacy shared between space and time is not so unprecedented, however. Something of the sort is evident in the way we treat the solstices and equinoxes. They are locations in the sky as well as the times when those places are occupied by the sun. When we analyzed the page from the Aztec *Codex Fejérváry-Mayer* and several similar calendrical diagrams, we discovered that the ancient Mesoamericans also associated the passage of time with the orientation of space. The interlocking symbolism of these diagrams reflects the reciprocity of space and time. Each year, for example, began with a year-bearer date, and each of these possible dates was identified with one of the cardinal directions. Intersecting crosses in the *Fejérváry-Mayer* codex map the cardinal and "intercardinal" (solstitial) directions, and time parades around the perimeter of the design as a series of dots that stand for the days.

To the ancient Mexicans, the earth was an island, floating on the sea. Its waters met the horizon and mingled there with the sky. The sky, too, was fluid, a fact confirmed by the fall of rain, but its waters were less clotted than those of the lakes, rivers, and sea. Divine sky-lifters or world trees held up the heavens and were located in the key horizon directions.

The Mesoamericans imagined the universe in metaphysical as well as physical terms as a hierarchy of realms, organized like an especially ambitious layer cake with 13 celestial levels and nine floors for the underworld. We have already seen how this concept of cosmic space was transformed into time by equating the two sets of levels with the passage of day and night. These same numbers were used to bundle nights together into groups of nine nights and the daylight periods into groups of 13 days. Separate counts for day and night were kept, then, with numbers that mirrored the structure of the cosmos.

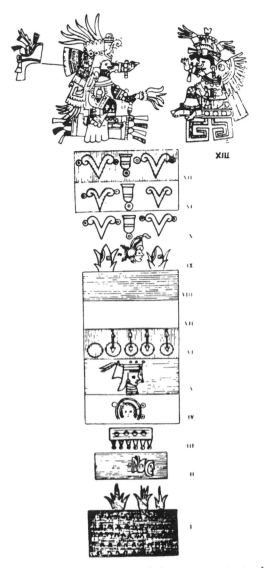

It may not look like it, but the Aztec idea of the cosmos is similar, in a way, to the Ptolemaic spheres. Both are layered universes. There, however, the resemblance ends. This drawing in the *Codex Vaticanus* shows the 13 layers of heaven and the nine layers of the underworld. In the middle (layer 1) is the earth. Above it are the zones of the moon, the stars, the sun, and the planet Venus, respectively. The important constellation of the Fire Drill occupied its own heaven in layer six. Above it rested the green heaven of winds and storms and the blue heaven of dust. Layer nine seems to have been the realm of thunder, and above it were three more colored zones: white, yellow, and red. Above all of these celestial realms, in the thirteenth and highest heaven, resided Ometeotl, the creator of space, time, and the world. (© *1971 by The University of Texas Press*)

Einstein's merger of space and time may not be exactly what the ancient Meso-americans had in mind, but they certainly realized one concept melts into the other, as long as we perceive the universe as an assembly of events. Einstein's handling of these ideas was considerably more abstract. Mesoamerican cosmology equates the location of a celestial event with the time at which it happens; Einsteinian relativity does not. Modern relativity is much more abstract, but we can still get a concrete feeling for the intimate relationship between space and time by giving some thought to what we really see overhead.

When we look at the moon, we don't see it as it is now, but as it was a little more than a second ago. The moon is about 240,000 miles away, but light travels at 186,000 miles per second. After bouncing off the face of the moon, it takes light 1.28 seconds to reach us. Similarly, the sun we see shining so dependably is the sun of 8 minutes and 19 seconds ago. If the sun were to blink out now, we would have 8.31 minutes of ignorant grace before being enveloped in darkness. Beyond the sun, the closest star is Proxima Centauri, one member of the triple system known as Alpha Centauri. An appalling distance of 25 trillion miles separates us from this nearest of stars, and its light travels 4⅓ years to reach us. It is convenient to specify such vast distances in light-years, the distance light travels in a year, and one light-year is about 6 trillion miles.

When Sung dynasty astronomers in China spotted a brilliant new star on July 4, 1054, in the constellation we call Taurus, the bull, they were actually observing the catastrophic explosion of a star that had occurred 6,000 years earlier, at the beginning of China's neolithic age. The star that burst was 6,000 light-years away, and news of its demise did not reach earth until the eleventh century after Christ. Today we see a chaotic, tangled knot of clouds and filaments of gas—the Crab Nebula. They are the still-expanding remains of the star as it appeared nearly a thousand years after its death. When we look out into space, we look back into time, and the remnant we see in Taurus today is really what the cloud was doing more than a thousand years before the first stones went up at Stonehenge.

Today, our telescopes see far beyond the Crab Nebula—far beyond, in fact the edges of our galaxy, with its 100,000-light-year diameter. Now the edge of the cosmos is no longer the surface of the celestial sphere but the limit of our instruments' abilities to collect the light of distant objects. The most distant things we see are 10 to 20 billion light-years from us in space and 10 to 20 billion years back in time.

Outdistanced by the Dark

Such distances and intervals of time give the universe grandiose proportions, and cosmology is an assessment of the grandest proportions by which we measure our lives. What the universe is like, how it got that way, and what's going to happen to it next—these don't sound like issues that affect our daily lives, but the answers affect our attitudes. And attitude, ultimately, is linked with survival.

We just want to feel at home in the universe. That is why we do cosmology, and that is why we ponder the sky. Cosmology and cosmic order have always been linked. To ignore cosmic mysteries is an invitation to be haunted by the real vacuum of space: loss of our sense of place. The universes we design have given us a footing in the avalanche of events that comprise the world.

And yet, our yardstick for plotting our location and course—the measurement of intergalactic distance—is so fraught with error that the entire history of modern cosmology sometimes seems like a tale of inadequate measurement.

Astronomers are still our intermediaries with the sky, and one of them, Dr. Allan Sandage, has been working at the cold end of a telescope for several decades to develop some sense of where we are in the universe. Back in 1970, Allan Sandage called cosmology "a search for two numbers." One of these is H, the Hubble constant. It is named after Dr. Edwin Hubble, the American astronomer who, in 1929, discovered that the universe was expanding. It still is, and the Hubble constant is the rate at which the "ordered whole" is getting larger.

The Hubble constant is an interesting number; it's related to the age of the universe. Evaluate H and we can guess how long things have been going on this way. The second interesting number is q, the deceleration constant, which specifies the rate at which the universe's expansion is slowing down. Essentially, q tells us how long all this will continue to go on.

But it is the Hubble constant that most interests us for the moment. To understand the universe, we have to understand H. Evaluated correctly, it tells us how the speeds of the distant galaxies, racing away from us, are related to their extravagant distances. And, of course, it was Hubble who first estimated its value.

Hubble is the kind of famous American who deserves to be on a postage stamp. He was certainly one of this century's greatest astronomers, and any one of the three major discoveries he made, all early in his career, would qualify him as one of the greatest astronomers of all time. The closest the U.S. Postal Service has come to commemorating him is a 1981 "Space Achievement" issue that portrays the Large Space Telescope superimposed upon a spiral galaxy and accompanied by the words "Comprehending the Universe." There is no reference to Hubble on the stamp, but the spirit is right. Just as Einstein's theories changed the way we look at space and time, Hubble's observations of the distant galaxies, or nebulae, as they were called in his day, transformed our comprehension of the universe.

Hubble was the first to demonstrate that other galaxies—"island universes" they were sometimes called—vast communities of stars, gas, and interstellar dust, drift in the dark realm beyond our own Milky Way galaxy. Hubble proved that other galaxies are farther away than the edges of the Milky Way. Distances to three relatively nearby galaxies expanded the universe into something larger than the Milky Way.

Hubble's estimates were based on a sound technique: comparison of a star's appearance to the known brightness of a star of the same type. A star grows dimmer as its distance increases, and Hubble stretched the technique to measure the dimmest and farthest stars he could measure with the telescope he had.

Once Hubble had distances at his disposal, and put them together with velocities he could measure, he chanced on a startling circumstance: The galaxies are all receding from each other. From our point of view, it looks as if the galaxies are deserting us, but the view from any other galaxy would be pretty much the same. From any galaxy, all but the nearest neighbors are rolling away, and the farthest are absconding with their respective portions of the cosmos at the fastest speeds. This antisocial behavior can be described in terms of the Hubble constant, which tells us how many miles per hour, say, the galaxies pick up with every quadrillion or so miles of additional distance.

Expansion of the universe is now accepted as the consequence of an explosion that took place long ago. It sent everything surfing on the breaking swell of a bubble of space-time, and the explosion is still going on. Because the farther we look into space, the earlier we see in time, really remote galaxies are our clue to what the universe was like when it and the explosion were still young.

It has become fashionable, particularly as we approach the end of a millennium and tune up our instincts for apocalypse, to call the time of the explosion the "creation of the universe." It is not as romantic to think of it as simply the "time of the explosion," but that is what it really is. We have no way of probing what went on, if anything, before the explosion, and our ignorance is transmuted into presumption by saying the "creation" is pinpointed. All we can really talk about is how long the universe has been behaving this way. The Hubble constant, by telling us the rate of expansion per space-time mile, can tell us, in turn, how long this cosmic performance piece has continued.

Our understanding of the universe is tied to knowing the distances, and despite Hubble's best effort, his estimates were short. His original determination pegged the universe at about 2 billion years old. Even in the 1930s we knew of rocks that were older than that. It was obvious the universe couldn't be younger than the rocks in it.

To understand what went wrong, we have to remember it is a very shaky series of measurements that puts us at the edge of the observable universe. Known objects calibrate the distances of nearby galaxies, and we then use the brightnesses of these galaxies to pull ourselves up to the more distant ones. Over decades, the sources of error in Hubble's distances were found, and, conveniently, the universe grew larger and older. The Hubble constant, for its part, had the courtesy to diminish in value. A series of refinements, many developed and imposed by Allan Sandage, improved the distances and gave the universe an age of 15 to 20 billion years.

Others besides Sandage have been active, however, and controversy has diluted our intimacy with the cosmos. Conventional distance measurements are still plagued, some claim, by errors that conceal the true geography of the universe. As recently as 1980, new work prompted some to argue that the universe isn't so old after all—

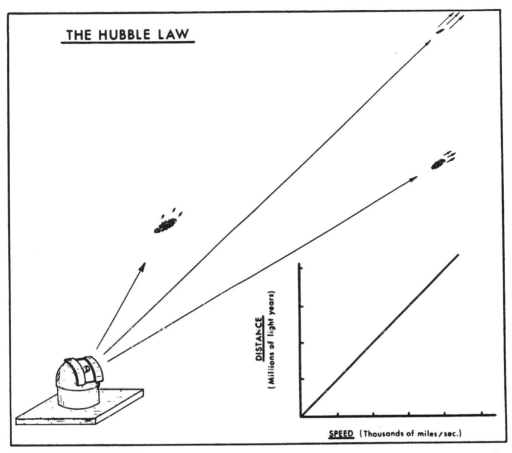

THE HUBBLE LAW

DISTANCE (Millions of light years)

SPEED (Thousands of miles/sec.)

By measuring the speeds of distant galaxies and by determining their distances as well, astronomer Edwin Hubble discovered that the universe is expanding. Most galaxies are traveling away from us, for the light they emit is reddened by the Doppler shift. Hubble's revolutionary observations demonstrated that the speed of the galaxies' departures depends on their distances. The farther away they are, the faster they are going. *(Griffith Observatory)*

perhaps only 10 billion years, a mere infant. Drs. Marc Aaronson of the Steward Observatory (near Tucson, Arizona), Jeremy Mould of nearby Kitt Peak National Observatory, and John Huchra, associated with the Harvard-Smithsonian Center for Astrophysics, tried something new in tagging distances. They looked at entire galaxies and classified their brightness in terms of how fast the galaxies spin. Rotation speed *can* be measured.

When Aaronson, Huchra, and Mould pulled their method together and looked at the big picture it provided, they found themselves sampling a universe of 10 billion years' vintage. This presents problems, for we think the universe contains stars and atoms older than this.

In a way, the new method seems cleaner, less likely to have errors than previous methods. If Aaronson, Huchra, and Mould are right, the galaxies are half as far, the universe is expanding twice as fast, and the explosion is half as old as we thought. But whether this be so or not, what will happen to the universe next?

We can say something about the future fate of the cosmos by knowing how quickly the explosion is settling down. With three assumptions—(1) The universe obeys Einstein's General Theory of Relativity; (2) mass and energy—taken together—can be neither created nor destroyed; and (3) from any point in the entire universe the overall picture would be the same—we can conclude that one of three possible destinies awaits us.

If the expansion of the universe is slowing down quickly, everything will reach its maximum separation from everything else, and then the whole process will reverse. The entire universe will collapse back upon itself. Alternatively, there may have been enough energy in the explosion to keep the galaxies expanding outward forever. Space in this kind of universe is curved to prohibit any recollapse, and the cosmos just grows larger and emptier. The third possibility provides the universe with enough energy to expand to infinity, but it will require infinite time to do so.

An intriguing variation on the first destiny also is sometimes considered—an oscillating universe. Explosion, expansion, and collapse: we may be in but one cycle in a string of such cosmic sighs. Every breath of space-time creates and destroys another cosmos.

We have squinted at the farthest and oldest celestial lights we can see to find out what kind of cosmos we inhabit, but the expanding universe and any of its possible outcomes have brought us right back to the old familiar myth of the cycle of cosmic order. The cosmos is created, it grows, and in some fashion it dies—either in emptiness or in suffocating collapse.

And, perhaps, it is created anew. We live in a strange universe. Its proportions are incredible, its destiny uncertain. It has no center, yet the view from anywhere gives the viewer the sense of being at the center, along with the knowledge he is not. Modern scientific cosmology has given us a universe but has provided no real inkling of what our role in it may be. And we find ourselves contending with the same old myth as we seek the signs of cosmic order.

We have arrived at an interesting juncture. We may not be able to sort out the universe, but our attempts to do so may help us sort out our minds. Cosmic order is a product of the processing that goes on in our brains. The human brain has a job to do: It must make sense of the world. In that respect, it does what a computer does. It processes information. To succeed, it must edit, organize, and interpret all the data our senses collect, and its ability to filter data out is as important as its capacity to integrate what it retains.

Imagine if we left it to our senses to deal with the world for us. They are unselective, and they would brutalize us with more facts and phenomena, more details and

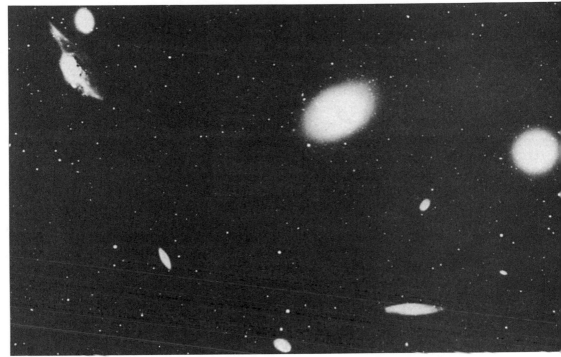

These are a few of the galaxies in the center of the Virgo cluster, the nearest large cluster of galaxies to us. Even though they are about 50 million light-years away, they are relatively close compared to the most distant objects we see, hundreds of times farther away. *(Kitt Peak National Observatory)*

distractions, than we could or should handle. The real genius of the brain is its ability to simplify the world, to reduce it to the ingredients essential for our survival, and to rescue us from the madness of information overload. Recent research on the neurochemistry of the brain suggests that evolution and natural selection determined long ago how the brain would excerpt the world—through chemistry—to enhance our ability to survive.

All this implies that there must be some shared structure to human thought and some order imposed on perception. Despite our individual and cultural differences this must be so, or communication—particularly through language—would be impossible. All of our conversations would be alternating non sequiturs.

Phosphenes—those multicolored, geometric luminous patterns the eye and brain generate by themselves—are shared by all people. Experts who have studied them associate them with the process of visual perception. In some way not yet understood, the phosphene patterns are thought to have something to do with the way in which information is actually transmitted along the neural pathways that participate in the experience of vision. These neural conduits determine what we see and how we perceive it. If the phosphenes are triggered by spontaneous electrochemical activity in the neurons, they may, in a sense, be showing us the structure of the neural circuitry. If so, phosphenes reflect a universal order that we impose upon the world around us.

Just because our interpretation of reality is subjective, it doesn't mean there isn't really any order out there. Our interaction with the sky tells us, however, that our brains specially emphasize the order. Why should they do this?

Organized, cooperative groups have an evolutionary advantage, but it takes intelligence—a big brain—for us to deal with each other. Intelligence, in part, is reflected in our ability to classify and to generalize—to create order. The patterns that we impose on the real world allow us to anticipate the events that conform to the patterns and respond with wit to those that don't. Those abilities help us cooperate and survive. Any structure that helps us communicate and reciprocate is going to be useful. This includes language, myth, calendars, clocks, a system of directions in an organized landscape, and many more tools that orient our lives.

We can be misled, however, by our consciousness of our own intelligence and believe that our big brains are really able—or will be able—to know the cosmos completely. We rarely think of the limits of intelligence. But each time we return to the myth of cosmic order—and realize we still use it—we should recognize our brains for what they are: tools that help us fill the environmental niche we so successfully occupy. The sky is one of the things that helped make the big brain work, but we are not detached, omniscient observers of the universe. We are participants.

In some strange way our participation includes awareness of the universe and an effort to find our place in it. And so we continue to try to snare the light of distant galaxies in the hope of comprehending the universe. The Hubble constant controversy is like an old-time Saturday matinee—"to be continued." All the concentrated work of 50 years at the telescope has failed, so far, to resolve the most fundamental information, the distances to the most distant things. Our ignorance in such matters is reflected, in turn, in the rest of the riddles that our reports, as eyewitnesses to the cosmos, provoke.

It is very hard to measure the distances of distant objects. It always has been. And all of our answers hinge on simple facts—how far away things are from us—in space and in time. We can see landmarks. We just don't know for certain where they are. Hubble's own assessment of what we knew in his day is accurate for our own:

> Thus the explorations of space end on a note of uncertainty. And necessarily so. We are, by definition, in the very center of the observable region. We know our immediate neighborhood rather intimately. With increasing distance, our knowledge fades. Eventually we reach the dim boundary—the utmost limits of our telescopes. There, we measure shadows, and we search among ghostly errors of measurements for landmarks that are scarcely more substantial.

The universe is 46 years larger than when Hubble wrote those words. The explosion is millions of miles older. It has lit candles of galaxies for us. It would be shabby to curse the darkness.

Comprehending the Universe

USA 18c

The way we design universes today, with the observational approach of modern science, may differ from the sacred metaphors of our ancestors, but we all do it for the same reason: to comprehend the universe in a way that lets us feel at home in it. *(U.S. Postal Service)*

Bibliography

(Many of the works cited below are available as Dover reprints.
Visit our website at www.doverpublications.com.)

General Archaeoastronomy

Aveni, Anthony F. "Archaeoastronomy." In *Advances in Archaeological Method and Theory,* vol. 4. Princeton, N.J.: Academic Press, 1981, pp. 1–77.

Aveni, Anthony F., ed. *Archaeoastronomy in the New World.* Cambridge: Cambridge University Press, 1982.

Aveni, Anthony F., and Gary Urton, ed. *Ethnoastronomy and Archaeoastronomy in the American Tropics.* New York: New York Academy of Sciences, 1982.

Baity, Elizabeth Chesley. "Archaeoastronomy and Ethnoastronomy So Far." *Current Anthropology* 14 (1973); 389–449.

Brecher, Kenneth, and Michael Feirtag, ed. *Astronomy of the Ancients.* Cambridge, Mass.: MIT Press, 1979.

Cornell, James. *The First Stargazers.* New York: Charles Scribner's Sons, 1981.

Hawkins, Gerald S. *Beyond Stonehenge.* New York: Harper & Row, 1973.

Hawkins, Gerald S. "Stargazers of the Ancient World." *1976 Yearbook of Science and the Future.* Chicago: Encyclopaedia Britannica, 1975, pp. 124–137.

Heggie, D. C., ed. *Archaeoastronomy in the Old World.* Cambridge: Cambridge University Press, 1982.

Hicks, Ronald. "Archaeoastronomy and the Beginnings of a Science," *Archaeology* 32, no. 2 (March/April, 1979): 46–52.

Hodson, F. R., ed. "The Place of Astronomy in the Ancient World." *Philosophical Transactions of the Royal Society of London* 276, no. 1257; 1–276.

Kern, Hermann. *Kalenderbauten.* Munich: Die Neue Sammlung, 1976.

Krupp, E. C. "Ancient Watchers of the Sky." *1980 Science Year.* World Book Science Annual. Chicago: World Book-Childcraft International, 1979, pp. 98–113.

Krupp, E. C., ed. *Archaeoastronomy and the Roots of Science.* Washington, D.C.: American Association for the Advancement of Science, 1983.

Krupp, E. C., ed. *In Search of Ancient Astronomies.* Garden City, N.Y.: Doubleday & Company, 1978 (and New York: McGraw-Hill Book Company, 1979 paperback reprint).

JOURNALS

Archaeoastronomy. Bulletin of the Center for Archaeoastronomy: Center for Archaeoastronomy, Space Sciences Building, University of Maryland, College Park, Maryland 20742.

Archaeoastronomy. Supplement to the *Journal for the History of Astronomy,* Science History Publications Ltd., Halfpenny Furze, Mill Lane, Chalfont St. Giles, Bucks., England, HP8 4NR.

General Astronomy

Abell, George O. *Exploration of the Universe.* 4th ed. New York: Holt, Rinehart and Winston, 1975.

Alter, Dinsmore, Clarence H. Cleminshaw, and John H. Phillips. *Pictorial Astronomy.* 4th revised ed. New York: Thomas Y. Crowell, 1974.

Cleminshaw, Clarence H. *The Beginner's Guide to the Skies.* New York: Thomas Y. Crowell, 1977.

Pasachoff, Jay M. *Contemporary Astronomy.* 2nd ed. Philadelphia: Saunders College Publishing, 1981.

Rudaux, Lucien, and G. de Vaucouleurs. *Larousse Encyclopedia of Astronomy.* New York: Prometheus Press, 1959.

Russell, Henry Norris, Raymond Smith Dugan, and John Quincy Stewart. *Astronomy, vol. 1: The Solar System.* Boston: Ginn & Company, 1926.

Zim, Herbert S., and Robert H. Baker. *Stars.* New York: Golden Press, 1975.

History of Astronomy

Berry, Arthur. *A Short History of Astronomy.* 1908. Reprint. New York: Dover Publications, 1968.

Coleman, James A. *Early Theories of the Universe.* New York: New American Library/Signet Books, 1967.

Dicks, D. R. *Early Greek Astronomy to Aristotle.* London: Thames and Hudson, 1970.

Dreyer, J. L. E. *A History of Astronomy from Thales to Kepler.* 1906. Reprint. New York: Dover Publications, 1956.

Heath, Sir Thomas. *Aristarchus of Samos, the Ancient Copernican.* 1913. Reprint. New York: Dover Publications, 1981.

Ley, Willy. *Watchers of the Skies.* New York: Viking Press, 1969.

Krupp, E. C. "Outdistanced by the Dark," *Griffith Observer* 45, no. 4 (April, 1981): 10–15 (also in *The Boogie Woogie Review & Scriblerus Papers* 2, no. 2 [March/April, 1980]: 1–2).

Pannekoek, A. *A History of Astronomy.* New York: Interscience Publishers, 1961.

Sarton, George. *A History of Science.* 1952. Reprint. 2 vols. New York: W. W. Norton & Company, 1970.

Toulmin, Stephen, and June Goodfield. *The Fabric of the Heavens.* New York: Harper & Row, 1961.

Star Lore and Ancient Calendars

Blake, John F. *Astronomical Myths.* London: Macmillan and Co., 1877.

de Santillana, Giorgio, and Hertha von Dechend. *Hamlet's Mill.* Boston: Gambit Incorporated, 1969.

O'Neil, W. M. *Time and the Calendars.* Sydney, Australia: Sydney University Press, 1975.

O'Neill, John. *The Night of the Gods, vol. 1.* London: Harrison & Sons and Bernard Quaritch, 1893.

Plunket, Emeline M. *Ancient Calendars and Constellations.* London: John Murray, 1903.

Porter, Jermain G. *The Stars in Song and Legend.* Boston: Ginn & Company, 1902.

Zinner, Ernst. *The Stars Above Us.* New York: Charles Scribner's Sons, 1957.

Mythology and Ancient Religion

Branston, Brian. *Gods of the North.* London: Thames and Hudson, 1955.

Branston, Brian. *The Lost Gods of England.* London: Thames and Hudson, 1957.

Burland, C. A. *Myths of Life & Death.* New York: Crown Publishers, 1974.

Campbell, Joseph. *The Hero with a Thousand Faces.* 1949. Reprint. Cleveland: World Publishing Company/Meridian Books, 1967.

Campbell, Joseph. *The Masks of God: Occidental Mythology.* New York: Viking Press, 1964.

Campbell, Joseph. *The Masks of God: Oriental Mythology.* New York: Viking Press, 1962.

Campbell, Joseph. *The Masks of God: Primitive Mythology.* New York: Viking Press, 1959.

Campbell, Joseph. *The Mythic Image.* Princeton, N.J.: Princeton University Press, 1975.

Cavendish, Richard. *Mythology: An Illustrated Encyclopedia.* New York: Rizzoli International Publications, 1980.

Cotterell, Arthur. *A Dictionary of World Mythology.* New York: G. P. Putnam's Sons, 1980.

Cumont, Franz. *The Mysteries of Mithra.* 1902. Reprint. New York: Dover Publications, 1956.

Dumézil, Georges. *The Destiny of the Warrior.* Chicago: University of Chicago Press, 1970.

Dumézil, Georges. *Gods of the Ancient Northmen.* Berkeley and Los Angeles: University of California Press, 1973.

Editor *et al. New Larousse Encyclopedia of Mythology.* London: Hamlyn Publishing Group, 1959.

Eliade, Mircea. *Cosmos and History (The Myth of the Eternal Return).* 1954. Reprint. New York: Harper & Row, 1959.

Eliade, Mircea. *The Forge and the Crucible.* 2d ed. Chicago: University of Chicago Press, 1978.

Eliade, Mircea. *A History of Religious Ideas. Vol. 1 (from the Stone Age to the Eleusinian Mysteries).* Chicago: University of Chicago Press, 1978.

Eliade, Mircea. *Myth and Reality.* 1963. Reprint. New York: Harper & Row, 1975.

Eliade, Mircea. *Myths, Dreams, and Mysteries.* 1960. Reprint. New York: Harper & Row, 1967.

Eliade, Mircea. *From Primitives to Zen.* New York: Harper & Row, 1977.

Eliade, Mircea. *The Quest (History and Meaning in Religion).* Chicago: University of Chicago Press, 1969.

Eliade, Mircea. *Rites and Symbols of Initiation (The Mysteries of Birth and Rebirth).* 1958. Reprint. New York: Harper & Row, 1975.

Eliade, Mircea. *The Sacred & the Profane (The Nature of Religion).* 1959. Reprint. New York: Harcourt Brace Jovanovich, no date.

Eliade, Mircea. *Zalmoxis, the Vanishing God.* Chicago: University of Chicago Press, 1972.

Ferm, Vergilius, ed. *Ancient Religions.* New York: Philosophical Library, 1950.

Frankfort, H., *et al. Before Philosophy (The Intellectual Adventure of Ancient Man).* 1946. Reprint. Harmondsworth, England: Penguin Books, 1968.

Frankfort, H. *Kingship and the Gods.* Chicago: University of Chicago Press, 1948.

Frazer, Sir James George. *The Illustrated Golden Bough* (ed. Mary Douglas). Garden City, N.Y.: Doubleday & Company, 1978.

Graves, Robert. *The White Goddess.* 1948. Reprint. New York: Vintage Books, no date.

Grimal, Pierre, ed. *Larousse World Mythology.* Secaucus, N.J.: Chartwell Books, 1973.

Hamilton, Edith. *Mythology.* Boston: Little, Brown & Company, 1942.

Hesiod and Theognis. *Theogony/Works and Days* and *Elegies.* Harmondsworth, England: Penguin Books, 1973.

Hultkrantz, Åke. *The Religions of the American Indians.* Berkeley and Los Angeles: University of California Press, 1979.

Huxley, Francis. *The Way of the Sacred.* Garden City, N.Y.: Doubleday & Company, 1974.

James, E. O. *The Ancient Gods.* New York: G. P. Putnam's Sons, 1960.

Jung, C. J. *Flying Saucers: A Modern Myth of Things Seen in the Sky.* 1959. Reprint. New York: New American Library, 1969.

Kramer, Samuel Noah, ed. *Mythologies of the Ancient World.* Garden City, N.Y.: Doubleday & Company/Anchor Books, 1961.

Krickeberg, Walter, *et al. Pre-Columbian American Religions.* London: Weidenfeld and Nicolson, 1968.

Krupp, E. C. "Sky Riders," *Griffith Observer* 46, no. 6 (June, 1982): 17–20.

Krupp, E. C. "Why God Is in His Heaven," contributed paper, Southwestern Anthropological Association symposium "Astronomy in Anthropology," in Santa Barbara, March 21, 1981. Abstract: *Archaeoastronomy* (Maryland) III, no. 4 (Oct.-Nov.-Dec., 1980): 7.

Leeming, David. *Mythology.* New York: Newsweek Books, 1976.

Perry, John Weir. *Lord of the Four Quarters* (Myths of the Royal Father). New York: George Braziller, 1966.

Pinsent, John. *Greek Mythology.* London: Hamlyn Publishing Group, 1969.

Shapiro, Max S., and Rhoda A. Hendricks. *Mythologies of the World.* Garden City, N.Y.: Doubleday & Company, 1979.

Vallee, Jacques. *The Invisible College.* New York: E. P. Dutton, 1975.

Vallee, Jacques. *Passport to Magonia (from Folklore to Flying Saucers).* Chicago: Henry Regnery Company, 1969.

Shamanism and Prehistoric Religion

Bean, Lowell John, and Sylvia Brakke Vane. "Shamanism: An Introduction." *Art of the Huichol Indians* (ed. Kathleen Berrin). New York: Harry N. Abrams, 1978, pp. 117–128.

Cook, Roger. *The Tree of Life (Image for the Cosmos).* New York: Avon Books, 1974.

Das, Prem. "Initiation by a Huichol Shaman." *Art of the Huichol Indians* (ed. Kathleen Berrin). New York: Harry N. Abrams, 1978, pp. 129–141.

Eliade, Mircea. *Shamanism (Archaic Techniques of Ecstasy).* Princeton, N.J.: Princeton University Press, 1964.

Furst, Peter T. *Hallucinogens and Culture.* San Francisco: Chandler & Sharp Publishers, 1976.

Gimbutas, Marija. *The Gods and Goddesses of Old Europe 7000–3500 B.C.* London: Thames and Hudson, 1974.

Holmberg, Uno. *The Mythology of All Races: Finno-Ugric/Siberian. Vol. IV.* Reprint. New York: Cooper Square Publishers, 1964.

James, E. O. *Prehistoric Religion.* New York: Frederick A. Praeger, 1957.

Levy, G. Rachel. *Religious Conceptions of the Stone Age.* 1948. Reprint. New York: Harper & Row, 1963.

Lommel, Andreas. *The World of the Early Hunters.* London: Evelyn, Adams and Mackay, 1967.

Mandell, Arnold J. "The Neurochemistry of Religious Insight and Ecstasy." *Art of the Huichol Indians* (ed. Kathleen Berrin). New York: Harry N. Abrams, 1978, pp. 71–81.

Maringer, Johannes. *The Gods of Prehistoric Man.* New York: Alfred A. Knopf, 1960.

Matossian, Mary Kilbourne. "Symbols of Seasons and the Passage of Time: Barley and Bees in the New Stone Age," *Griffith Observer*, 44, no. 11 (November, 1980): 9–17.

Mellaart, James. *Çatal Hüyük.* New York: McGraw-Hill Book Company, 1967.

Schultes, Richard Evans, and Albert Hofmann. *Plants of the Gods.* New York: McGraw-Hill Book Company, 1979.

Human Evolution

Leakey, Richard E. *The Making of Mankind.* New York: E. P. Dutton, 1981.

Leakey, Richard E., and Roger Lewin. *The People of the Lake.* Garden City, N.Y.: Doubleday & Company/Anchor Press, 1978.

Leakey, Richard E., and Roger Lewin. *Origins.* New York: E. P. Dutton, 1977.

The Ice Age Artists

Hadingham, Evan. *Secrets of the Ice Age.* New York: Walker and Company, 1979.

Leroi-Gourhan, A. *Treasures of Prehistoric Art.* New York: Harry N. Abrams, no date.

Lewin, Roger. "An Ancient Cultural Revolution," *New Scientist* 83, no. 1166 (August 2, 1979): 352–355.

Marshack, Alexander. "The Art and Symbols of Ice Age Man," *Human Nature* 1, no. 9 (September, 1978): 32–41.

Marshack, Alexander. "Exploring the Mind of Ice Age Man," *National Geographic* 147, no. 1 (January, 1975): 64–89.

Marshack, Alexander. "Ice Age Art," *Explorers Journal* 59, no. 2 (June, 1981): 50–57.

Marshack, Alexander. *The Roots of Civilization.* New York: McGraw-Hill Book Company, 1972.

Sieveking, Ann. *The Cave Artists.* Ancient People and Places, vol. 93. London: Thames and Hudson, 1979.

Ucko, Peter J., and Andrée Rosenfeld. *Paleolithic Cave Art.* New York: World University Library/McGraw-Hill Book Company, 1967.

Megaliths

Atkinson, R. J. C. *Stonehenge.* Harmondsworth, England: Penguin Books, 1979.

Atkinson, R. J .C. *Stonehenge and Neighboring Monuments.* London: Her Majesty's Stationery Office, 1978.

Atkinson, R. J. C. "Some New Measurements on Stonehenge," *Nature* 275 (September 7, 1978): 50–52.

Balfour, Michael. *Stonehenge and Its Mysteries.* London: Macdonald & Jane's, 1979.

Barber, John. "The Orientation of the Recumbent-Stone Circles of the South-West of Ireland," *Journal of Kerry Archaeological and Historical Society* 6 (1973): 26–39.

Brennan, Martin. *The Boyne Valley Vision.* Portlaoise, Ireland: Dolmen Press, 1980.

Brown, Peter Lancaster. *Megaliths and Masterminds.* London: Robert Hale, 1979.

Brown, Peter Lancaster. *Megaliths, Myths and Men.* Poole, Dorset, England: Blanford Press, 1976.

Burgess, Colin. *The Age of Stonehenge.* London: J. M. Dent & Sons, 1980.

Burl, Aubrey. "Dating the British Stone Circles," *American Scientist* 61 (March-April, 1973): 167–174.

Burl, Aubrey. *Prehistoric Avebury.* New Haven: Yale University Press, 1979.

Burl, Aubrey. *Prehistoric Stone Circles.* Aylesbury, Bucks., England: Shire Publications, 1979.

Burl, Aubrey. "The Recumbent Stone Circles of Scotland," *Scientific American* 245, no. 6 (December, 1981): 66–72.

Burl, Aubrey. *Rings of Stone.* London: Frances Lincoln Publishers, 1979.

Burl, Aubrey. *Rites of the Gods.* London: J. M. Dent & Sons, 1981.

Burl, Aubrey. "Science or Symbolism: Problems of Archaeo-astronomy," *Antiquity* LIV (1980): 191–200.

Burl, Aubrey. *The Stone Circles of the British Isles.* New Haven: Yale University Press, 1976.

Coffey, George. *New Grange and Other Incised Tumuli in Ireland.* 1912. Reprint. Poole, Dorset, England: Dolphin Press, 1977.

Daniel, Glyn. *The Megalith Builders of Western Europe.* 1958. Reprint. Harmondsworth, England: Penguin Books, 1962.

Daniel, Glyn. "Megalithic Monuments," *Scientific American* 243, no. 1 (July, 1980): 78–90.

Ellegård, Alvar. "Stone Age Science in Britain," *Current Anthropology* 22, no. 2 (April, 1981): 99–125.

Fowles, John. *The Enigma of Stonehenge.* New York: Summit Books, 1980.

Gelling, Peter, and Hilda Ellis Davidson. *The Chariot of the Sun.* New York: Frederick A. Praeger, 1969.

Giot, P. R. *Brittany.* Ancient Peoples and Places, vol. 13. New York: Frederick A. Praeger, 1960.

Hadingham, Evan. "Carnac Revisited," *Archaeoastronomy* (Maryland) III, no. 3 (July-August-September, 1980): 10–13.

Hadingham, Evan. *Circles and Standing Stones.* New York: Walker and Company, 1975.

Hawkins, Gerald S. *Beyond Stonehenge.* New York: Harper & Row, 1973.

Hawkins, Gerald S., in collaboration with John B. White. *Stonehenge Decoded.* Garden City, N.Y.: Doubleday & Company, 1965.

Heggie, Douglas C. "Highlights and Problems of Megalithic Astronomy," *Archaeoastronomy* (*JHA* Supplement 3, 1981): S17–S37.

Heggie, Douglas C. *Megalithic Science.* London: Thames and Hudson, 1981.

Herity, Michael. *Irish Passage Graves.* Dublin: Irish University Press, 1974.

Hoddinott, R. F. *The Thracians.* Ancient Peoples and Places, vol. 98. New York: Thames and Hudson, 1981.

Hoyle, Fred. *On Stonehenge.* San Francisco: W. H. Freeman and Company, 1977.

Hoyle, Fred. *From Stonehenge to Modern Cosmology.* San Francisco: W. H. Freeman and Company, 1972.

Hutchinson, G. Evelyn. "Long Meg Reconsidered," *American Scientist* 60 (January-February, 1972): 24–31.

Hutchinson, G. Evelyn. "Long Meg Reconsidered, Part 2," *American Scientist* 60 (March-April, 1972): 210–219.

Krupp, E. C. "Upon the Blue Horizon," contributed paper, January 12, 1981, meeting of the Historical Astronomy Division, American Astronomical Society, Albuquerque, N.M. Abstract: *Archaeoastronomy* (Maryland) III, no. 4 (October-November-December, 1980): 5.

Laing, Lloyd, and Jennifer Laing. *The Origins of Britain.* New York: Charles Scribner's Sons, 1980.

Lockyer, Sir J. Norman. *Stonehenge and Other British Stone Monuments Astronomically Considered.* 2d ed. London: Macmillan and Co. 1909.

Lynch, B. M., and L. H. Robbins. "Namoratunga: The First Archeoastronomical Evidence in Sub-Saharan Africa," *Science* 200 (May 19, 1978): 766–768.

MacKie, Euan. *The Megalith Builders.* Oxford: Phaidon Press, 1977.

MacKie, Euan W. *Science and Society in Prehistoric Britain.* London: Elek Books, 1977.

McCreery, T. "The Kintraw Stone Platform," *Kronos* 5, no. 3 (April, 1980): 71–79.

Megaw, J. V. S., and D. D. A. Simpson. *Introduction to British Prehistory.* Leicester: Leicester University Press, 1979.

Michell, John. *A Little History of Astro-archaeology.* London: Thames and Hudson, 1977.

Michell, John. *Megalithomania.* London: Thames and Hudson, 1982.

Morrison, L. V. "Analysing Lunar Sightlines," *Archaeoastronomy* (*JHA* Supplement 2, 1980): S78–S89.

Newall, R. S. *Stonehenge, Wiltshire.* London: Her Majesty's Stationery Office, 1959.

Newham, C. A. *The Astronomical Significance of Stonehenge.* Gwent, Wales: Moon Publications, 1972.

O'Kelly, Claire. *Illustrated Guide to Newgrange and the Other Boyne Monuments.* Cork, Ireland: C. O'Kelly, 1978.

Ó Ríordáin, Sean P., and Glyn Daniel. *New Grange and the Bend of the Boyne.* Ancient Peoples and Places, vol. 40. London: Thames and Hudson, 1964.

Pitts, Michael W. "Stones, Pits and Stonehenge," *Nature* 290 (1981): 46–47.

Robinson, Jack H. "Sunrise and Moonrise at Stonehenge," *Nature* 225 (March 28, 1970): 1236–1237.

Roy, A. E., McGrail, and R. Carmichael. "A New Survey of the Tormore Circles," *Transactions of the Glasgow Archaeological Society,* New Series, XV, part II (1963): 59–67.

Ruggles, Clive. "Prehistoric Astronomy: How Far Did It Go?" *New Scientist* 90, no. 1258 (June 18, 1981): 750–753.

Ruggles, C. L. N., and A. W. R. Whittle, eds. *Astronomy and Society in Britain during the Period 4000–1500 B.C.* (B.A.R. British Series 88). Oxford: BAR, 1981.

Service, Alastair, and Jean Bradbery. *Megaliths and Their Mysteries.* London: Weidenfeld and Nicolson, 1979.

Stover, Leon E., and Bruce Kaig. *Stonehenge: The Indo-European Heritage.* Chicago: Nelson-Hall, 1978.

Thom, A. *Megalithic Lunar Observatories.* Oxford: Oxford University Press, 1971.

Thom, A. *Megalithic Sites in Britain.* Oxford: Oxford University Press, 1967.

Thom, A., and A. S. Thom. *Megalithic Remains in Britain and Britanny.* Oxford: Oxford University Press, 1978.

Thom, A., and A. S. Thom. "A New Study of All Lunar Sightlines," *Archaeoastronomy* (*JHA* Supplement 2, 1980): S78–S89.

Thom, A., and A. S. Thom. "The Standing Stones in Argyllshire," *Glasgow Archaeological Journal* 6 (1979): 5–10.

Thom, A., and A. S. Thom, with A. Burl. *Megalithic Rings* (B.A.R. British Series 81). Oxford: BAR, 1980.

Tyler, Larry. "Megaliths, Medicine Wheels, and Mandalas," *The Midwest Quarterly* XXI, no. 3 (spring, 1980): 290–305.

Wood, John Edwin. *Sun, Moon and Standing Stones.* Oxford: Oxford University Press, 1978.

The Celts and the Druids

Chadwick, Nora K. *Celtic Britain*. Ancient Peoples and Places, vol. 34. London: Thames and Hudson, 1963.

Chadwick, Nora. *The Celts*. Harmondsworth, England: Penguin Books, 1970.

Chadwick, Nora K. *The Druids*. Cardiff, Wales: University of Wales Press, 1966.

Dillon, Myles, and Nora Chadwick. *The Celtic Realms*. 1967. Reprint. London: Cardinal Books, 1973.

Hickey, Elizabeth. *The Legend of Tara*. Dundalk, Ireland: Dundalgan Press (W. Tempest), 1969.

Kendrick, T. D. *The Druids*. 1927. Reprint. London: Frank Cass & Co. 1966.

Laing, Lloyd. *Celtic Britain*. New York: Charles Scribner's Sons, 1979.

Long, George. *The Folklore Calendar*. 1930. Reprint. East Ardsley, Wakefield, England: EP Publishing, 1977.

Macalister, R. A. S. *Tara*. New York: Charles Scribner's Sons, 1931.

Macbain, Alexander. *Celtic Mythology and Religion*. Stirling, Scotland: Eneas Mackay, 1917.

MacCana, Proinsias. *Celtic Mythology*. London: Hamlyn Publishing Group, 1973.

MacGowan, Kenneth. *The Hill of Tara*. Dublin: Kamac Publications, 1979.

Ó Ríordáin, Sean P. *Tara, the Monuments on the Hill*. Dundalk, Ireland: Dundalgan Press (W. Tempest), 1974.

Owen, A. L. *The Famous Druids*. Oxford: Clarendon Press, 1962.

Piggott, Stuart. *The Druids*. Ancient Peoples and Places, vol. 63. 1968. New edition. New York: Praeger Publishers, 1975.

Powell, T. G. E. *The Celts*. Ancient Peoples and Places, vol. 6. 1958. New edition. London: Thames and Hudson, 1980.

Rees, Alwyn, and Brinley Rees. *Celtic Heritage*. 1961. Reprint. London: Thames and Hudson, 1975.

Rolleston, T. W. *Myths and Legends of the Celtic Race*. Boston: David D. Nickerson, 191-.

Ross, Anne. *Pagan Celtic Britain*. 1967. Reprint. London: Cardinal Books, 1973.

Squire, Charles. *Celtic Myth & Legend, Poetry & Romance*. London: Gresham Publishing Company, 191-.

California Indians

Benson, Arlene. "California Sun-Watching Site," *Archaeoastronomy* (Maryland) III, no. 1 (winter, 1980): 16–19.

Blackburn, Thomas C. *December's Children—a Book of Chumash Oral Narratives.* Berkeley and Los Angeles: University of California Press, 1975.

Grant, Campbell. *The Rock Paintings of the Chumash.* Berkeley and Los Angeles: University of California Press, 1965.

Hedges, Ken. "Winter Solstice Observatory Sites in Kumeyaay Territory, San Diego County, California." *Archaeoastronomy in the Americas* (ed. Ray A. Williamson). Los Altos, Cal.: Ballena Press/Center for Archaeoastronomy, 1981, pp. 151–156.

Hudson, Travis, and John B. Carlson. *Visions of the Sky: Archaeological and Ethnographic Studies of California Indian Astronomy.* Ramona, Cal.: Acoma Press/Center for Archaeoastronomy, 1983.

Hudson, Travis, Georgia Lee, and Ken Hedges. "Solstice Observers and Observatories in Native California," *Journal of California and Great Basin Anthropology* 1, no. 1 (summer, 1979): 39–63.

Hudson, Travis, and Ernest Underhay. *Crystals in the Sky: An Intellectual Odyssey Involving Chumash Astronomy, Cosmology and Rock Art.* Socorro, N.M.: Ballena Press, 1977.

Krupp, E. C. "Emblems of the Sky." *Ancient Images: Rock Art of the Californias* (ed. Jo Anne Van Tilburg). Los Angeles: Rock Art Archive, Institute of Archaeology, University of California, 1982.

Librado, Fernando. *Breath of the Sun* (notes of John P. Harrington, ed. Travis Hudson). Banning, Cal.: Malki Museum Press, 1979.

Librado, Fernando. *The Eye of the Flute* (notes of John P. Harrington, ed. Travis Hudson, et al.). Santa Barbara, Cal.: Santa Barbara Museum of Natural History, 1977.

Other North American Indians

Alexander, Hartley Burr. *The Mythology of All Races: North American, vol. X.* 1916. Reprint. New York: Cooper Square Publishers, 1964.

Alexander, Hartley Burr. *The World's Rim.* Lincoln, Neb.: University of Nebraska Press, 1953.

Bahti, Tom. *Southwestern Indian Ceremonials.* Las Vegas, Nev.: KC Publications, 1970.

Brown, Joseph Epes. *The Sacred Pipe.* Norman, Okla.: University of Oklahoma Press, 1953.

Brown, Lionel A. "The Fort Smith Medicine Wheel," *Plains Anthropologist* 8 (1963): 225–230.

Chamberlain, Von Del. *When the Stars Came Down to Earth: Cosmology of the Skidi Pawnee Indians of North America.* Los Altos, Cal.: Ballena Press, 1982.

Chamberlain, Von Del. "The Skidi Pawnee Chart of the Heavens," *Sky and Telescope* 62 (July, 1981): 23–28.

Eddy, John A. "Astronomical Alignment of the Big Horn Medicine Wheel," *Science* 184 (June 7, 1974): 1035–1043.

Eddy, John A. "Medicine Wheels and Plains Indian Astronomy." *Native American Astronomy* (ed. Anthony F. Aveni). Austin, Tex.: University of Texas Press, 1977, pp. 147–169.

Eddy, John A. "Medicine Wheels and Plains Indian Astronomy." *Astronomy of the Ancients* (ed. Kenneth Brecher and Michael Feirtag). Cambridge, Mass.: MIT Press, 1979, pp. 1–24.

Eddy, John A. "Probing the Mystery of the Medicine Wheels," *National Geographic* 151, no. 1 (January, 1977): 140–146.

Ellis, Florence Hawley. "A Thousand Years of the Pueblo Sun-Moon-Star Calendar." *Archaeoastronomy in Pre-Columbian America* (ed. Anthony F. Aveni). Austin, Tex.: University of Texas Press, 1975, pp. 59–87.

Farrer, Claire R. "Mescalero Apaches and Ethnoastronomy," *Archaeoastronomy* (Maryland) III, no. 1 (winter, 1980): 20.

Farrer, Claire R., and Bernard Second. "Living the Sky: Aspects of Mescalero Apache Ethnoastronomy." *Archaeoastronomy in the Americas* (ed. Ray A. Williamson). Los Altos, Cal.: Ballena Press/Center for Archaeoastronomy, 1981, pp. 137–150.

Ferguson, Erna. *Dancing Gods.* Albuquerque, N.M.: University of New Mexico Press, 1931.

Fletcher, Alice C., and Francis La Flesche. *The Omaha Tribe, vol. 1.* 1905–1906. Twenty-Seventh Annual Report of the Bureau of American Ethnology. Reprint. Lincoln, Neb.: University of Nebraska Press, 1972.

Fowler, Melvin L. "A Pre-Columbian Urban Center on the Mississippi," *Scientific American* 233 (August, 1975): 92–101.

Frazier, Kendrick. "The Anasazi Sun Dagger," *Science 80* 1, no. 1 (November-December, 1979): 56–67.

Frazier, Kendrick. "Solstice Watchers of Chaco," *Science News* 114, no. 9 August 26, 1978): 148–151.

Frazier, Kendrick. "Stars, Sky and Culture," *Science News* 116, no. 5 (August 4, 1979): 90–93.

Frazier, Kendrick. "Western Horizons." *Fire of Life: The Smithsonian Book of the Sun* (ed. Joe Goodwin *et al.*). Washington, D.C.: Smithsonian Exposition Books, 1981, pp. 168–175.

Fries, Allan G. "Vision Quests at the Big Horn Medicine Wheel and Its Date of Construction," *Archaeoastronomy* (Maryland) III, no. 4 (October-November-December, 1980): 20–24.

Green, Jesse, ed. *Zuñi (Selected Writings of Frank Hamilton Cushing).* Lincoln, Neb.: University of Nebraska Press, 1979.

Hudson, Dee T. "Anasazi Measurement Systems at Chaco Canyon, New Mexico," *The Kiva* 38, no. 1 (1972): 27–42.

Kehoe, Alice, B., and Thomas F. Kehoe. *Solstice-Aligned Boulder Configurations in Saskatchewan.* Ottawa: National Museums of Canada, 1979.

Krupp, E. C. "Cahokia: Corn, Commerce, and the Cosmos," *Griffith Observer* 41, no. 5 (May, 1977): 10–20.

Krupp, E. C. "Sun and Stones on Medicine Mountain," *Griffith Observer* 38, no. 11 (November, 1974): 9–20.

Lister, Robert H., and Florence C. Lister. *Chaco Canyon.* Albuquerque, N.M.: University of New Mexico Press, 1981.

Mansfield, Victor N. "The Big Horn Medicine Wheel as a Site for the Vision Quest," *Archaeoastronomy* (Maryland) III, no. 2 (April-May-June, 1980): 26–29.

McCluskey, Stephen C. "The Astronomy of the Hopi Indians," *Journal for the History of Astronomy* 8, part 3, no. 23 (October, 1977): 174–195.

Neihardt, John G. *Black Elk Speaks.* Lincoln, Neb.: University of Nebraska Press, 1961.

Norrish, Dick. "This Priest-Astronomer, This Genius," *Cahokian* (February, 1978): 1–11.

O'Kane, Walter Collins. *The Hopis: Portrait of a Desert People.* Norman, Okla.: University of Oklahoma Press, 1953.

Pfeiffer, John E. "Indian City on the Mississippi." *Nature/Science Annual 1974* (ed. Jane D. Alexander). New York: Time-Life Books, 1973, pp. 124–139.

Reyman, Jonathan E. "The Emics and Etics of Kiva Wall Niches," *Journal of the Steward Anthropological Society* 7, no. 1 (1976): 107–129.

Robinson, Jack H. "Archaeoastronomical Alignments at the Fort Smith Medicine Wheel," *Archaeoastronomy* (Maryland) IV, no. 3 (July-August-September, 1981): 14–23.

Robinson, Jack H. "Fomalhaut and Cairn D at the Big Horn and Moose Mountain Medicine Wheels," *Archaeoastronomy* (Maryland) III, no. 4 (October-November-December, 1980): 15–19.

Simmons, Leo W., ed. *Sun Chief.* New Haven: Yale University Press, 1942.

Sofaer, Anna, Volker Zinser, and Rolf M. Sinclair. "A Unique Solar Marking Construct," *Science* 206, no. 4416 (October 19, 1979): 283–291.

Stevenson, Matilda Coxe. *The Zuñi Indians: Their Mythology, Esoteric Societies, and Ceremonies.* 1901–1902. Twenty Third Annual Report of the Bureau of American Ethnology. Washington, D.C.: Government Printing Office, 1904.

Tyler, Hamilton. *Pueblo Gods and Myths.* Norman, Okla.: University of Oklahoma Press, 1964.

Waters, Frank. *Book of the Hopi.* 1963. Reprint. Harmondsworth, England: Penguin Books, 1979.

Waters, Frank. *Masked Gods.* 1950. Reprint. New York: Ballantine Books. 1975.

Wedel, Waldo R. "Native Astronomy and the Plains Caddoans." *Native American Astronomy* (ed. Anthony F. Aveni). Austin, Tex.: University of Texas Press, 1977, pp. 131–146.

Wedel, Waldo R. *Prehistoric Man on the Great Plains.* Norman, Okla.: University of Oklahoma Press, 1961.

Weltfish, Gene. *The Lost Universe.* New York: Basic Books, 1965.

Williamson, Ray A., ed. *Archaeoastronomy in the Americas.* Los Altos, Cal.: Ballena Press/Center for Archaeoastronomy, 1981.

Williamson, Ray A. "Native Americans Were Continent's First Astronomers," *Smithsonian* 9, no. 7 (October, 1978): 78–85.

Williamson, Ray A. "Pueblo Bonito and the Sun," *Archaeoastronomy Bulletin* (Maryland) I, no. 2 (February, 1978): 5–7.

Williamson, Ray A., Howard J. Fisher, and Donnel O'Flynn. "Anasazi Solar Observatories." *Native American Astronomy* (ed. Anthony F. Aveni). Austin, Tex.: University of Texas Press, 1977, pp. 203–217.

Wilson, Michael, Kathie L. Road, and Kenneth J. Hardy. *Megaliths to Medicine Wheels: Boulder Structures in Archaeology.* Calgary, Alberta: The University of Calgary Archaeological Association, 1981.

Wittry, Warren L. *Summary Report on 1978 Investigations of Circle No. 2 of the Woodhenge, Cahokia Mounds State Historic Site.* Chicago: Department of Anthropology, University of Illinois at Chicago Circle, March, 1980.

Mesoamerica

Alexander, Hartley Burr. *The Mythology of All Races: Latin American, vol. XI.* 1920. Reprint. New York: Cooper Square Publishers, 1964.

Arochi, Luis E. *La Pirámide de Kukulcan, Su Simbolismo Solar.* 3d ed. Mexico City: Editorial Orion, 1981.

Aveni, Anthony F. "Archaeoastronomy in the Maya Region: A Review of the Past Decade." *Archaeoastronomy* (*JHA* Supplement 3, 1981): S1–S37.

Aveni, Anthony F., ed. *Archaeoastronomy in Pre-Columbian America.* Austin, Tex.: University of Texas Press, 1975.

Aveni, Anthony F., ed. *Native American Astronomy.* Austin, Tex.: University of Texas Press, 1977.

Aveni, Anthony F. "Old and New World Naked Eye Astronomy." *Astronomy of the Ancients* (ed. Kenneth Brecher and Michael Feirtag). Cambridge, Mass.: MIT Press, 1979, pp. 61–89.

Aveni, Anthony F. *Skywatchers of Ancient Mexico.* Austin, Tex.: University of Texas Press, 1980.

Aveni, Anthony F. "Tropical Archaeoastronomy," *Science* 213 (July 10, 1981): 161–171.

Aveni, Anthony F. "Venus and the Maya," *American Scientist* 67, no. 3 (May-June, 1979): 274–285.

Aveni, Anthony F., and S. Gibbs. "On the Orientation of Pre-Columbian Buildings in Central Mexico," *American Antiquity* 41 (1976): 510–517.

Aveni, Anthony F., Sharon L. Gibbs, and Horst Hartung. "The Caracol Tower at Chichén Itzá: An Ancient Astronomical Observatory?" *Science* 188 (June 6, 1975): 977–985.

Aveni, Anthony F., and Horst Hartung. "The Cross Petroglyph: An Ancient Mesoamerican Astronomical and Calendrical Symbol." *Indiana 6.* Berlin: Gebr. Mann Verlag, 1981.

Aveni, Anthony F., and Horst Hartung. "The Observation of the Sun at the Time of Passage through the Zenith in Mesoamerica," *Archaeoastronomy* (*JHA* Supplement 3, 1981): S51–S70.

Aveni, Anthony F., and Horst Hartung. "Some Suggestions About the Arrangements of Buildings at Palenque." *The Art, Iconography, and Dynastic History of Palenque. Part IV* (ed. Merle Greene Robertson). Monterey, Cal.: Pre-Columbian Art Research, The Robert Louis Stevenson School, 1978, pp. 173–178.

Aveni, Anthony F., and Horst Hartung. "Three Round Towers in the Yucatán Peninsula," *Interciencia* 3 (1978): 136–143.

Aveni, Anthony F., Horst Hartung, and Beth Buckingham, "The Pecked Cross in Ancient Mesoamerica," *Science* 202 (October 20, 1978): 267–279.

Aveni, Anthony F., Horst Hartung, and J. Charles Kelley. "Alta Vista, Chalchihuites, a Mesoamerican Ceremonial Outpost at the Tropic of Cancer: Astronomical Implications," *American Antiquity* 47, no. 2 (1982): pp. 316–335.

Aveni, Anthony F., and R. Linsley. "Mound J, Monte Albán: Possible Astronomical Orientation," *American Antiquity* 37 (1972): 528–531.

Benson, Elizabeth, ed. *Mesoamerican Sites and World-Views.* Washington, D.C.: Dumbarton Oaks, 1981.

Broda, Johanna. "La Fiesta Azteca del Fuego y el Culto de las Pleyades." *Space and Time in the Cosmovision of Mesoamerica* (ed. Franz Tichy and Anthony F. Aveni). Nuremberg: University of Erlangen-Nuremberg, 1982. (also in *Homenaje a R. Girard: La Antropología Americanista en la Actualidad. Tomo 2.* Mexico City: Editores Mexicanos Unidos, 1980, pp. 283–304.)

Brotherston, Gordon. "Huitzilopochtli and What Was Made of Him." *Mesoameri-*

can Archaeology, New Approaches (ed. Norman Hammond). Austin, Tex.: University of Texas Press, 1974, pp. 155–166.

Brotherston, Gordon. *Image of the New World.* London: Thames and Hudson, 1979.

Brundage, Burr Cartwright. *The Fifth Sun.* Austin, Tex.: University of Texas Press, 1979.

Brundage, Burr Cartwright. *The Phoenix of the Western World—Quetzalcóatl and the Sky Religion.* Norman, Okla.: University of Oklahoma Press, 1982.

Burland, C. A. *The Gods of Mexico.* New York: G. P. Putnam's Sons, 1967.

Carlson, John B. "Astronomical Investigations and Site Orientation Influences at Palenque." *The Art, Iconography, and Dynastic History of Palenque. Part III* (ed. Merle Greene Robertson). Pebble Beach, Calif.: Pre-Columbian Art Research, The Robert Louis Stevenson School, 1976, pp. 107–117.

Carlson, John B., and Linda Landis. "Bands, Bicephalic Dragons, and Other Beasts: The Skyband in Maya Art and Iconography." Paper presented at the Cuarta Mesa Redonda de Palenque, June 8–14, 1980.

Caso, Alfonso. *The Aztecs, People of the Sun.* Norman, Okla.: University of Oklahoma Press, 1958.

Caso, Alfonso. *Los Calendarios Prehispánicos.* Mexico City: Universidad Nacional Autónoma de Mexico, 1967.

Caso, Alfonso. "Calendrical Systems of Central Mexico." *Handbook of Middle American Indians. Vol. 10. Archaeology of Northern Mesoamerica, Part One* (ed. Gordon F. Ekholm and Ignacio Bernal). Austin, Tex.: University of Texas Press, 1971, pp. 333–348.

Chiu, B. C., and Philip Morrison. "Astronomical Origin of the Offset Street Grid at Teotihuacán," *Archaeoastronomy* (*JHA* Supplement 2, 1980): S55–S64.

Coe, Michael D. *The Maya.* Ancient Peoples and Places, vol. 52. Revised and enlarged edition. London: Thames and Hudson, 1980.

Coe, Michael D. *The Maya Scribe and His World.* New York: Grolier Club, 1973.

Coe, Michael D. *Mexico.* Ancient Peoples and Places, vol. 29. New York: Frederick A. Praeger, 1962.

Collea, Beth A. "The Celestial Bands in Maya Hieroglyphic Writing." *Archaeoastronomy in the Americas* (ed. Ray A. Williamson). Los Altos, Calif.: Ballena Press/Center for Archaeoastronomy, 1981, pp. 215–232.

Cook, Ange Garcia, and Raul M. Arana A. *Rescate Arqueológico del Monolito Coyolxauhqui.* Mexico City: Instituto Nacional de Antropología e Historia, 1978.

Dow, J. W. "Astronomical Orientations at Teotihuacán, a Case Study in Astro-archaeology," *American Antiquity* 32, 1967: 326–334.

Durán, Fray Diego. *Book of the Gods and Rites* and *The Ancient Calendar.* Norman, Okla.: University of Oklahoma Press, 1971.

Elzey, Wayne. "The Nahua Myth of the Suns," *Numen* XXIII, fasc. 2 (1976): 114–135.

Elzey, Wayne. "Some Remarks on the Space and Time of the 'Center' in Aztec Religion," *Estudios de Cultura Nahuatl, Vol. 12.* Mexico City: Instituto de Investigaciones Históricos/Universidad Nacional Autónoma de Mexico, 1976, pp. 315–334.

Gossen, Gary H. "A Chamula Calendar Board from Chiapas." *Mesoamerican Archaeology, New Approaches* (ed. Norman Hammond). Austin, Tex.: University of Texas Press, 1974, pp. 217–254.

Gossen, Gary H. *Chamulas in the World of the Sun.* Cambridge, Mass.: Harvard University Press, 1974.

Hartung, Horst. "Alte Stadt in Mexico: Monte Albán," *Deutsche Bauzeitung,* 2 (1974): 152–159.

Hartung, Horst. "An Ancient 'Astronomer' on a Relief of Monte Albán?" *Griffith Observer* 45, no. 6 (June, 1981): 11–20.

Hartung, Horst. "Ancient Maya Architecture and Planning: Possibilities and Limitations for Astronomical Studies," in *Native American Astronomy* (ed. Anthony F. Aveni). Austin, Tex.: University of Texas Press, 1977, pp. 111–130.

Hartung, Horst. "Astronomical Signs in the Codices Bodley and Selden," in *Native American Astronomy* (ed. Anthony F. Aveni). Austin, Tex.: University of Texas Press, 1977, pp. 38–41.

Hartung, Horst. "Bauwerke der Maya weisen zur Venus." *Umschau 76,* Heft 16 (1976): 526–528.

Hartung, Horst. "Copan—Raum, Kunst und Astronomie in einem Maya-Zeremonial-zentrum," *Das Altertum Heft 1,* Bd. 25 (1979): 5–15.

Hartung, Horst. "El Ordenamiento Espacial en los Conjuntos Arquitectónicos Mesoamericanos—El Ejemplo de Teotihuacán," *Comunicaciones Proyecto Puebla-Tlaxcala,* 16 (1979): 89–103.

Hartung, Horst. "Pre-Columbian Settlements in Mesoamerica," *Ekistics* 45, no. 271 (July/August, 1978): 326–330.

Hartung, Horst. "A Scheme of Probable Astronomical Projections in Mesoamerican Architecture." *Archaeoastronomy in Pre-Columbian America* (ed. Anthony F. Aveni). Austin, Tex.: University of Texas Press, 1975, pp. 191–204.

Hartung, Horst. *Die Zeremonialzentren der Maya.* Graz, Austria: Akademische Druckung Verlagsanstalt, 1971.

Henderson, John S. *The World of the Ancient Maya.* Ithaca, N.Y.: Cornell University Press, 1981.

Hunt, Eva. *The Transformation of the Hummingbird.* Ithaca, N.Y.: Cornell University Press, 1977.

Ivanoff, Pierre. *Monuments of Civilization: Maya.* New York: Grosset & Dunlap, 1973.

Kelly, Joyce. *The Complete Visitor's Guide to Mesoamerican Ruins.* Norman, Okla.: University of Oklahoma Press, 1982.

Kendall, Timothy. *Patolli, a Game of Ancient Mexico.* Belmont, Mass.: Kirk Game Company, 1980.

Krupp, E. C. "An Aztec 'Calendar' Stone and Its Celestial Seal of Approval," *Griffith Observer* 45, no. 7 (July, 1981): 1–8.

Krupp, E. C. "The 'Binding of the Years,' the Pleiades, and the Nadir Sun," *Archaeoastronomy* (Maryland) 5, no. 1 (Jan.-Mar., 1982): 10–13.

Krupp, E. C. "The Observatory of Kukulcan," *Griffith Observer* 41, no. 9 (September, 1977): 1–20.

Krupp, E. C. "The Serpent Descending," *Griffith Observer* 46, no. 9 (September, 1982): 10–20.

Lamb, Weldon. "The Sun, Moon and Venus at Uxmal," *American Antiquity* 45, no. 1 (1980): 79–86.

León-Portilla, Miguel, *Aztec Thought and Culture.* Norman, Okla.: University of Oklahoma Press, 1963.

León-Portilla, Miguel. *Mexico-Tenochtitlán: Su Espacio y Tiempo Sagrados.* Mexico City: Instituto Nacional de Antropología e Historia, 1978.

León-Portilla, Miguel. *Pre-Columbian Literatures of Mexico.* Norman, Okla.: University of Oklahoma Press, 1975.

Lhuillier, Alberto Ruz. *The Tomb of Palenque.* Mexico City: Instituto Nacional de Antropología e Historia, 1974.

Lounsbury, Floyd G. "Astronomical Knowledge and Its Uses at Bonampak, Mexico." *Archaeoastronomy in the New World* (ed. A. F. Aveni). Cambridge: Cambridge University Press, 1982, pp. 143–168.

Lounsbury, Floyd G. "Maya Numeration, Computation, and Calendrical Astronomy." *Dictionary of Scientific Biography, vol. XV, Suppl. I.* New York: Charles Scribner's Sons, 1978, pp. 706–727.

Marquina, Ignacio. *El Templo Mayor de Mexico.* Mexico City: Instituto Nacional de Antropología e Historia, 1960.

Marquina, Ignacio. *Templo Mayor de Mexico Official Guide.* Mexico City: Instituto Nacional de Antropología e Historia, 1968.

Marshack, Alexander. "The Chamula Calendar: An Internal and Comparative Analysis." *Mesoamerican Archaeology, New Approaches* (ed. Norman Hammond). Austin, Tex.: University of Texas Press, 1974, pp. 255–270.

Milbrath, Susan. "Star Gods and Astronomy of the Aztecs." *La Antropología Americanista en la Actualidad,* Tomo I. Mexico City: Editores Mexicanos Unidos, 1980, pp. 289–303.

Millon, Rene. "Teotihuacán." *Pre-Columbian Archaeology* (ed. Gordon R. Willey and Jeremy A. Sabloff). San Francisco: W. H. Freeman and Company, 1980 (originally published in *Scientific American,* (June, 1967), pp. 107–117.

Musser, Curt. *Facts and Artifacts of Ancient Middle America.* New York: E. P. Dutton, 1978.

Nicholson, Irene. *Mexican and Central American Mythology.* London: Hamlyn Publishing Group, 1967.

Nicholson, H. B. "Religion in Prehispanic Central Mexico." *Handbook of Middle American Indians* (ed. Robert Wauchope). Austin, Tex.: University of Texas Press, 1971, pp. 395–446.

Ordoño, César Macazaga. *Coyolxauhqui, la Diosa Lunar.* Mexico City: Editorial Cosmos, 1978.

Ordoño, César Macazaga. *Mito y Simbolismo de Coyolxauhqui.* Mexico City: Editorial Cosmos, 1978.

Ordoño, César Macazaga. *Ritos y Esplendor del Templo Mayor.* Mexico City: Editorial Innovación, 1978.

Palacios, Enrique Juan. *The Stone of the Sun and the First Chapter of the History of Mexico.* Bulletin VI. Chicago: University of Chicago Press, 1971.

Remington, Judith A. "Current Astronomical Practices Among the Maya." *Native American Astronomy* (ed. Anthony F. Aveni). Austin, Tex.: University of Texas Press, 1977, pp. 75–88.

Sahagún, Fray Bernardino de. *Florentine Codex: General History of the Things of New Spain* (ed. Arthur J. O. Anderson and Charles E. Dibble). Santa Fe, N.M.: The School of American Research and the University of Utah. *Book 2, The Ceremonies,* 1981 (2d ed.); *Books 4 and 5, The Soothsayers, the Omens,* 1957; *Book 7, The Sun, Moon, and Stars, and the Binding of the Years,* 1953.

Schele, Linda. "Palenque: The House of the Dying Sun." *Native American Astronomy* (ed. Anthony F. Aveni). Austin, Tex.: University of Texas Press, 1977, pp. 42–56.

Séjourné, Laurette. *Burning Water: Thought and Religion in Ancient Mexico* New York: Vanguard Press, 1956.

Teeple, John E. "Maya Astronomy." *Contributions to American Anthropology and History* 1, no. 2 (Carnegie Institution of Washington, November, 1931): 29–116. Reprint: New York: Johnson Reprint Corporation, 1970.

Thompson, J. Eric S. *A Commentary on the Dresden Codex.* Philadelphia: American Philosophical Society, 1972.

Thompson, J. Eric S. "Maya Astronomy." *Philosophical Transactions of the Royal Society of London* 276 ("The Place of Astronomy in the Ancient World, ed. F. R. Hodson, 1974): 83–98.

Thompson, J. Eric S. *Maya History and Religion.* Norman, Okla.: University of Oklahoma Press, 1970.

Thompson, J. Eric S. *The Rise and Fall of Maya Civilization.* 2d ed. Norman, Okla.: University of Oklahoma Press, 1966.

Townsend, Richard Fraser. *State and Cosmos in the Art of Tenochtitlán.* "Studies in Pre-Columbian Art and Archaeology Number Twenty." Washington, D.C.: Dumbarton Oaks, 1979.

Westheim, Paul. *The Art of Ancient Mexico.* Garden City, N.Y.: Anchor Books/ Doubleday & Company, 1965.

Willey, Gordon R. "Maya Archaeology," *Science* 215 (January 15, 1982): 260–267.

Peru

Aveni, Anthony F. "Horizon Astronomy in Incaic Cuzco." *Archaeoastronomy in the Americas* (ed. Ray A. Williamson). Los Altos, Cal.: Ballena Press/Center for Archaeoastronomy, 1981, pp. 305–318.

Bingham, Hiram. *Lost City of the Incas.* New York: Duell, Sloan and Pearce, 1948.

Cobo, Father Bernabé. *History of the Inca Empire.* 1653. Austin, Tex.: University of Texas Press, 1979.

Dearborn, D. S., and R. E. White. "Archaeoastronomy at Machu Picchu," *Ethnoastronomy and Archaeoastronomy in the American Tropics.* (ed. A. F. Aveni and G. Urton). New York: New York Academy of Sciences, 1982, pp. 249–259.

de la Vega, Garcilaso. *The Incas (The Royal Commentaries of the Inca).* 1609. New York: Orion Press, 1961.

Frost, Peter. *Exploring Cuzco.* Lima: Lima 2000, 1979.

Gasparini, Graziano, and Luise Margolies. *Inca Architecture.* Bloomington, Ind.: Indiana University Press, 1980.

Guidoni, Enrico, and Robert Magni. *Monuments of Civilization: The Andes.* New York: Grosset and Dunlap, 1977.

Isbell, William H. "The Prehistoric Ground Drawings of Peru." *Pre-Columbian Archaeology* (ed. Gordon R. Willey and Jeremy A. Sabloff). San Francisco: W. H. Freeman and Company, 1980, 188–196 (originally published in *Scientific American,* October, 1978).

Kendall, Ann. *Everyday Life of the Incas.* London: B. T. Batsford, 1973.

Morrison, Tony (incorporating the work of Gerald S. Hawkins). *Pathways to the Gods.* Salisbury, Wilts., England: Michael Russell, 1978.

Müller, Rolf. *Sonne, Mond und Sterne über dem Reich der Inka.* Berlin: Springer-Verlag, 1972.

Poma de Ayala. Don Felipe Huamán. *Letter to a King.* 1584-1614. New York: E. P. Dutton, 1978.

Reiche, Maria. *Mystery on the Desert.* Stuttgart: Maria Reiche, 1968.

Urton, Gary. *At the Crossroads of the Earth and the Sky.* Austin, Tex.: University of Texas Press, 1981.

Zuidema, R. T. "The Inca Calendar." *Native American Astronomy* (ed. Anthony F. Aveni). Austin, Tex.: University of Texas Press, 1977, pp. 219–259.

Zuidema, R. T. "Inca Observations of the Solar and Lunar Passages through Zenith and Anti-Zenith at Cuzco." *Archaeoastronomy in the Americas* (ed. Ray A. Williamson). Los Altos, Cal.: Ballena Press/Center for Archaeoastronomy, 1981, pp. 319–342.

South American Indians

Hugh-Jones, Stephen. *The Palm and the Pleiades.* Cambridge, England: Cambridge University Press, 1979.

Levi-Strauss, Claude. *From Honey to Ashes.* New York: Harper & Row, 1973.

Levi-Strauss, Claude. *The Origin of Table Manners.* New York: Harper & Row, 1978.

Levi-Strauss, Claude. *The Raw and the Cooked.* New York: Harper & Row, 1969.

Reichel-Dolmatoff, Gerardo. *Amazonian Cosmos.* Chicago: University of Chicago Press, 1971.

Reichel-Dolmatoff, Gerardo. "Astronomical Models of Social Behavior Among Some Indians of Colombia," *Ethnoastronomy and Archaeoastronomy in the American Tropics.* New York: New York Academy of Sciences, 1982, pp. 165–181.

Reichel-Dolmatoff, Gerardo. *Beyond the Milky Way.* Los Angeles: U.C.L.A. Latin American Center Publications, 1978.

Reichel-Dolmatoff, Gerardo. "Brain and Mind in Desana Shamanism," *Journal of Latin American Lore* (U.C.L.A. Latin American Center) 7, no. 1 (1981): 73–98.

Reichel-Dolmatoff, Gerardo. "Desana Animal Categories, Food Restrictions, and the Concept of Color Energies," *Journal of Latin American Lore* (U.C.L.A. Latin American Center) 4, no. 2 (1978); 243–291.

Reichel-Dolmatoff, Gerardo. "Desana Curing Spells: An Analysis of Some Shamanistic Metaphors," *Journal of Latin American Lore* (U.C.L.A. Latin American Center) 2, no. 2 (1976); 157–219.

Reichel-Dolmatoff, Gerardo. "Desana Shamans' Rock Crystals and the Hexagonal Universe," *Journal of Latin American Lore* (U.C.L.A. Latin American Center) 5, no. 1 (1979), 117–128.

Reichel-Dolmatoff, Gerardo. "The Loom of Life: A Kogi Principle of Integration," *Journal of Latin American Lore* (U.C.L.A. Latin American Center) 4, no. 1 (1978): 5–27.

Reichel-Dolmatoff, Gerardo. *The Shaman and the Jaguar.* Philadelphia: Temple University Press, 1975.

Reichel-Dolmatoff, Gerardo. "Templos Kogi: introducción al simbolismo y a la astronomía del espacio sagrado," *Revista Colombiana de Antropología* 19 (1977); 199–246.

Reichel-Dolmatoff, Gerardo. "Training for the Priesthood among the Kogi of Colombia." *Enculturation in Latin America: An Anthology* (ed. Johannes Wilbert). Los

Angeles: U.C.L.A. Latin American Center Publications, University of California, 1977, pp. 265–288.

Trupp, Fritz. *The Last Indians, South America's Cultural Heritage.* Wörgl, Austria: Perlinger Verlag, 1981.

Wilbert, Johannes. "Eschatology in a Participatory Universe: Destinies of the Soul among the Warao Indians of Venezuela." *Death and the Afterlife in Pre-Columbian America* (ed. Elizabeth P. Benson). Washington, D.C.: Dumbarton Oaks, 1973, pp. 163–189.

Wilbert, Johannes. "The House of the Swallow-Tailed Kite: Models of Shamanic Symbolism." In press.

Wilbert, Johannes. *Survivors of El Dorado.* New York: Praeger Publishers, 1972.

Wilbert, Johannes. "Warao Cosmology and Yekuana Roundhouse Symbolism," *Journal of Latin American Lore* (U.C.L.A. Latin American Center) 7, no. 1 (1981): 37–72.

Wilbert, Johannes. "The Warao Lords of the Rain." *The Shape of the Past: Studies in Honor of Franklin D. Murphy.* Los Angeles: Institute of Archaeology and Office of the Chancellor, University of California, Los Angeles, 1982.

Egypt

Antoniadi, Eugene Michel. *L'Astronomie Egyptienne Depuis les Temps le Plus Recules.* Paris: Gauthiers-Villars, 1934.

Badawy, Alexander. *A History of Egyptian Architecture, Vol. 1.* Giza, Egypt: Alexander Badawy, 1954.

Badawy, Alexander. *A History of Egyptian Architecture: The Empire (the New Kingdom).* Berkeley and Los Angeles: University of California Press, 1968.

Badawy, Alexander. "The Stellar Destiny of Pharaoh and the So-Called Air-Shafts of Cheops' Pyramid," *Mitteilungen des Instituts für Orientforschung,* Band X (1964): 189–206.

Baines, John, and Jaromír Málek. *Atlas of Ancient Egypt.* New York: Facts on File Publications, 1980.

Barguet, Paul. *Le Temple d'Amon-Rê à Karnak.* Cairo: L'Institut Français d'Archeologie Orientale du Caire, 1962.

Barocas, C. *Monuments of Civilization: Egypt.* New York: Grosset and Dunlap, 1972.

Breasted, James H. *Development of Religion and Thought in Ancient Egypt.* 1912. Reprint. Philadelphia: University of Pennsylvania Press, 1972.

Brugsch, Heinrich. *Astronomical and Astrological Inscriptions on Ancient Egyptian Monuments.* 1883. English translation serialized in *Griffith Observer,* published by Griffith Observatory, 2800 East Observatory Road, Los Angeles, Cal., 90027, 1978–80.

Budge, E. A. Wallis. *The Book of the Opening of the Mouth.* 1909. Reprint. New York: Arno Press, 1980.

Budge, E. A. Wallis. *From Fetish to God in Ancient Egypt.* 1934. Reprint. New York: Benjamin Blom, 1972.

Budge, E. A. Wallis. *The Gods of the Egyptians.* 1904. Reprint, 2 vols. New York: Dover Publications, 1969.

Budge, E. A. Wallis. *Osiris and the Egyptian Resurrection.* 1911. Reprint, 2 vols. New York: Dover Publications, 1973.

Clark, R. T. Rundle. *Myth and Symbol in Ancient Egypt.* London: Thames and Hudson, 1978.

Cole, John. *A Treatise on the Circular Zodiac of Tentyra.* London: Longmans & Co., 1824.

Cooke, Harold P. *Osiris, a Study in Myths, Mysteries and Religion.* London: C. W. Daniel Company, 1931.

Davis, Virginia Lee. "Pathways to the Gods." *Ancient Egypt: Discovering Its Splendors* (ed. Jules B. Bellard). Washington, D.C.: National Geographic Society, 1978, pp. 154 201.

Edwards, I. E. S. *The Pyramids of Egypt.* Harmondsworth, England: Penguin Books, 1961.

Fagan, Cyril. *Zodiacs Old and New.* London: Anscombe & Company, 1951.

Fagan, Cyril. *Astrological Origins.* St. Paul, Minn.: Llewellyn Publications, 1971.

Fairman, H. W. *The Triumph of Horus.* Berkeley and Los Angeles: University of California Press, 1974.

Fakhry, Ahmed. *The Pyramids.* Chicago. University of Chicago Press, 1969.

Frankfort, Henri. *Ancient Egyptian Religion.* 1948. Reprint. New York: Harper & Row, 1961.

Gleadow, Rupert. *The Origin of the Zodiac.* New York: Castle Books, 1968.

Goyon, Georges. *Le Secret des Batisseurs des Grandes Pyramides "Kheops."* Paris: Editions Pygmalion, 1977.

Grinsell, Leslie V. *Barrow, Pyramid and Tomb.* London: Thames and Hudson, 1975.

Grinsell, Leslie V. *Egyptian Pyramids.* Gloucester, England: John Bellows Limited, 1947.

Habachi, Labib. *The Obelisks of Egypt.* New York: Charles Scribner's Sons, 1977.

Hawkins, Gerald S. "Astroarchaeology: The Unwritten Evidence." *Archaeoastronomy in Pre-Columbian America* (ed. Anthony F. Aveni). Austin, Tex.: University of Texas Press, 1975, pp. 131–162.

Hawkins, Gerald S. "Astronomical Alignments in Britain, Egypt, and Peru." *Philosophical Transactions of the Royal Society of London* 276 ("The Place of Astronomy in the Ancient World," ed. F. R. Hodson, 1974): 157–167.

Hawkins, Gerald S. *Beyond Stonehenge.* New York: Harper & Row, 1973.

Ions, Veronica. *Egyptian Mythology.* London: Hamlyn Publishing Group, 1968.

Krupp, E. C. "Ancient Watchers of the Sky." *1980 Science Year.* World Book Science Annual. Chicago: World Book-Childcraft International, 1979, pp. 98–113.

Krupp, E. C. "Astronomers, Pyramids, and Priests." *In Search of Ancient Astronomies* (ed. E. C. Krupp). Garden City, N.Y.: Doubleday & Company, 1978, pp. 203–239.

Krupp, E. C. "Egyptian Astronomy: The Roots of Modern Timekeeping," *New Scientist* 85 (January 3, 1980): 24–27.

Krupp, E. C. "Great Pyramid Astronomy," *Griffith Observer* 42, no. 3 (March, 1978): 1–18.

Krupp, E. C. "Recasting the Past: Powerful Pyramids, Lost Continents, and Ancient Astronauts." *Science and the Paranormal* (ed. George O. Abell and Barry Singer). New York: Charles Scribner's Sons, 1981, pp. 253–295.

Krupp, E. C. "The Sun Gods." *Fire of Life: The Smithsonian Book of the Sun.* (ed. Joe Goodwin *et al.*) Washington, D.C.: Smithsonian Exposition Books, 1981, pp. 160–167.

Lauer, Jean-Philippe. *Le Mystère des Pyramides.* Paris: Presses de la Cité, 1974.

Lockyer, J. Norman. *The Dawn of Astronomy.* London: Cassell and Company, 1894.

MacKenzie, Donald. *Egyptian Myth and Legend.* London: Gresham Publishing Company, 191–.

Macnaughton, Duncan. *A Scheme of Egyptian Chronology.* London: Luzac & Co., 1932.

Mendelssohn, Kurt. *The Riddle of the Pyramids.* New York: Praeger Publishers, 1974.

Mercer, Samuel A. B. *Earliest Intellectual Man's Idea of the Cosmos.* London: Luzac & Co., 1957.

Morenz, Siegfried. *Egyptian Religion.* Ithaca, N. Y.: Cornell University Press, 1973.

Müller, W. Max, and Sir James George Scott. *The Mythology of All Races: Vol. XII. Egyptian/Indo-Chinese.* Boston: Marshall Jones Company, 1918.

Neugebauer, O. *The Exact Sciences in Antiquity.* 2d ed. 1957. Reprint. New York: Harper & Row, 1962.

Neugebauer, O. *A History of Ancient Mathematical Astronomy, Part 2.* New York: Springer-Verlag, 1975.

Neugebauer, O., and R. A. Parker. *Egyptian Astronomical Texts I. The Early Decans.* Providence, R. I.: Brown University Press, 1960.

Neugebauer, O., and R. A. Parker. *Egyptian Astronomical Texts II. The Ramesside Star Clocks.* Providence, R. I.: Brown University Press, 1964.

Neugebauer, O., and R. A. Parker. *Egyptian Astronomical Texts III. Decans, Planets, Constellations and Zodiacs.* 2 vols. Providence, R. I.: Brown University Press, 1969.

Parker, R. A. "Ancient Egyptian Astronomy." *Philosophical Transactions of the Roy-*

al Society of London 276 ("The Place of Astronomy in the Ancient World," ed. F. R. Hodson, 1974): 51–65.

Parker, R. A. *The Calendars of Ancient Egypt.* Chicago: University of Chicago Press, 1950.

Parker, R. A. "Egyptian Astronomy, Astrology and Calendrical Reckoning." *Dictionary of Scientific Biography, vol. XV, Suppl. I.* New York: Charles Scribner's Sons, 1978, pp. 706–727.

Petrie, W. M. Flinders. *The Pyramids and Temples of Gizeh.* London: Field & Tuer, et al., 1885.

Petrie, W. M. Flinders. *Wisdom of the Egyptians,* vol. LXIII. London: Bernard Quaritch Ltd., 1940.

Piankoff, Alexandre. *Mythological Papyri.* Princeton, N. J.: Princeton University Press, 1957.

Piankoff, Alexandre. *The Shrines of Tut-Ankh-Amon.* Princeton, N. J.: Princeton University Press, 1955.

Piankoff, Alexandre. *The Tomb of Ramesses VI.* Princeton, N. J.: Princeton University Press, 1954.

Piazzi-Smyth, C. *Our Inheritance in the Great Pyramid.* 5th ed. London: Charles Burnet & Co., 1890.

Plumley, J. M. "The Cosmology of Ancient Egypt." *Ancient Cosmologies* (ed. Carmen Blacker and Michael Loewe). London: George Allen & Unwin, 1975, pp. 17–41.

Poole, Reginald Stuart. *Horae Aegyptiacae, or the Chronology of Ancient Egypt.* London: John Murray, 1851.

Proctor, Richard A. *The Great Pyramid: Observatory, Tomb and Temple* London: Longmans, Green, 1888.

Reymond, E. A. E. *The Mythological Origin of the Egyptian Temple.* Manchester, England: Manchester University Press, 1969.

Ruffle, John. *The Egyptians.* Ithaca, N. Y.: Cornell University Press, 1977.

St. Clair, George. *Creation Records.* London: David Nutt, 1898.

Shorter, Alan W. *The Egyptian Gods.* London: Routledge & Kegan Paul, 1937.

Spence, Lewis. *The Myths of Ancient Egypt.* London: George G. Harrap & Co., 1915.

Trimble, Virginia. "Astronomical Investigation Concerning the So-Called Air-Shafts of Cheops' Pyramid," *Mitteilungen des Instituts für Orientforschung,* Band X (1964): 183–187.

Wainwright, Gerald A. *Sky Religion in Egypt: Its Antiquity & Effects.* 1938. Reprint. Westport, Conn.: Greenwood Press, 1971.

Žába, Zbyněk. *L'Orientation Astronomique dans l'Ancienne Égypte, et la Précession de l'Axe du Monde.* Prague: Éditions de l'Académie Tchécoslovaque des Sciences, 1953.

Mesopotamia

Aaboe, A. "Scientific Astronomy in Antiquity." *Philosophical Transactions of the Royal Society of London* 276 ("The Place of Astronomy in the Ancient World," ed. F. R. Hodson, 1974): 21–42.

Burney, Charles. *The Ancient Near East*. Ithaca, N.Y.: Cornell University Press, 1977.

Contenau, Georges. *Everyday Life in Babylon and Assyria*. New York: W. W. Norton & Company, 1966.

Gaster, Theodor H. *Thespis—Ritual, Myth, and Drama in the Ancient Near East*. Garden City, N.Y.: Anchor Books/Doubleday & Company, 1961.

Gray, John. *Near Eastern Mythology*. London: Hamlyn Publishing Group, 1969.

Handcock, Percy S. P. *Mesopotamian Archaeology*. New York: G. P. Putnam's Sons, 1912.

Hartner, Willy. "The Earliest History of the Constellations in the Near East and the Motif of the Lion-Bull Combat," *Journal of Near Eastern Studies* XXIV, nos. 1 & 2 (Jan.-Apr., 1965): 1–16.

Heidel, Alexander. *The Babylonian Genesis*. 2d ed. Chicago: University of Chicago Press, 1951.

Hinke, William J. "A New Boundary Stone of Nebuchadnezzar I from Nippur." *The Babylonian Expedition of the University of Pennsylvania*. Series D: Researches and Treatises (ed. H. V. Hilprecht). Philadelphia: University of Pennsylvania, 1907.

Hooke, S. H. *Babylonian and Assyrian Religion*. Norman, Okla.: University of Oklahoma Press, 1963.

Hooke, S. H. *Middle Eastern Mythology*. Harmondsworth, England: Penguin Books, 1963.

Jacobsen, Thorkild. *Toward the Image of Tammuz and Other Essays on Mesopotamian History and Culture* (ed. William L. Moran). Cambridge, Mass.: Harvard University Press, 1970.

Jacobsen, Thorkild. *The Treasures of Darkness*. New Haven: Yale University Press, 1976.

James, E. O. *Myth and Ritual in the Ancient Near East*. New York: Frederick A. Praeger, 1958.

Jastrow, Morris. *Aspects of Religious Belief and Practice in Babylonia and Assyria*. 1911. Reprint. New York: Benjamin Blom, 1971.

Jastrow, Morris. *The Civilization of Babylonia and Assyria*. Philadelphia: J. B. Lippincott Company, 1915.

Jastrow, Morris. *The Religion of Babylonia and Assyria*. Boston: Ginn & Company, 1898.

Kramer, Samuel Noah. *Sumerian Mythology*. Revised edition. New York: Harper & Row, 1961.

Kramer, Samuel Noah. *The Sumerians: Their History, Culture and Character*. Chicago: University of Chicago Press, 1963.

Krupp, E. C. "Astronomical Symbolism in Mesopotamian Religious Imagery," contributed paper, January 8, 1979, meeting of the American Astronomical Society, Mexico City. Abstract: *Archaeoastronomy Bulletin* (Maryland) II, no. 1 (November, 1978): 5.

Lambert, W. G. "The Cosmology of Sumer and Babylon." *Ancient Cosmologies* (ed. Carmen Blacker and Michael Loewe). London: George Allen & Unwin, 1975, pp. 42–65.

Langdon, Stephen Herbert. *The Mythology of All Races: Vol. V. Semitic*. 1931. Reprint. New York: Cooper Square Publishers, 1964.

Laroche, Lucienne. *Monuments of Civilization: The Middle East*. New York: Grosset & Dunlap, 1974.

Lloyd, Seton. *The Archaeology of Mesopotamia*. London: Thames and Hudson, 1978.

Mackenzie, Donald A. *Myths of Babylonia and Assyria*. London: Gresham Publishing Company, 191–.

Neugebauer, O. *The Exact Sciences in Antiquity*. 2d ed. 1957. Reprint. New York: Harper & Row, 1962.

Oppenheim, A. Leo. *Ancient Mesopotamia*. Chicago: University of Chicago Press, 1964.

Oppenheim, A. Leo. "Man and Nature in Mesopotamian Civilization." *Dictionary of Scientific Biography, vol. XV. Suppl. I* New York: Charles Scribner's Sons, 1978, pp. 634–666.

Parrot, André. *The Arts of Assyria*. New York: Golden Press, 1961.

Parrot, André. *Sumer, the Dawn of Art*. New York: Golden Press, 1961.

Pinches, Theophilus G. *The Religion of Babylonia and Assyria*. London: Archibald Constable & Co., 1906.

Pritchard, James B., ed. *The Ancient Near East, vol. 1 and 2*. Princeton, N.J.: Princeton University Press, 1958 and 1975.

Sachs, A. "Babylonian Observational Astronomy." *Philosophical Transactions of the Royal Society of London* ("The Place of Astronomy in the Ancient World," ed. F. R. Hodson, 1974), pp. 43–50.

Saggs, H. W. F. *The Greatness That Was Babylon*. 1962. Reprint. New York: New American Library/Mentor Books, 1968.

Sandars, N. K., trans. *Poems of Heaven and Hell from Ancient Mesopotamia*. Harmondsworth, England: Penguin Books, 1971.

Sayce, A. H. *Astronomy and Astrology of the Babylonians*. 1874. Reprint. San Diego: Wizards Bookshelf, 1981.

Smith, George. *The Chaldean Account of Genesis.* 1876. Reprint. Minneapolis: Wizards Bookshelf, 1977.

Spence, Lewis. *Myths and Legends of Babylonia and Assyria.* London: George Harrap & Company, 1916.

Thompson, R. Campbell. *The Reports of the Magicians and Astrologers of Nineveh and Babylon in the British Museum. Vol. 1 and 2.* 1900. Reprint. New York: AMS Press Inc., 1977.

van der Waerden, B. L. "Mathematics and Astronomy in Mesopotamia." *Dictionary of Scientific Biography. Vol. XV. Suppl. I.* New York: Charles Scribner's Sons, 1978, pp. 667–680.

China and Japan

Anon. *New Archaeological Finds in China.* Beijing: Foreign Languages Press, 1974.

Anon. *New Archaeological Finds in China (II).* Beijing: Foreign Languages Press, 1978.

Bredon, Juliet, and Igor Mitrophanow. *The Moon Year.* Shanghai: Kelly & Walsh, 1927.

Chang, Kwang chih. *Shang Civilization.* New Haven: Yale University Press, 1980.

Chien, Szuma. *Selections from Records of the Historian.* Beijing: Foreign Languages Press, 1979.

Christie, Anthony. *Chinese Mythology.* London: Hamlyn Publishing Group, 1968.

Chung-kuo she hui k'o-hsueh yuan (Institute of Archaeology, Chinese Academy of Social Sciences, ed.). *Chung-kuo ku-tai t'ien-wen wen-wu t'u chi* ("Illustrations of Ancient Chinese Astronomical Artifacts: Monographs in Archaeology, B.17"). Beijing: Wen-wu ch'u-pan-she, 1980.

Cotterell, Arthur. *The First Emperor of China.* New York: Holt, Rinehart and Winston, 1981.

Cottrell, Leonard. *The Tiger of Ch'in.* New York: Holt, Rinehart and Winston, 1962.

Davis, F. Hadland. *Myths and Legends of Japan.* Boston: David D. Nickerson & Company, 191–.

Dupree, Nancy Hatch. "T'ang Tombs in Chien County, China," *Archaeology* 32, no. 4 (July/August, 1979): 34–44.

Ferguson, John C., and Masaharu Anesaki. *The Mythology of All Races: Vol. VIII. Chinese/Japanese.* 1928. Reprint. New York: Cooper Square Publishers, 1964.

Hackin, J., *et al. Asiatic Mythology.* New York: Crescent Books (Crown Publishers,), no date.

Hall, Alice. "A Lady from China's Past," *National Geographic* 145, no. 5 (May, 1974): 660–681.

Hay, John. *Ancient China.* New York: Henry Z. Walck, 1973.

Hearn, Maxwell. "An Ancient Chinese Army Rises from Underground Sentinel Duty," *Smithsonian* 10, no. 8 (November, 1979): 38–51.

Krupp, E. C. "Earthquakes and Mooncakes," *Griffith Observer* 37, no. 5 (May, 1973): 12–16.

Krupp, E. C. "The Mandate of Heaven," *Griffith Observer* 46, no. 6 (June, 1982), 8–17.

Krupp, E. C. "Shadows Cast for the Son of Heaven," *Griffith Observer* 46, no. 8 (August, 1982): 8–18.

Krupp, E. C. "Tombs That Touched the China Sky," *Griffith Observer* 46, no. 7 (July, 1982): 9–17.

Loewe, Michael. *Ways to Paradise, the Chinese Quest for Immortality.* London: George Allen & Unwin, 1979.

MacFarquar, Roderick. *The Forbidden City.* New York: Newsweek, 1972.

Needham, Joseph. "Astronomy in Ancient and Medieval China." *Philosophical Transactions of the Royal Society of London* 276 ("The Place of Astronomy in the Ancient World," ed. F. R. Hodson, 1974): 67–82.

Needham, Joseph. "The Cosmology of Early China." *Ancient Cosmologies* (ed. Carmen Blacker and Michael Loewe). London: George Allen & Unwin, 1975, pp. 87–109.

Needham, Joseph. *Science & Civilisation in China,* vol. 3. Cambridge, England: Cambridge University Press, 1959.

Picken, Stuart D. B. *Shinto: Japan's Spiritual Roots.* Tokyo: Kodansha International, 1980.

Piggott, Juliet. *Japanese Mythology.* London: Hamlyn Publishing Group, 1969.

Pirazzoli-T'Serstevens, Michèle. *Living Architecture: Chinese.* New York: Grosset & Dunlap, 1971.

The Purple Mountain Observatory and the Beijing Planetarium. "The Star Map Found in the Northern Wei Tomb in Loyang," *Wen Wu* ("Cultural Relics"), no. 12 (1974): 56–60.

Qian, Hao, Chen Heyi, and Ru Suichu. *Out of China's Earth.* New York: Harry N. Abrams, 1981.

Schafer, Edward H. "Astral Energy in Medieval China," *Griffith Observer* 46, no. 7 (July, 1982): 18–20.

Schafer, Edward H. *Ancient China.* New York: Time-Life Books, 1967.

Schafer, Edward H. *Pacing the Void.* Berkeley and Los Angeles: University of California Press, 1977.

Schafer, Edward H. "From the Tombs of China." *Discovery of Lost Worlds* (ed. Joseph J. Thorndike, Jr.). New York: American Heritage Publishing Company, 1979, pp. 316–344.

Stephenson, F. Richard. "Chinese Roots of Modern Astronomy," *New Scientist* 86 (June 26, 1980): 380–383.

Stephenson, F. Richard, and David H. Clark. "Ancient Astronomical Records from the Orient," *Sky and Telescope* 53, no. 2 (February, 1977): 84–91.

Topping, Audrey. "China's Incredible Find," *National Geographic* 153, no. 4 (April, 1978): 440–459.

Wen, Fong., ed. *The Great Bronze Age of China.* New York: Alfred A. Knopf, 1980.

Williams, C. A. S. *Outlines of Chinese Symbolism and Art Motives.* 1941. Reprint. New York: Dover Publications, 1976.

Yutang, Lin. *Imperial Peking.* New York: Crown Publishers, 1961.

Index